Astrobiology

Scrivener Publishing
100 Cummings Center, Suite 541J
Beverly, MA 01915-6106

Astrobiology Perspectives on Life of the Universe

Series Editors: Richard Gordon and Joseph Seckbach

In his 1687 book *Principia*, Isaac Newton showed how a body launched atop a tall mountain parallel to the ground would circle the Earth. Many of us are old enough to have witnessed the realization of this dream in the launch of Sputnik in 1957. Since then our ability to enter, view and understand the Universe has increased dramatically. A great race is on to discover real extraterrestrial life, and to understand our origins, whether on Earth or elsewhere. We take part of the title for this new series of books from the pioneering thoughts of Svante Arrhenius, who reviewed this quest in his 1909 book *The Life of the Universe as Conceived by Man from the Earliest Ages to the Present Time*. The volumes in **Astrobiology Perspectives on Life of the Universe** will each delve into an aspect of this adventure, with chapters by those who are involved in it, as well as careful observers and assessors of our progress. Guest editors are invited from time to time, and all chapters are peer-reviewed.

Publishers at Scrivener
Martin Scrivener (martin@scrivenerpublishing.com)
Phillip Carmical (pcarmical@scrivenerpublishing.com)

Astrobiology

Science, Ethics, and Public Policy

Edited by

Octavio Alfonso Chon Torres,
Universidad de Lima, Peru

Ted Peters,
Graduate Theological Union, Berkeley, CA, USA

Joseph Seckbach
The Hebrew University of Jerusalem, Israel

and

Richard Gordon
*Gulf Specimen Marine Laboratory & Aquarium, Panacea, FL, USA
and Wayne University, Detroit, MI, USA*

Scrivener
Publishing

WILEY

This edition first published 2021 by John Wiley & Sons, Inc., 111 River Street, Hoboken, NJ 07030, USA
and Scrivener Publishing LLC, 100 Cummings Center, Suite 541J, Beverly, MA 01915, USA
© 2021 Scrivener Publishing LLC
For more information about Scrivener publications please visit www.scrivenerpublishing.com.

Wiley Global Headquarters
111 River Street, Hoboken, NJ 07030, USA

For details of our global editorial offices, customer services, and more information about Wiley products visit us at www.wiley.com.

Limit of Liability/Disclaimer of Warranty
While the publisher and authors have used their best efforts in preparing this work, they make no representations or warranties with respect to the accuracy or completeness of the contents of this work and specifically disclaim all warranties, including without limitation any implied warranties of merchantability or fitness for a particular purpose. No warranty may be created or extended by sales representatives, written sales materials, or promotional statements for this work. The fact that an organization, website, or product is referred to in this work as a citation and/or potential source of further information does not mean that the publisher and authors endorse the information or services the organization, website, or product may provide or recommendations it may make. This work is sold with the understanding that the publisher is not engaged in rendering professional services. The advice and strategies contained herein may not be suitable for your situation. You should consult with a specialist where appropriate. Neither the publisher nor authors shall be liable for any loss of profit or any other commercial damages, including but not limited to special, incidental, consequential, or other damages. Further, readers should be aware that websites listed in this work may have changed or disappeared between when this work was written and when it is read.

Library of Congress Cataloging-in-Publication Data

ISBN 978-1-119-71116-2

Cover image: The editors
Cover design by Russell Richardson

Set in size of 11pt and Minion Pro by Manila Typesetting Company, Makati, Philippines

Printed in the USA

Contents

4   Who Goes There? When Astrobiology Challenges Humans   79
Jacques Arnould

5   Social and Ethical Currents in Astrobiological Debates   95
Kelly C. Smith

Foreword

The science of astrobiology may be understood as a book with four chapters: the origin, evolution, distribution and destiny of life in the universe. Astrobiology's still unfinished first chapter emerged mainly from the work of Alexander Oparin (1894–1980) and other organic chemists. They gave rise to the subdiscipline of astrobiology that was called chemical evolution, a scientific approach to the origin of life on Earth. NASA was established in 1958. Since then, the young space agency encouraged space exploration of the Solar System: their efforts, together with the space agencies that came after them, could lead to at least a single additional example of life in our cosmic neighborhood. This would be the beginning of a second chapter of astrobiology—the evolution of life in the universe. A preliminary development, a third chapter of astrobiology, was due to the molecular biologist and Nobel Laureate Joshua Lederberg (1925–2008). He raised the question of the origin of life, not as a terrestrial phenomenon, but rather as a cosmic distribution of life. A fourth chapter, the destiny of life in the universe, is a different inspiring topic. For getting off the ground, it will need interdisciplinary interactions at the frontier of astrobiology and humanism.

The eighteen chapters of *Astrobiology: Science, Ethics and Public Policy* attempt to fill a gap in the current literature on the continuing growth of this new science of life in the cosmos. Even though astrobiology has made remarkable progress, the humanistic neighbors across its cultural frontiers are only at the beginning of confronting the problem of other life. A specific neighboring humanistic area is a main concern of the present book. It has been called alternatively *astroethics*, or *astrobioethics*. We will adopt the latter denomination, following the suggestion of the 2016 International Working Group on Astrobioethics.

As suggested by the present book, there are two time-honored philosophic subdisciplines that will be relevant for progress in astrobiology. Firstly, ethics, which goes back to Aristotle's *Nicomachean Ethics* (c. 340 BC). The other one, political philosophy, has its roots in the best known of Plato's dialogues, *Republic* (c. 375 BC). Ethics covers questions such as culture, religion and human-nonhuman relations. But our main interest regards especially human-nonhuman relations, since they concern not only public policy, but the future of astrobiology itself. To be fair to the society we live in, we need the assistance of government to ensure that justice is implemented, so that our rights, and those of others, are respected. Then, we should return to political philosophy to guide us in enquiries on public policy:

> What are the right policies for implementing public power in order to respect, preserve, and improve the quality of life on Earth, and elsewhere?

Governments have the vocation to face difficult decisions concerning the distribution of limited public funds that are available to the State. One aspect of this obligation is the support of big science. The main example goes back to the middle of the last century. It involves physics of high energies with their large accelerators. More recently, astrobiology has been inserted into this restricted group, whose most urgent expenses are due to Solar System exploration. Once again, political philosophy comes to our aid regarding the enormous long-term decisions that our expenses force upon public offices. For instance, if we commit ourselves to terraforming in the Red Planet, this activity presents us with a clear-cut question that begs for a political answer. Even closer to the present, though, governments will face the economic exploitation of the Moon, Mars and the asteroids. For these activities we may profit from an earlier analogous multinational experience that has already been addressed with the exploitation of the Antarctic.

Similarly, we are becoming aware that spacefaring nations, with their corresponding space agencies, will need to take possession of new resources pacifically, according to the UN's Outer Space Treaty. Consequently, political agreement is necessary within the United Nations Organization. All space agencies, which are capable of space exploration, should respect UN agreements: the European Union, the United States of America, Russia, Japan, China and India. More recently, other national agencies have come to the foreground, including Israel and the United Arab Emirates. Clearly, political philosophy may come again to our aid.

In a different line of thinking, philosophical studies of morality serve as a basis for extending ethics into considerations that are forced upon us by the eventual understanding of the distribution of life in the universe. In this case the term "neighbor" takes a new, deeper, inspiring, unexpected and unprecedented philosophical significance. We generally accept the principle of equality as a proper ethical basis for relations with other human beings. But with Peter Singer in *Practical Ethics* (1979), we are aware that the principle of equality is also a proper ethical basis for the more restricted question of human-nonhuman relations on Earth. A very remarkable example of an animal that we should keep in mind—the dolphin—was singled out by the neuroscientist Lori Marino: she found that the rate of encephalization (variation in relative brain size) in the hominid line may have been matched by this marine mammal's encephalization as recently as only one million years ago. But independent of this special evolutionary factor, all nonhuman animals should be encompassed, without exception, in our ethical codes. But our search for other manifestations of the phenomenon of life ranges from microbial evolution in the Solar System to the evolution of intelligence in worlds elsewhere in our galaxy. Thus, a bigger, inevitable and evident question in morality cannot be avoided:

> *With nonhuman species, on Earth and elsewhere,*
> *how far should we extend our ethical codes?*

In agreement with Edward Osborne Wilson in *Consilience* (1998), the origin of ethics is not a religious debate between believers and non-believers, but rather between "transcendentalists," those of us who believe that ethical precepts (such as justice and human rights) are independent of human experience, and "empiricists," who believe that ethical principles are human inventions. In what follows we shall understand how, for astro-bioethics, both sides of this debate are fruitfully complementary.

Even though we have already underlined that independent of any theological consideration, the main debate on ethics is between transcendentalists and empiricists, nevertheless we must not exclude, but instead we should pay special attention to some religious aspects both of morality and public policy. Independent of any ethical system, our Judeo-Christian traditions contain writings that are remarkable from an ethical point of view, as they address fundamental questions. An outstanding example is Jesus' *The Sermon on the Mount* (Mathew, 5,1-14, written c. 85 AD), which is inserted in a long biblical tradition (Psalm 1 and Jeremiah 17,7).

On the other hand, as astrobiologists we are mainly concerned with an empirical approach to ethics. Its insertion in science goes back to Charles Darwin in *The Descent of Man* (1875). This work offers a rationalization of the origin of ethics. Since the second half of the last century, the application of Darwinian theory to social behavior—sociobiology—has taught us how ethical behavior, as well as astrobioethics, can be given solid scientific bases. Consequently, under empiricism, progress in the search for life in the universe is bound to induce us to abandon the idea that ethics is uniquely human.

However, we should keep in mind the other major approach to ethics. In philosophy, from Socrates to Singer, there is a long history of transcendentalism. The following short selection of outstanding contributions clearly illustrates this remark: John Locke's *Second Treatise on Civil Government* (1689), David Hume's *A Treatise of Human Nature* (1739), Immanuel Kant's *The Categorical Imperative* (1785), Georg Wilhelm Friedrich Hegel's *The Philosophy of Right* (1831), George Edward Moore's *Principia Ethica* (1903), and John Rawls' *A Theory of Justice* (1971).

With these major philosophical contributions, we are once again in the satisfactory position that has characterized progress: When empirical bases have been identified, rationalism arises as its inevitable complement. In science, from Democritus to Darwin, the concert between empiricism and rationalism has been the general rule. For example, in classical mechanics, early empirical observations of Galileo were later rationalized by Newton's theory of gravitation. Exceptionally, in the astrobiological context, empiricism arose long after rationalization had preceded it in the form of transcendentalism. Fortunately, both sides of the current debate on ethics, and *a fortiori* on astrobioethics, provide solid bases for a consensus. We are ready to face astrobiology's most pressing objective due to the programs on exploration of the Solar System: our eventual interaction with life beyond our own horizons.

Julian Chela-Flores
The Abdus Salam International Centre for Theoretical Physics,
Trieste, Italy
May 2021

Preface

Science is awesome. Well, actually, it's not the science that's awesome. It's the natural world that science helps uncover, expose, reveal. My friend and Nobel Prize winning physicist, the late Charles Townes, once averred to some students visiting in our home, "science is a form of revelation."

In this sense, astrobiology is awesome. Among the revelations over the last quarter century are exoplanets, more than 4,000 confirmed with thousands more waiting in line for confirmation. Our space spectroscopists are watching and our SETI scientists are listening for biosignatures that could reveal extraterrestrial intelligence. Meanwhile, our solar system spelunkers energetically mine the subsurface of Titan and the atmosphere of Venus, looking for possible microbial neighbors closer to home. Astrobiology offers video game adventure for grown-ups.

Astrobiology provides one component to the more comprehensive matrix of science and technology that comprise space research. Just standing in awe at the numinosity of the cosmos is only part of the picture. Funding launch and orbital technology in the midst of geopolitical competition and tension occupy private entrepreneurs and governmental leaders alike. A competition has arisen between stockholders investing in off-Earth mining, on the one hand, and scientists wishing to maintain pristine off-Earth laboratories for their research, on the other hand.

This competition provides honest work for philosophers who then ask: would critters living in an off-Earth biosphere have intrinsic value? And, if so, would the imputation of intrinsic value protect them from terrestrial profiteering? Regardless of how we respond to these ethical quandaries, the answers should rise to the level of public policy formulation to guide the next generation of space explorers.

We need a book. We need a book that looks at *Astrobiology: Science, Ethics, and Public Policy.* You are now reading this book. Yet, as we delve into the details of reading this book with our eyes focused on the pages, we dare not forget the awesome beauty of the cosmos that can be glimpsed only when we turn to look in the direction of the stars.

Let me alert you to some subtleties of vocabulary. With the term, *astro-biology*, we work with standard definitions summarized as: Astrobiology is the scientific study of the origin and evolution of life on Earth and beyond Earth that draws upon a host of disciplines such as astronomy, physics, planetary science, geology, chemistry, biology.

What about ethics? In general, the term, *ethics*, refers to the theoretical work undergirding standards of value and moral responsibility. Be alert to overlaps and distinctions in various chapters. The panoptic terms, *astroethics* and *space ethics*, are inclusive. They include reflection on the broad scope of ethical concerns arising from concrete procedures in space exploration as well as speculations regarding extraterrestrial life. A more focused term is *astrobioethics*, which concentrates on matters having to do specifically with *bios*, life. Astrobioethics, you will read in Octavio Chon-Torres' chapter, is a branch of philosophy and astrobiology that studies the moral implications of the search for life in space.

The construction of future public policy can be built on a solid foundation laid already in 1958 and 1967. As Jacques Arnould at France's *Centre National d'Etudes Spatiales* (CNES) reminds us, the United Nations Committee on Space Research was established in 1958 on the occasion of the International Geophysical Year. The result is a perduring UN mandate to develop recommendations that form the basis of what is now known as planetary protection.

To this foundational principle of planetary protection was added some superstructure in the 1967 UN Treaty on Principles Governing the Activities of States in the Exploration and Use of Outer Space, Including the Moon and Other Celestial Bodies. This prescient document stipulated:

> "§ 1. The exploration and use of outer space, including the moon and other celestial bodies, shall be carried out for the benefit and in the interest of all countries, irrespective of their degree of economic or scientific development, and shall be the province of all mankind.

> § 2. Outer space, including the moon and other celestial bodies, shall be free for exploration and use by all States without discrimination of any kind, on a basis of equality and in accordance with international law, and there shall be free access to all areas of celestial bodies."

In many quarters, this moral foundation for public policy has been forgotten. New debates have broken out over weaponization of space, selling off-Earth real estate, competition for planting national flags, establishing colonies, and property rights for yet-to-be-discovered precious resources.

When we remind ourselves of the foundation laid in 1958 and 1967, we are inspired to see how the awesome magnificence of outer space revealed by the astro-sciences has been bolted to steel moral girders, one of which is to support the notion of a single Earthly society of moral deliberation. When we turn from staring at the stars and look back to Earth, we can perceive a oneness that might have been overlooked in previous centuries. The space sciences reveal something about the cosmos, and something about ourselves as well.

Ted Peters
Berkeley, California, USA
January 1, 2021

Astrobioethics: Epistemological, Astrotheological, and Interplanetary Issues

Octavio A. Chon Torres[1,2]

[1]*Programa de Estudios Generales, Universidad de Lima, Lima, Perú*
[2]*Asociación Peruana de Astrobiología, Lima, Perú*

Abstract

The themes that arise as we enter the philosophical discussion on astrobiology are many and diverse. Of all these, ethics is presented as a rather complex one. Therefore, astrobioethics is the branch of philosophy and astrobiology that is responsible for studying the moral implications of the search for life in space. In this chapter I will analyze three fundamental aspects: epistemological, astrotheological, and interplanetary issues. Each has its own field of discussion and questions that need to be addressed, so that our new small step for mankind does not end up crushing the life we find in the universe.

Keywords: Astrobioethics, astrotheology, interplanetary, teloempathy, transdisciplinary

1.1 Introduction

For a long time, humans have wondered if we are alone in the universe. This has manifested itself in culture, in religion, in philosophy, and in a variety of forms as different as human groups can be on Earth. Although we have not found empirical evidence that we are not alone in the cosmos, eventually this can happen. We do not know exactly when this will happen. However, that is not an impediment to the question of what we should

Email: ochon@ulima.edu.pe

Octavio Alfonso Chon Torres, Ted Peters, Joseph Seckbach and Richard Gordon (eds.) Astrobiology: Science, Ethics, and Public Policy, (1–16) © 2021 Scrivener Publishing LLC

morally do about it. Science and technology advance by leaps and bounds in the search for extraterrestrial life.

Every so often we see news about habitable exoplanets being detected. We have the disciplinary nature of Astrobiology, which brings together several disciplines made up of different specialists whose modus operandi is to work in an orchestrated and coordinated way. But what about the humanities, specifically ethics? Can we have a breakthrough that is matched with the Natural Sciences? To make the comparison would not do justice to either of them, since the nature of both respond to forms of knowledge with their own characteristics.

No. Ethics is not a science that gives us answers like mathematical formulas or experiments in a laboratory or astronomical observations. Ethics is a branch of philosophy that studies the moral dimension of human actions and thinking and, as such, since it does not have a unified methodology in which all experts agree and whose proposal is immutable in time, there are no universal moral laws. However, thanks to reflections on morality we can realize and reflect on our actions and thoughts, on their consequences and implications. That is why it is much more difficult to establish a moral system with coherence and adequate sustenance. And if that is so for earthly matters, for matters that go beyond life on Earth this could become a great mental exercise which will take time and the results of which will not be available every few months as if they were the product of the latest technological advances. To be able to engage in the thinking of astrobioethics, one must approach ethics as a branch of philosophy in addition to astrobiology, because astrobioethics was born in conjunction with moral reflection on issues expressly related to extraterrestrial life and, unlike astroethics, it deals with aspects that are more broad and general such as the responsibility of taking care of space junk or the right to property in an interplanetary context [1.6] [1.10] [1.11] [1.24].

The first time the word astrobioethics was used was in 2016 at two international events: the 35th International Geological Congress in Cape Town, South Africa [1.20] and the 12th Rencontres du Vietnam in Quy Nhon [1.21]. The first academic article that directly addressed this issue was published in the *International Journal of Astrobiology* under the title "Astrobioethics." It states that

> "Astrobioethics is an interdisciplinary field of astrobiology and ethics; it studies the ethical implications of astrobiological research. However, astrobioethics must have transdisciplinary practices in order to enrich itself and propose a broader judgement according to the context where it is applied [1.8]."

The concept of the scientific discipline of astrobiology should include a humanities perspective. In addition, three fundamental axes need to be analyzed in astrobioethics research, which are: the legal aspect, the ethical aspect, and the social aspect. One of the ideas reflected upon is the moral relevance of the Planetary Protection Policy and the need to make it more impactful on ethical discussions concerning any space exploration involving astrobiological components.

Currently, the Working Group on Astrobioethics, a working team belonging to the International Association for Geoethics led by Jesús Martínez-Frías, is the first official international working group on astrobioethics. "One of the main tasks of the WG will be to analyze the potential societal and ethical implications related to astrobiology..." [1.19]. Considering the above, this work is the first book that addresses the astrobiological theme from its ethical aspect.

Different issues will be presented below related to the astrobioethics discussion to determine its importance and opportunities for reflection from different starting points. All of these have to do with whether we are alone in the universe, but also with what would happen if a discovery of life in the universe happens. Here we will look at the discussion involving epistemological, astrotheological, and interplanetary aspects.

1.2 Epistemological Issue

When we deal with moral problems that relate to life in the universe we are presented with a great challenge in relation to knowledge. Considering that astrobiology itself is a transdisciplinary form of knowledge [1.7], the way it connects disciplines must be done in a way that avoids any kind of reductionism. It is interesting to note that the NASA Astrobiology Strategy [1.23] encompasses a diversity of disciplines and each is given a place in this new scenario.

It does not put biology above (and here we are differentiating ourselves from exobiology) other scientific disciplines, but also involves planetary sciences, law, epistemology, etc. A goal of the NASA Astrobiology Strategy is to foster interdisciplinary science. "Astrobiology is multidisciplinary in its content and interdisciplinary in its execution. Its success depends critically upon the close coordination of diverse scientific disciplines and programs, including space missions" [1.28]. Astrobioethics, as well as Astrobiology, requires the participation of each discipline according to the need that emerges in the study process.

In this sense, astrobioethics, working from different areas of knowledge, reminds us that reality is complex and that we must live up to the circumstances. Perhaps the emergence of the need to have these new perspectives is not an isolated phenomenon regarding astrobiology and astrobioethics alone but is part of a much broader context. This would not be strange; after all, science is a product culture and obeys the demands that each era requires. We see today that it is increasingly crucial to address different aspects of reality as an interconnected whole, and not only to know it in theory, but also to know how to apply it. The recent SARS-CoV-2 pandemic is an example of this. What happens to one country is no stranger to what happens to other countries, and it is also foolish to worry more about the economy than public health, because in the end one affects the other. In addition, when implementing health measures in different countries, it is not only the recommendation to prevent a disease that counts, it is also necessary to consider the idiosyncrasies that each context possesses. Reality teaches us what is not taught in college.

Astrobioethics must learn from the lessons we already have on Earth because, after all, whether we want it or not, astrobiological ethics is conditioned by a biogeocentric approach [1.1] [1.4] [1.5]; that is, the paradigmatic dependence we have where we only use as a reference the knowledge we have about life on Earth.

That any moral way of proceeding against extraterrestrial life will necessarily have to be linked to what we have learned on Earth represents the greatest epistemological challenge of astrobioethics. This in turn represents an ontological limitation since we cannot escape our human point of view. But we are no stranger to it in relation to the living beings that inhabit this planet. Anthropocentric conditioning could be overcome if a thinking being alien to the human species could establish a system of values by which to compare it to ours. At present, this is not the case and does not appear to be so anytime soon. On the other hand, we also have the reason-centric conditioning [1.27], on which we base our morals from reason. If not, will it be possible to conceive of other moral systems that do not depend on reason?

The attempt to develop an "inclusive" astrobiological ethic must face this epistemological problem which has no solution for now, but this does not mean that we stop working on it. Even if we do the mental experiment of assuming that a way of life has already been discovered, we could raise four scenarios:

- The first: discovery of microbial life forms.
- The second: discovery of primitive life forms, like those that inhabit the Earth (such as dolphins, dogs, giraffes, lions, ants, etc.).

- The third: discovery of intelligent life forms that are unable to communicate with us.
- The fourth: discovery of intelligent life forms with interplanetary communication capacity.

The first case is more likely to happen in our Solar System, on Mars or below the surface of Europa, Jupiter's Moon. The moral approach in this aspect is obvious, since these microbial life forms would not have a moral system of their own, it would be us who would establish the ethical system of action. Although the second case is not too far away in terms of a moral approach, at least we could identify a certain type of protoconsciousness as happens in some terrestrial animal species, or understand how they behave in order to respect and evaluate how to deal with it.

It also depends on how we might disrupt their habitat or interact with those extraterrestrial life forms. The ethical normative aspect would have greater weight whenever there is a possibility of intervening in their ecosystem. It would not be the same to detect extraterrestrial life forms on exoplanets that we cannot visit to detect it on a nearby planet and be able to have some degree of impact on it.

The third case would have more of an impact on us than on them, in the case of detecting intelligent life forms that may not be able to communicate with us, but we do detect them. This scenario does not seem highly likely now (and here the convenience of the mental experiment, because we can imagine it). However, if that were the case, it would have a significant impact on how we understand life in the universe. In this scenario, since we cannot communicate with them much less visit them, we would not have the possibility to exchange moral systems to establish an authentic astrobiocentric way of behavior. We would be limited to rearticulate what we are as intelligent beings sharing existence with other intelligent beings, but at the moral level we would still be conditioned by the ethical perspective based on the biogeocentric approach, so that even if there are considerable implications both for the natural and social sciences, we would still be watching with eyes anthropocentric to our peers' smarts.

The fourth scenario is closer to a science fiction one, but sometimes science fiction gets ahead of the facts and helps us imagine scenarios and develop interesting arguments. If a communication with intelligent life forms were to take place, the epistemological gap would be more affected than in the other cases since it represents a rather rich cognitive element.

In this exchange of information, the elaboration of an astrobiological code of ethics with an astrobiocentric approach would find its most developed form, since it will be based not only on one reference, but on two. Assuming, of course, that in this communication there is a minimum of mutual understanding and there is no threat presented by it. This scenario is hard to imagine and less likely to happen. What most likely could be expected to happen is scenario one, considering the current state of the search for life in the universe. That being the case, our problem on how to establish an astrobioethics criterion with an appropriate epistemological foundation will have to be limited to an inevitable anthropo-bio-geocentric bias.

1.3   Astrotheological Issue

It is interesting to think about the implications that the discovery of life in other worlds would have on theology. It would not be the first time that a scientific discovery would modify or infuse the spiritual world's perspective on the universe. The antecedent of this can be seen in Darwin's theory of evolution and the passage of the geocentric model to a heliocentric one in the Copernican turn.

The positivism of Auguste Comte thought that after passing through the theological state and the metaphysical state, you would follow the state science, where the hubris of humanity would finally surpass the other forms of knowledge to give way to one in which science leads the way (this is known as the law of the three states). In the theological state, human beings explain the world through supernatural beings; in the metaphysical state, reason supplants these beings to give an explanation of your environment; and finally assumes the positive state, in which the explanation based on scientific evidence predominates, which is the one that would eventually be established as the one form of knowledge that leads us to truth, freeing us from any form of theological thought [1.16].

Religion still exists and, in some countries, is even stronger today. Perhaps analyzing the etymological origin of the word religion will help us to understand it in another way and not as positivist thought was trying to do.

> "The word religion comes from the Latin *religare* which means 'to bind together.' Religion in this sense would be the construct that for a long time has allowed us to unite our world, giving shape and meaning, giving us a character of teleological beings, or beings that seek a purpose, which is not given *a priori* but is rather developed [1.3]."[1]

---

[1]Own translation.

If we understand religion as giving meaning, then its function can be better understood, since it is not a question of worshipping certain gods or one in particular, but of offering the meaning that human beings need to guide their lives. Religious experience is a dimension of the human being that cannot be left out. Religion can be understood as a natural phenomenon, which is a cultural product that society has needed to give meaning to its environment, and has been so for thousands of years, leaving an undeniable imprint on the human mind [1.17].

On the other hand, the universality of religious experience can also be understood as a form of perennial wisdom, as Aldous Huxley explains:

> "To this the fully developed Perennial Philosophy always has and, in all places, given fundamentally the same answer. The divine Ground of all existence is a spiritual Absolute, ineffable in terms of discursive thought, but (in certain circumstances) susceptible of being directly experienced and realized by the human being. This Absolute is the God-without-form of Hindu and Christian mystical phraseology [1.18]."

If we understand it from this perspective, religion is not a secondary accessory for humanity. For a person it can be dispensable, since you can be an atheist and use the sense of religare in other activities such as science itself (an extreme version of this can be seen in the Religion of Humanity by Auguste Comte, where science was an important part of the cult).

If we have this perspective, it is important not to forget the theological aspect in astrobiology, regardless of whether one is a believer or not, because it is a relevant issue to consider and has an impact on the way we see the world of millions of people on Earth. The mere discovery of a second Genesis would potentially have significant implications for the way we see the world. Based on this we could speculate and say that religions will adapt. Perhaps there are religions more flexible to this type of new scenario than others.

But we must also consider another factor. We have religion whose Praxis is institutionalized, and we have the personal way in which people live their religious lives. In accordance with their belief system, perhaps for the average person it is not exceedingly difficult to assume that life exists on other worlds, but another scenario may be that the religious institution makes adjustments in the short term. The truth is that religion will not disappear because there is life on other worlds.

The theological explanation of Christian creation will find ways to contextualize in this new scenario, Buddhism will be able to share existence with other forms of life, and ancestral spiritual practices, such as in the

Andes and its cult of Pachamama, could contemplate the fact that Mother Earth (in a cosmic context) is always generous in giving life even in other lands.

Considering the complexity involved in having a religious discussion in relation to the discovery of life in other worlds, it is important to talk about astrotheology, which is the discipline that studies the theological implications related to the results of astrobiological research. The word astrotheology was coined by Ted Peters, and he defines it as:

> "Astrotheology is that branch of theology that provides a critical analysis of contemporary space sciences combined with an explanation of classical doctrines such as creation and Christology for the purpose of building a comprehensive and meaningful understanding of our human situation within an astonishingly immense cosmos [1.25]."

In astrotheology we could discuss religious aspects in a context of astrobiological discoveries. Traditionally accepted questions in some religions can be seen in a renewed way in astrotheology. We could take as a starting point of reflection the four fundamental axes of astrotheology proposed by Tom Peters:

1. To reflect from different religious traditions the issue of creation and geocentrism.
2. Discuss the parameters on the debate of the person of Christ and the work of Christ.
3. Analyze and discuss astrobiology and related sciences from within, exposing extra-scientific assumptions, interpreting the high value of scientific enterprise.
4. Cooperate between scientists and religious leaders to prepare for possible extraterrestrial contact [1.25].

At least the second axis corresponds more to a reflective aspect of Christian astrotheology. We could include other axes that also represent important aspects of different religions, but this could increase them unnecessarily. The other option would be to encompass the more general notions that have to do with this form of *perennial philosophy* of all religions and reflect them in the framework of astrobiological discoveries.

One issue that is extremely important is that the astrotheological discussion should come from religious representatives who have adequate knowledge of the working form of science and astrobiology. Point 3 is essential in order to not fall into absolute biases that blind the assimilation of new results that arise, which will ultimately influence point 4.

The most fundamental aspect we could infer from these four axes is that of our place in the universe from a religious perspective. What is our place in the universe considering the presence of life outside the Earth within the framework of religion? are we moving towards new religious forms that contemplate "other creations" as well as a second astrobiological Genesis? This will push us to rethink several of the religious concepts we currently handle [1.2].

However, when discussing axis 4 we would have to take into account the following:

> "Astrotheology should take seriously that most astrobiologists are searching for life that is far from an ETI with which we might have conversations over metaphysics. It is microbial life that is being imagined and that seems like it potentially could be found [1.26]."

There are important issues worth discussing about astrotheology and the discovery of life outside the Earth. To what extent is it relevant for religious studies to find microbial life compared to extraterrestrial intelligence (ETI)? Would discovering non-intelligent and microbial life forms imply a sort of preference for "intelligent creation" on Earth over other planets from a biogeocentric point of view? These and other issues deserve proper discussion to reach a consensus.

1.4 Interplanetary Issue

What if a fact-finding mission involves extraterrestrial life? What would happen if we accidentally caused damage to these life forms? Of course, I mean non-intelligent life, which presents us with a situation where we must decide for ourselves what is the right way to proceed.

These kinds of questions differ from those that might be asked in astroethics. In the latter, there would be more of a focus on issues regarding territorial conflicts, such as: What would happen if a nation under the pretext of staying in an area, the moon, for example, for the purpose of research, intends for second intentions to "appropriate" that land? What would prevent a country from indefinitely occupying a place on Mars under the pretext of continuing its research, considering that no celestial object can be the object of militarization?

We cannot deny that eventually any issues involving the presence of life and the militarization of space or space mining will have to intersect. In the future it is most likely that they will, and to date there is no normative document that addresses all the implications and responsibilities involving all the variables to consider, especially since it is one thing to discuss it around a table, and another to have the problem directly in front of us.

But coming back to the issue of Interplanetary, in astrobioethics it will have to be limited to our role in relation to these other forms of life. This puts us in the position of thinking about defining the axiological dimension that extraterrestrial life will have for us. Should we consider that these forms of life have value in themselves or just an instrumental value? Therefore, we could look at extraterrestrial life from two perspectives, which, since it is always changing, could be more or even interconnect. The perspective of value in itself—taking Kant as a reference—and that of instrumental value—considering the utilitarian perspective.

It should be emphasized that the idea that the human being has value in himself, according to Kant's argument, comes mainly from the fact of his being a rational being. We do not expect in the short term to meet beings from other planets with intelligence and use of reason, so we will subtract from this thought only the notion of value in itself, since an authentic use of Kantian morality has no place here. Taking into consideration that we think that non-intelligent extraterrestrial life forms cannot defend themselves or have a way of communicating with us in any way, to what extent can we give them some value? The value we could give a non-intelligent extraterrestrial life forms is perhaps as the only galactic companions we know, and that neglecting them or not treating them properly, would then make us alone in the known universe. Here you can enter the word coined by Charles Cockell, which is that of "teloempathy." What does it mean? Basically, to have empathy for other terrestrial and non-terrestrial life forms because they have interests. What is that interest? That of not being destroyed [1.12] [1.13] [1.15].

However, when we start talking about interests, we are moving away from Kantian morality, so it is more convenient to talk about an astrobioethics with utilitarian rather than Kantian tendencies,

> "…if we really want to consider a 'universal' ethic in the most real sense possible, it should be based on the experience and the cases studied, so that we avoid a metaphysical attempt that can bring us difficulties rather than solutions [1.7]."

Considering a form of universal ethics would be inconvenient if we put it in front of the transdisciplinary nature of astrobioethics discussed ages ago. The *a posteriori* form of work that involves the process of astrobiological discovery does not have a universal common theoretical framework to establish a univocal criterion, so our moral reflection is more about the interest that an object may have for us, or, to the contrary, the value that the object of study may have in itself; in this case, an extraterrestrial life form.

If the extraterrestrial life we find has extrinsic or instrumental value, is it really a moral act? If we think of it in terms of the value to humanity that results from research, perhaps yes, but the status of extraterrestrial life places it on an axiological plane different from that of experimenting with other forms of terrestrial life. While teloempathy is an extension of the ecological morality existing on Earth, dealing with the intrinsic value of an extraterrestrial life form can place it in a different position. What happens if in space mining we also find extraterrestrial life forms and that, in addition, the removal of the minerals present in them is of great importance for the Earth? Perhaps it would be more appropriate if this thought was taken to the extreme: What would happen if, after having studied it enough, this way of life does not represent greater scientific value for humanity and there is pressure to use its environment for space mining?

If we only entertain the idea that protecting an extraterrestrial microbial life form has value because it is important for research, then when that value no longer exists, we will have no fear of intervening. Perhaps we could think of places of planetary protection as large reserves, special places where it is forbidden to intervene [1.14]. We should not repeat the ecological damage we have caused on planet Earth in other planetary environments where life may exist. May this new small step for mankind not be the start of trampling life in the universe, as we did with nature on Earth. To avoid this, our value system must be rethought not only on a theoretical but also practical level.

For this reason, an instrumental ethic that only values extraterrestrial microbial life for its usefulness to science is a pseudomorph since in turn it will only be a matter of time for that value to cease to exist. Then, we will have no other reason not to intervene and give it up to be used for the interests that some company or state might have for that place. It is an issue that sooner or later will occur, if not in this generation, the next or subsequent generations. Therefore, it must be considered with the seriousness necessary to achieve a consensus.

However, we could press the question of what to do in a crucial situation where it is imperative to intervene in an environment which has the potential for life or where life has already been confirmed. If it is a potential environment, the pressure might be less than if it is a place with the confirmed presence of life. Let us not forget that we are thinking of an extreme situation, where all other options to avoid landing there have been exhausted and we cannot prolong the decision any longer. In this case we could evoke the third principle of deep ecology, as presented in the book *The Basic Principles of Deep Ecology* written by Naess and Sessions [1.22], which states that humans have no right to reduce the richness and diversity of natural life except to meet vital needs. How could this be understood in the framework of astrobioethics and special exploration?

> "...it does not matter how safe we are that we have established plots of land where we can land and settle, or promote some form of terrestrial life, there is always the difficulty of not being totally sure of avoiding some kind of collateral damage, at least with the current technology [1.9]."

In a situation where we must make a decision that involves some risk to native life forms, the third principle adapted in this scenario may be our guideline for how we should behave. Having exhausted all possible resources to avoid compromising extraterrestrial microbial life, what can be done is to safeguard it in the best possible way. Thus, the genuine value that native life forms have is always respected.

On the other hand, if our moral maxim is that we should not interfere or land where native life forms are under any circumscription, then there will be no exception to justify our actions. It is difficult to think of such an absolutist circumstance, and that is why I have decided to develop the option of the extreme situation to be able to highlight what our options would be. However, we must not forget that the decision we take must be in consensus with the countries involved. It is possible that what is considered inviolable for one country, may not be for another, so a common voice is necessary to be able to make such decisions.

1.5   Conclusions

From an epistemological aspect, it can be inferred that the fact of being able to elaborate an ethical theoretical framework involving other forms of life represents a gnoseological challenge. Therefore, it is also a transdisciplinary challenge to be able to connect different disciplines and thus to develop an appropriate normative moral framework. The implication that so many disciplines have from social to natural, makes astrobioethics an authentic exercise between disciplines. However, as much as we want, the ontological frontier is always present, reminding us that we cannot leave our human brains and our carbon-based nature.

From an astrotheological aspect, it is interesting to think that perhaps there may be religions more likely to accept life on other worlds. It is important to have a discussion-based perspective among different religious leaders. It is inevitable that the subject of religion will be dealt with once life is found in other worlds; and yet, depending on the type of discovery, we might see different reactions from religion. However, it seems obvious that religion will have to adapt its worldview to the new component: life on other worlds.

Regarding the aspect of Interplanetary issue, we see that instrumental ethics is not sustainable over time since the value of extraterrestrial life depends on its usefulness for scientific research. On the contrary, if we assume a teloempathic position, this changes because we will try as much as possible to look after the interests of these discovered beings—always assumed to be microbial forms—and generate large planetary reserve parks. However, there is always the extreme imaginary challenge where we are asked if it is necessary to intervene in an inhabited place, and it is here that the third principle of deep ecology comes into play. This principle allows exceptions to the rule of respect for diversity of life. It is important to keep this in mind so that we do not have to spoil an ecosystem that we may never discover again, and not repeat the disaster we have caused in nature on Earth.

References

[1.1]   Aretxaga, R., Astrobiology and biocentrism, in: *Life in the Universe*, J. Seckbach, J. Chela-Flores, T. Owen, F. Raulin (Eds.), pp. 345–348, Kluwer Academic, Dordrecht, The Netherlands, 2004.

[1.2]   Arnould, J., Astrotheology, astroethics, and the new challenges. *Theol. Sci.*, 16, 380–381, 2018.

[1.3]   Atlan, H. and Thompson, W.I., *Gaia*, 5th ed, pp. 107–123, Kairós, Barcelona, 2009.

[1.4]   Chela-Flores, J., *The New Science of Astrobiology from Genesis of the Living Cell to Evolution of Intelligent Behavior in the Universe*, Kluwer Academic, Dordrecht, The Netherlands, 2001.

[1.5]   Chela-Flores, J., *A Second Genesis: Stepping Stones Towards the Intelligibility of Nature*, World Scientific, Singapore, 2009.

[1.6]   Chon Torres, O., Astrobiology and its influence on the renewal of the way we see the world from the teloempathic, educational and astrotheological perspective. *Int. J. Astrobiology*, 19, 4, 1–5, 2020a.

[1.7]   Chon-Torres, O., Astrobioethics: A brief discussion from the epistemological, religious and societal dimension. *Int. J. Astrobiology*, 19, 1, 61–67, 2020b.

[1.8]   Chon-Torres, O., Astrobioethics. *Int. J. Astrobiology*, 17, 1, 51–56, 2018.

[1.9]   Chon-Torres, O., Moral challenges of going to Mars under the presence of non-intelligent life scenario. *Int. J. Astrobiology*, 19, 1, 1–4, 2020.

[1.10] Chon-Torres, O.A., Disciplinary nature of astrobiology and astrobioethic's epistemic foundations. *Int. J. Astrobiology*, 1–8, 2018a.

[1.11] Chon-Torres, O.A., Astrobioethics. *Int. J. Astrobiology*, 17, 51–56, 2018b.

[1.12] Cockell, C., Planetary protection – a microbial ethics approach. *Space Policy*, 21, 281–292, 2005a.

[1.13] Cockell, C., Duties to extraterrestrial microscopic organisms. *J. Br. Interplanet. Soc.*, 58, 367–373, 2005b.

[1.14] Cockell, C. and Horneck, G., A planetary park system for Mars. *Space Policy*, 20, 291–295, 2004.

[1.15] Cockell, C., Landfester, U., Remuss, N.-L., Schrogl, K.-U., Worms, J.-C., *Humans in Outer Space – Interdisciplinary Perspectives*, pp. 80–114, Springer, Berlin, 2011.

[1.16] Comte, A., *The Positive Philosophy of Auguste Comte*, translation by H. Martineau, Batoche Books, Canada, 2000.

[1.17] Dennett, D., *Breaking the Spell*, Penguin Random House, United States, 2007.

[1.18] Huxley, A., *The Perennial Philosophy*, Chatto & Windus, London, 1947.

[1.19] IAGETH Working Group on Astrobioethics, *Astrobioethics*, International Association for Geoethics, 2020.

[1.20] Martínez Frías, J. and Gargaud, M., Does Astrobiology require an astrobioethical approach? *35th International Geological*, 2016.

[1.21] Martínez Frias, J., Ethics and space exploration: from geoethics to astrobioethics. Search for life: from early earth to exoplanets, *XII The rencontres du Vietnam - Quy Nhon*, 2016.

[1.22] Naess, A. and Sessions, G., The Deep Ecology Platform, in: *Foundations for Deep Ecology*, 1984, Available at http://www.deepecology.org/platform.htm.

[1.23] Hays, L. (Ed.), *The Astrobiology Strategy 2015*, NASA, EE.UU, 2015, Available at https://astrobiology.nasa.gov/nai/media/medialibrary/2016/04/NASA_Astrobiology_Strategy_2015_FINAL_041216.pdf.

[1.24] Peters, T., Does extraterrestrial life have intrinsic value? An exploration in responsibility ethics. *Int. J. Astrobiology*, 18, 4, 304–310, 2019.

[1.25] Peters, T., Astrotheology: a constructive proposal. *Zygon*, 49, 443–457, 2014.

[1.26] Pryor, A., It's a great big universe: astrobiology and future trends for an astrotheology. *Dialog*, 57, 5–11, 2018.

[1.27] Smith, K.C., The trouble with intrinsic value: an ethical primer for astrobiology, in: *Exploring the Origin, Extent, and Future of Life: Philosophical, Ethical, and Theological Perspectives*, C. Bertka (Ed.), Cambridge University Press, Cambridge, 2009.

[1.28] Des Marais, D.J., Nuth, J.A. III, Allamandola, L.J., Boss, A.P., Farmer, J.D., Hoehler, T.M., Jakosky, B.M., Meadows, V.S., Pohorille, A., Runnegar, B., Spormann, A.M., The NASA astrobiology roadmap. *Astrobiology*, 8, 715–730, 2008.

2

Astroethics for Earthlings: Our Responsibility to the Galactic Commons

Ted Peters

Graduate Theological Union in Berkeley, California, USA

Abstract

The Astroethics of Responsibility proposed here is founded on a substructure of quandary-responsibility ethics, supported by a theological notion of the common good plus a naturalistic justification for response and care. Within the sphere of the solar neighborhood, ten already articulated quandaries are addressed: (1) planetary protection; (2) intrinsic value of off-Earth biospheres; (3) application of the Precautionary Principle; (4) space debris; (5) satellite surveillance; (6) weaponization of space; (7) scientific versus commercial space exploration; (8) terraforming Mars; (9) colonizing Mars; and (10) anticipating natural space threats. Within the sphere of the Milky Way metropolis in which the "galactic common good" becomes the astroethical norm, engagement with intelligent extraterrestrials is analyzed within three categories: (1) ETI less intelligent than Earth's *Homo sapiens*; (2) ETI equal in intelligence; and (3) ETI superior in intelligence. Superior ETI may come in both biological and postbiological forms. Our ethical mandate: respond with care.

Keywords: Astrobiology, astroethics, astrobioethics, quandary-responsibility ethics, intrinsic value, dignity, common good, galactic commons

2.1 Introduction

Astrobiologists along with astronomers and astrophysicists are ready to take the next step in refining astroethics, ethics for space exploration. Might

---

Email: tedfpeters@gmail.com

Octavio Alfonso Chon Torres, Ted Peters, Joseph Seckbach and Richard Gordon (eds.) Astrobiology: Science, Ethics, and Public Policy, (17–56) © 2021 Scrivener Publishing LLC

we consider thinking of space beyond Earth as a commons, a domain to be shared by both earthlings and space neighbors? Might we lift up a vision of the common good that includes yet transcends our home planet?

Apollo 11 astronaut, Buzz Aldrin, says the time has come in which "Space offers us, or rather has allowed us to adopt for ourselves, a new dimension of freedom, which we must use for the benefit of humanity, to enrich and not degrade our lives" [2.1]. The time has come because space explorers need policies, policies that are ethically informed and formed. SETI astrobiologist Margaret Race identifies this need to add ethics to our science. "There are no specific policies or statements regarding ethical considerations or the broader impacts of human activities, particularly in relation to ET life and environments. Moreover, there is no guiding framework for considering any non-scientific issues" [2.69]. The need for foundational ethical—especially astrobioethical—deliberation has arisen [2.13].

In this chapter we will entertain a series of quandaries in "astroethics," sometimes called *space ethics*. We will divide the universe into two concentric spheres of moral concern, our solar neighborhood and the Milky Way metropolis. Because of the untraversable distances between galaxies, and because we have virtually no hope of ever devising a technology by which we could communicate faster than the speed of light, the largest sphere of moral concern we can seriously consider is the Milky Way. We will rely on the term "astroethics" to encompass the full scope of space ethics; the term "astrobioethics" will be employed when dealing specifically with off-Earth life forms.[1]

In addition to applying the common good to the Milky Way galaxy, this astroethical investigation will require identifying the moral agent. Here I nominate for the role of moral agent: Earth as a single planetary society. However, this raises a problem. Designating all the peoples of Earth functioning together as a moral agent is problematic because a single planetary society does not currently exist. Our planet Earth is governed formally by independent nations or, more materially, by competing economic and social interest groups. The very formation of a planetary society to deal with off-Earth matters is itself an ethical ideal and goal to pursue. As we pursue this vision of a cooperative human race, we must overcome the presumption of national sovereignty and the individuated interests

---

[1]"Astrobioethics is an emerging discipline that studies, evaluates, and analyzes the moral, legal, and social issues of the search for life in the Universe" [2.13] [2.15].

competing in the economic sector.[2] In addition. this vision of a single planetary community of moral deliberation must draw into the center of participation those who today are marginalized due to poverty, race, or gender. Astroethics is first of all earthling ethics, and then it becomes galactic ethics [2.59].

In what follows we will lay the foundation and begin the superstructure for a refined astroethics to aid in formulating public policy. In laying the foundation we will pose three quandary questions: Who are we? What do we value? What should we do? The cumulative answers to these three questions will lead to a foundation we will label "Astroethics of Responsibility" [2.63].

Atop this foundation we will frame a superstructure of quandaries regarding specific astroethical issues. The load-bearing vertical supports will include: (1) the moral agent: Earth; (2) the moral norm: the galactic common good; (3) the moral spheres: the solar neighborhood and the Milky Way metropolis; (4) the moral justification: a theological apprehension of the common good combined with a naturalistic apprehension of the Golden Rule.

The floor plan will designate a conference room for each of thirteen previously formulated ethical quandaries. Ten of these quandaries lie within the sphere of the solar neighborhood: (1) planetary protection; (2) intrinsic value of off-Earth biospheres; (3) application of the Precautionary Principle; (4) space debris; (5) satellite surveillance; (6) weaponization of space; (7) scientific versus commercial space exploration; (8) terraforming Mars; (9) establishing human settlements on Mars; and (10) anticipating natural space threats [2.63]. Three quandaries lie within the sphere of the Milky Way metropolis: engagement with intelligent extraterrestrials who are (1) less intelligent than Earth's *Homo sapiens*; (2) ETI equal in intelligence; and (3) ETI superior in intelligence in both biological and postbiological forms [2.58].

---

[2] I rely on philosopher Eric Voegelin to grasp the notion of a universal humanity both as an ideal and promise. "Universal mankind is not a society existing in the world, but a symbol which indicates [human] consciousness of participating, in this earthly existence, in the mystery of a reality that moves toward its transfiguration. Universal mankind is an eschatological index." [2.90]. There is a proleptic component to the vision of a universal humanity. The biblical symbol of the imminent Kingdom of God enjoins us in the present moment to anticipate the eschatological consummation which will entail a new creation unified in justice and love.

2.2   Laying the Foundation for an Astroethics of Responsibility

When NASA launched the New Horizons probe to Pluto in 2006, the Earth-relative launch speed was 36,000 miles per hour. After its sling from Earth's orbital motion, it sailed toward the edge of our Solar System at 100,000 miles per hour. By 2018 the NASA launch of the Parker Solar Probe included an orbital velocity of 430,000 miles per hour. Science moves fast. Can ethics keep up?

"Space ethics appear today as a new *terra incognita,* an unknown country," writes Jacques Arnould, astroethicist at France's *Centre National d'Etudes Spatiales* (CNES). For this reason Arnould likens space ethicists to pioneers. As pioneers, space ethicists should begin their journey with humility, seeking first to learn the new territory. "That is the reason too why the first challenge is not to organize, to legalize and to reduce ethics to its repressive aspect. At the present time, we need to explore the field of space ethics. We need to determine the responsibilities; and to debate them. Major decisions about space cannot remain in the hands of individual leaders or the property of politic, scientific or financial lobbies" [2.2]. Here I intend to "determine the responsibilities" by constructing an Astroethics of Responsibility.

Astroethicists are pioneers. While astrobiologists are exploring the heavens, the ethicists are exploring the astrobiologists. Astrobiology needs more than science to explore astroethics. Astrophysicist Neil deGrasse Tyson puts a fence around science. "The methods of science have little or nothing to contribute to ethics, inspiration, morals, beauty, love, hate, or aesthetics. These are vital elements of civilized life, and are central to the concerns of every religion. What it all means is that for many scientists there is no conflict of interest" [2.84]. For ethical inspiration, the astroethicist must draw upon extrascientific sources. In this case, we will draw on the religious notion of the common good and the naturalistic notion of responsibility.

The core maxim derived from a naturalistic contribution to responsibility ethics is this: *respond with care.* I rely for this core maxim on the late Hans Jonas, a Jewish philosopher who parses the role of the "ought" when envisioning what "ought" to be done. Responsibility responds to the "forward determination of what is to be done," he observes. "First comes the ought-to-be of the object, second the ought-to-do of the subject who, in virtue of his power, is called to care" [2.36]. As living creatures, we are called by the natural domain to care. We are to care for all that lives. "Only what is alive, in its constitutive indigence and fragility, *can* be an object of responsibility" [2.36].[3] In this treatment I explore the question: can Jonas' notion of responsibility help us lay the foundation for an Astroethics of Responsibility [2.63]? My answer is affirmative.

2.2.1   First Foundational Question: Who Are We?

How do we ground our ethics when we earthlings are looking at the sky? By grounding I mean justifying. Ethics, as the theory underlying moral action, cannot simply ride the winds of whim or personal preference. Its foundation needs to be cemented down. How are we going to do this?

We can begin the planting process by asking three fundamental questions: Who are we? What do we value? What should we do? Let us address each of these in turn; and then we will address the issues already on our list.

Who are we? Evolution has made us into responders. According to theologian H. Richard Niebuhr, responsibility ethics is grounded in human nature. "What is implicit in the idea of responsibility is the image of man-the-answerer, man engaged in dialogue, man acting in response to action upon him" [2.54]. An astroethics of responsibility could be grounded in the responsive trait belonging to our human nature.

Who are *we*? By *we* here I mean the entire human race on planet Earth. Inherent in asking about astroethics for earthlings is the question: Who speaks for Earth? [2.83]. How could we justify a moral agent that does not build on responsibility to humanity and of humanity in the form of a single planetary society?

A single planetary society becomes a community of moral deliberation when addressing the relationship between Earth and what is off-Earth. Our solar neighborhood or the Milky Way metropolis is not the private property of one nation; nor is an off-Earth site the claim of whichever team of astronauts arrives first. The competition and rivalry that plague our everyday territorial claims on Earth must be superseded by a global community about to enter the space environment which surrounds all of us.

This single Earth community does not exist yet, even though the United Nations has been working with this concept of the *we* at least since 1967. The 1967 UN Treaty on Principles Governing the Activities of States in the Exploration and Use of Outer Space, Including the Moon and Other

---

[3]Human nature derives via evolution from pre-human nature, and pre-human life already exhibits the "ought" imperative. According to Jonas, because living creatures struggle to survive and thrive, they presuppose it is "worth the effort." If it is worth the effort, then this "must mean that the object of the effort is *good*, independent of the verdict of my inclinations. Precisely this makes it the source of an 'ought'" [2.36].

Celestial Bodies, stipulated: "§ 1. The exploration and use of outer space, including the moon and other celestial bodies, shall be carried out for the benefit and in the interest of all countries, irrespective of their degree of economic or scientific development, and shall be the province of all mankind. § 2. Outer space, including the moon and other celestial bodies, shall be free for exploration and use by all States without discrimination of any kind, on a basis of equality and in accordance with international law, and there shall be free access to all areas of celestial bodies" [2.85]. After looking toward the heavens, we earthlings look back at each other and recognize a newly founded unity. Octavio Chon-Torres notes: "The treaty of outer space is a good example of how our expansion in the Universe should help us to conceive [of] ourselves as a united humankind" [2.13].

Our first moral responsibility is to work toward the establishment of a single planetary society, which may in time become expanded into a galactic moral community. Here is the warrant: virtually every decision regarding what earthlings do in space will have repercussions for every resident of Earth. Therefore, the concept of planetary ethics includes, among other things, representative participation. We can envision a future replete with a single universal humanity; and we can incarnate that vision proleptically by acting now out of that vision. *Our first obligation is to become who we are: the one people of Earth, diverse in the past but united in the future.*

2.2.2   Second Foundational Question: What Do We Value?

Life. Like cream in ol' fashioned milk bottles, life floats to the top of the astrobiologist's value bottle. "I suggest that the long-term goal for astrobiology and society is to enhance the richness and diversity of life in the Universe," avers NASA's Christopher McKay [2.47].

No one debates whether or not to value life. Yet, just how we value life has become a quandary. "Whether secular or theological, the most important question about the foundations for ethical value turns on the distinction between intrinsic value—value independent of a valuing agent—and instrumental value—value in relation to something else like the needs of humans" [2.26]. Steven Dick, former holder of the Baruch S. Blumberg NASA/Library of Congress Chair in Astrobiology, turns on this distinction to develop a cosmocentric ethic. The cosmos has intrinsic value, concludes Dick. Or, perhaps more modestly, a cosmocentric ethic will stretch the scope of terrestrial human valuing beyond anthropocentrism or even geocentrism to incorporate extraterrestrial life in both microbial and intelligent forms. Dick's cosmocentric ethic "establishes the universe and all or part of its life as a priority rather than just humans or even terrestrial life in general" [2.26]. Where Dick puts cosmocentrism, I put the galactic common good.

Is it possible for anthropocentric and geocentric earthlings to transcend their own myopia? Yes, according to astrotheologian Andreas Losch, "We cannot avoid some anthropocentric bias, but we humans are also the ones who can speculate beyond the bounds of our experience" [2.41]. The shift toward a galactic or even cosmocentric perspective will require a realistic respect for the tension existing within our human nature: our proclivity toward self-centered myopia in tension with our capacity to speculate broadly and altruistically.

Dick's proposal of a cosmocentric ethic—in conjunction with my proposal for a galactic common good—compels us to ask: What do we already value? Do we actually value the safety, welfare, and future health of Planet Earth? Our ecoethicists say, no. They complain bitterly that *de facto* the human race values its home planet too little. Even with enlightened self-interest as a motive, we planetary citizens have befouled our terrestrial home. One might reasonably ask: If we terrestrials have befouled our own planetary nest, might we do the same for every off-Earth site we visit? [2.9].

Geocentric values are constantly assaulted by rival greeds. Even high-minded Enlightenment values—freedom, equality, justice, dignity, peace—are left orphaned by the vicious competition for economic survival if not domination. Arnould, using the metaphor of evolution, fears that what has happened on Earth may be repeated in space [2.4] [2.5]. The human attitude of domination of the fittest (or, sometimes also, of survival) leads to growing terrestrial pollution, toxic waste, even climate change which will modify, in a few decades, the level of the oceans, the rain pattern, the distribution of the deserts and the cultivatable zones.

To avoid the same polluting of space with earthling myopia, Arnould draws on the equivalent of intrinsic value and proposes that we "santuarize" outer space. By recognizing that space "transcends all our actual economic motivations .... It is probably the role of national and international space agencies to devise and introduce rules of effective control, and create conditions that would govern any form of exploitation still to come from space" [2.2]. In short, Arnould recommends that, by "sanctuarizing" space, our policy-setting transcends the vested interest of nations and businesses.

If "sanctuary" here connotes precedents already set on Earth for wildlife sanctuaries or protected parks, then I concur that setting a policy of sanctuarization of space would be consistent with our fundamental values. Yet, it could mean more. It could imply communion between what is natural and what is distinctively human; and this communion spawns moral power. Ecoethicist Cynthia Moe-Lobeda testifies how "human moral power flows *primarily* from deep communion between God, human creatures, and the broader community of life" [2.51]. Or, "the sacramental *communion*—God incarnate in us and among us as human communities and as a planetary or even cosmic community of life—is a locus of moral power" [2.51]. In a moment I will lift up the concept of a "galactic commons," one way in which such values can empower us and lead to policy formation.

2.2.2.1 Science and Value

The question—What do we value?—takes on complexity and nuance when drawing science into the picture. The place of science raises two challenges. First, if science is value-neutral, then all the ethicist can do is paint values over science, distorting science. Second, whichever moral color the ethicist selects will seem to be arbitrary, merely the color of the moral painter's subjective choice. Let us look at these two challenges in turn.

First, should we paint science with an ethical brush? If we do, would ethical deliberation distort value-free science? Or, should we ask a bit critically: Can science actually claim to be value-neutral in the first place? Scientists certainly strive for objectivity, employing multiple tests by blind referees to confirm or disconfirm a hypothesis. Such honest rigor is to be commended, even applauded. Yet, beneath the value-free patina, scientific research is always guided by either a worldview or by someone's vested interests. Big science as practiced today requires funding, and funding is supplied only by funders who are expressing their agendas.

Money talks. Power speaks. Space Studies researcher Mark Bullock alerts us to the force of financial influence. "The role science will play in determining the quality of life for every human being on the planet is of course determined by the elite that funds science. In this way all scientific enterprise is embedded in the greater moral problem of how individuals and groups should conduct themselves" [2.10].[4] In short, follow the money.

However, following the money is not enough. All moral issues cannot be reduced to money. A more subtle concern must be addressed. We must ask whether values or morals or obligations are inherent to scientific research and technological development, or if the ethicist must add it. Alternatively, we might ask whether values are inherent to the subject matter science studies—namely, nature—so that, even if science is value-neutral, nature

---

[4]We are not working here specifically with scientific ethics; rather, we are working with terrestrial social and ecological ethics with the contributions of scientific research to formulate the issues. "The term *scientific ethics* may refer to the ethics of doing science" [2.11].

is not?[5] Might we find value concerns rising up either from nature itself or from the science we employ to explain nature? If the answer is affirmative in either case, then science and value come together in a single package; and to separate science from value would constitute an abstraction.

To say it another way: human life is fundamentally and inextricably embedded in nature; and this embeddedness is already value-laden. Therefore, when the scientific method excises only objective data from our already value-rich experience with nature—drawing a picture of nature as valueless—this amounts to an abstraction. The value-free conclusion of science is actually an assumption; it is a circular argument that does not account for our fundamental relationship of the human within nature.

Despite the abstractive component to this method, we will operate here with the hypothesis that ethics and what science learns about nature are co-original; they belong together in the relationship between nature, science, and the wider culture.

When nature herself emits value, the ethicist does not simply paint values over an otherwise neutral physical world. The ethicist needs to demonstrate that the values already at work in scientific discovery can be subjected to analysis, their presuppositions exposed and made available for ethical critique. With existing value assumptions then out in the light, the ethicist can coax the researcher toward self-conscious realism, authenticity, and care. Research scientists, in large part, concur. UNESCO rejoices in that "the world of scientific and technical research now regards ethical reflection as an integral part of the development of its own domain" [2.30]. An Astroethics of Responsibility will rely on the hypothesis that science and ethics belong intrinsically together; and we will see just how illuminating this exercise might be.

2.2.2.2   Religious Reliance on the Common Good

In addition to hypothesizing that ethics is inherent within the scientific interpretation of nature, should the scientific enterprise allow itself to be painted with a religious brush? Do ethicists working within a specific religious orientation have any right to speak to the direction taken by science?

Durham University astrotheologian and astronomer David Wilkinson [2.92] relies on the hypothesis that science and its subject matter, the natural world, already emit moral valence. With the prospect of meeting extraterrestrial life on an exoplanet, Wilkinson reminds us that "theology will want to stress

---

[5]Hans Jonas listens as nature speaks. Life says that it seeks to live; and Jonas hears this cry as evidence of a supreme good. "Without the concept of good, one cannot even begin to approach the subject of behavior. Whether individual or social, intentional action is directed toward a good [even a] highest good, the *summum bonum*." [2.35].

the importance of an ethical dimension in any contact with life-forms else-where in the Universe" [2.67]. Note that Wilkinson will not simply paint religious values over an otherwise value-free scientific discovery. Rather, as a theologian-scientist hybrid, he will recognize values as they arise from the new situation. It is inherent to the theological task to engage in ethical spec-ulation and moral commitment and, in this case, theological speculation responds to what we will learn through science about the natural realm.

With such a theological perspective, a Christian moral posture would be erected on God's plan for a promised new creation [2.67]. That new creation entails an important moral norm, namely, the "common good." A middle axiom that connects God's promised new creation to the present moment is our vision of the common good, symbolized biblically as the eschatological Kingdom of God.

Pope Paul VI defined the common good as "the sum of those conditions of social life which allow social groups and their individual members rel-atively thorough and ready access to their own fulfillment" [2.89]. Saint Pope John XXIII previously said clearly that "the attainment of the com-mon good is the sole reason for the existence of civil authorities" [2.94]. The Lutherans have extended the common good to embrace all of life with-out directly specifying extraterrestrial life.

> "Today, the meaning of 'common good' or 'good of all' must include the community of all living creatures. The meaning also should extend beyond the present to include consideration for the future of the web of life. The sphere of moral consideration is no longer limited to human beings alone" [2.29].

What the Holy See and the Lutherans make explicit is already implicit within naturally derived ethical insights, namely, the moral responsibility of governments and all persons of good will is to serve the common good.

Let me entertain two likely objections to this theological input into astroethics. First, one might object that this is only what Christians think. Christianity is only one religion among many; and adherents to other religious beliefs have their own ethical groundings and norms. To make matters more complicated, moral values are relative to one's culture and context and personal preferences. A Christian has no right to superimpose sectarian morality on those who do not freely affirm the same basic com-mitment. To put it another way: the ethical ground on which a Christian stands is shifting sand for the non-Christian.

Let me respond. Christians live in neighborhoods with Muslims, Jews, Hindus, Buddhists, agnostics, and atheists, among others. Despite this diversity, however, we all live in a shared planetary community, or at least within a worldview that makes a shared planetary community plausible. At some level, shared ethical commitments and responsible moral behavior are indispensable for any communal cooperation in pursuit of the common good.

When it comes to space ethics, we must think of the entire human race on the third planet from the sun as a single community of moral deliberation. "The world common good demands the existence of a world community, a planetary public consisting of all those affected by the actions of others on earth," writes Georgetown University political scientist, Victor Ferkiss [2.31]. What this implies is that public ethical reasoning—the ethical reasoning by Christian ethicists as well as ethicists coming from any other tradition—must be sufficiently transparent as to make sense for a collection of diverse perspectives while lifting up a vision of the common good. This is what I am attempting here, even if briefly.

A second objection might come from postcolonial or postmodern critics who decry the ideal of a single global community of moral deliberation. Why do they object? Because such an ideal of a universal humanity is a distinctively "modern" and, therefore, outdated idea. The problem is this: Such a global ethic would rely upon universal thin values apart from any tradition-specific thick values. The thick traditional practices of specific moral communities might become marginalized if we develop a single moral community and an astroethic based on the lowest common denominator. "Rather," comments Kenyan theologian Gavin D'Costa, "I would suggest that it is important to allow different communities to advance their own thick descriptions, and then to work with what arises at that point" [2.23]. The assumption at work here is that thin values unite while thick values divide. Is this hopeless?

In response to D'Costa, I suggest that an Astroethics of Responsibility should enlist each existing tradition-rich moral community into a common planetary endeavor. This shared endeavor would then deal with a matter that concerns all earthlings equally, namely, our planet's relation to what is beyond our planet. Given the existing competition between moral groups on Earth, and given the urgency of a unitary approach, perhaps a thin yet shared set of moral values would be the best we could practically ask for. A distinctively modern ethic that avoids top-down tyranny and encourages bottom-up participation would provide the requisite foundation for a planetary approach to space ethics [2.25].

2.2.2.3   A Secular Grounding for Astroethics?

We have just reviewed a distinctively Christian approach to ethics based upon a vision of the common good; and we have reviewed some religious alternatives. What about an avowedly non-religious or secular public agenda for astroethics? One admirable attempt is offered by Kelly C. Smith of Clemson University.

How does Smith answer our grounding questions? Who are we? According to Smith, we are human beings who have evolved into social creatures and have constructed culture. What do we value? We value reason. Smith describes the human race as "ratiocentric." This leads to a "package deal" in which the "sociality-reason-culture triad (SRCT) is the proper basis for intrinsic moral value" [2.78]. What should we do? We should search for the extraterrestrial equivalent of ourselves—for SRCT aliens—and then expect that both the aliens and ourselves would share the same basic ethical structure. "The SRCT linkage may be so strong that it constitutes a universal property of other intelligent species in the universe" [2.78] [2.68].

Rather than ol' fashioned manifest destiny, Smith advocates "manifest complexity," a doctrine according to which rational intelligence should come to rule in the universe. Our purpose as rational beings is to "maximize universal complexity" [2.78], something SRCT aliens would understand and share with us.

This is a secular ethic grounded in a selected human trait, rational capacity or intelligence. Reason is lifted up as morally foundational, not an uncommon move in our post-Enlightenment era. In principle, any ethicist could lift up any trait to serve as the ground for intrinsic value. Christian ethicists are not likely to embrace Smith's ratiocentrism. Much higher on the Christian value scale than reason are affections such as compassion, love, service, and justice. For the Christian, mentally disabled or socially inept or culturally marginalized persons each have dignity; each has intrinsic value and should be treated as a moral end and not merely a means. Smart humans are not more valuable than dumb ones on the Christian scale of values. This commitment to universal human dignity is grounded in God's conferral of dignity on us in the incarnation. In sum, Smith's astroethic may seem persuasive only to that segment of society which is already ratiocentric.

Despite this demur, Smith has provided a laudably coherent argument, in my judgment. The fact that Smith's program seems arbitrary is by no means his fault. It is due to the fact that we live in a pluralistic world with moral relativity. It is difficult to ground any ethic that could be universal in our situation. Therefore, what a theological ethicist should do is promulgate a religiously grounded point of departure and then look for overlap with others representing differing perspectives toward the end of seeking sufficient common grounds to launch a single global community of moral deliberation. Smith's naturalistic justification for valuing life provides an enticing overlap if not partnership.

As just suggested, a secular astroethics of responsibility is likely to seek some level of grounding in one or another form of naturalism. The ethicist will ask nature to provide foundational insights for human valuing. This method risks committing the naturalistic fallacy—that is, drawing a moral "ought" out of an already existing "is." Yet, without a transcendent grounding in the divine, nature is about all that is left to undergird an ethical vision.

Kelly Smith and Steven Dick, along with Hans Jonas, bring to the astroethical construction site a welcome supply of nature's building materials. When we pour an aggregate of naturalistic gravel and theological stones into the concrete, the resulting foundation will be firm. "While theology can provide potentially universal principles such as compassion and dignity that will be useful in the context of astroethics, the problematic naturalistic fallacy should not stand in the way of secular ethics playing an important and perhaps predominant role" [2.26]. In short, theological and secular ethicists should link arms in constructing a superstructure of quandaries that lead to fitting moral responses.

2.2.3   Third Foundational Question: What Should We Do?

2.2.3.1   From Quandary to Responsibility

What should we do? We have already asked: Who are we? And we asked: What do we value? Now, to answer this third question—What should we do?—we draw on a pair of concepts: quandary and responsibility. No doubt communities and traditions find themselves frequently confronted with a quandary accompanied by a sense of responsibility. Such communities must work through the quandary by drawing practical applications out of their fundamental ethical orientation. Perhaps we can build on this common phenomenon as a foundation for multicultural ethics. We will call it the "Quandary-Responsibility method" within an Astroethics of Responsibility.

First, to demonstrate, today we are confronted by a quandary: How should we think ethically about the prospect of sharing the cosmos with space neighbors? Roman Catholic ethicist Charles Curran provides a framework. "Quandary ethics deal with concrete, objective human situations. In addition, it is here that human reason, science, and human experience predominate" [2.22]. This quandary regarding space exploration and ETI is not religion-specific. It does not begin with dogma and then seek application; rather, it begins with an astroethical question and then surfs the tradition for a helpful answer.

Second, we approach the quandary with a sense of moral responsibility. As the etymology of the Latin, *respondere,* meaning *to answer,* suggests, responsibility ethics answers questions raised by our changing situation [2.37]. Responding or answering belongs to the "primordial experience of the Judeo-Christian-Islamic tradition: a call from God that human beings accept or reject" [2.50]. In the present discussion, what calls to us for a response or answer is the quandary posed by new space knowledge.

The Quandary-Responsibility method provides a keeled river raft to navigate the rapids, a stable boat to ride the rushing whitewater of science, technology, and social change. When the quandary is prompted by a situation that lies beyond Earth and more than likely will affect our entire planet, then perhaps we need to think of a single planetary moral agent. It is our global community that should be morally responsive and responsible. To work with the notion of responsibility for earthlings and to work with the idea of a shared commons in space, we will need to commit ourselves to a vision of universality. The universe requires universality.

2.2.3.2   *From Space Sanctuary to Galactic Commons*

Earlier we introduced Arnould's proposal to treat outer space as a sanctuary; and we introduced the Roman Catholic notion of the common good. In extending these ideas [2.66] we ask: Is it fitting to think of circumterrestrial space as a commons, as belonging to us all and not to any person or nation in particular? [2.62]. Boston University theologian John Hart would answer in the affirmative. "The sacred cosmic commons is a communion of commonses cosmically interrelated and integrated. It is stardust become spirit; it is atoms become life and thought, all in the presence of a transcendent-immanent Being-Becoming, creating Spirit" [2.33].

Let's return for a moment to what the United Nations has said. "The exploration and use of outer space, including the moon and other celestial bodies, shall be carried out for the benefit and in the interests of all countries, irrespective of their degree of economic or scientific development, and shall be the province of all mankind" [2.85]. This has come to be known as the Common Heritage of Mankind Principle or CHP. The principle "confers on a region the designation of *domino util* or beneficial domain that should be legally defined as a *res communis humanitatis,* a common heritage that is not owned by any nation, but from which all nations may garner profits and benefits" [2.71]. Rather than national interests, the UN works with planetary interests. What about the interests of extraterrestrials?

Hart lifts up for us the ethical norm of a cosmic commons which takes into account the interest of the extraterrestrials.

"The *cosmic commons* is the spatial and local context of interactions among corporeal members of integral being who are striving to meet their material, spiritual, social, and aesthetic needs, and to satisfy their wants....The cosmic commons includes the aggregate of goods which, beyond their intrinsic value, have instrumental value in universe dynamics or as providers for the well-being of biotic existence. In the cosmic commons, goods that will eventually be accessible on the moon, asteroids, meteors, or other planets should prove useful to humankind, to other intelligent life, and to biokind collectively" [2.32].

Might Hart's notion of a cosmic commons help us move forward from quandary to responsibility? Yes.

Perhaps some elements of the Roman Catholic concept of the "common good" could bleed over into our concept of the cosmic commons. "The common good as sum of the goods possessed by many and directed toward the utility of individuals," writes Sergio Bastianel, "will be the *common reaching out* to realize a way of living together that can be accurately called *communion*" [2.6]. Or, the common good "indicates an *ultimate goal* of society, its *utopia*, in such a way that the intermediate aims will be critically evaluated in their being conformed toward such an aim of communion" [2.6]. For the near future, the commons will be shared by all of us who live on planet Earth. Perhaps in the more distant future, after we will have encountered extraterrestrial life and incorporated that life into our commons, the community of moral discernment will broaden. In short, our ethical vision directs our gaze toward a future communion shared by earthlings and spacelings.

Hart is not alone in proffering the idea of a cosmic commons as an ethical category. Like Hart and Dick, Mark Lupisella at NASA's Goddard Space Flight Center proposes a "cosmocentric ethic," which he contends "may be helpful in sorting through issues regarding the moral considerability of primitive extraterrestrial life as well as other ethical issues that will confront humanity as we move into the solar system and beyond" [2.42].

Arnould uses the term "Greater Earth" to communicate the same basic notion, although perhaps more limited in space. "Greater Earth defines the area, the space territory that surrounds the Earth and where most future space activities could take place" [2.3]. The Greater Earth within our solar system would host economic activities and provide the sphere of terrestrial moral responsibility. What we see here is a growing convergence toward the vision of a cosmic commons—whether called "Greater Earth" or a "cosmocentric ethic"—that makes the entire human community on Earth responsible for ethical deliberation and includes in our sphere of moral responsibility everything in space we can influence.

My preferred term, as I've already indicated, is "galactic commons" or "galactic common good."

2.3   Astroethical Quandaries Arising Within the Solar Neighborhood

The discipline of astroethics responds to earthlings going to space whether or not we meet new extraterrestrial neighbors. "Clearly, the 'holy grail' of astrobiology would be the actual discovery of life elsewhere in the Universe, and such a discovery would have profound scientific and very likely also philosophical and societal implications," exclaims Ian Crawford; "Needless-to-say, there will also be significant scientific and philosophical implications if extraterrestrial life is not discovered, despite ever more sophisticated searches for it" [2.20]. If "astrobioethics" within "astroethics" would focus directly on quandaries arising from engaging off-Earth life, then the more encompassing category of "astroethics" would include matters that may or may not involve off-Earth life forms.

We turn now to ten astroethical quandaries arising within our solar neighborhood, quandaries already familiar to astrobiologists: (1) planetary protection; (2) intrinsic value of off-Earth biospheres; (3) application of the Precautionary Principle; (4) space debris; (5) satellite surveillance; (6) weaponization of space; (7) scientific versus commercial space exploration; (8) terraforming Mars; (9) colonizing Mars; and (10) anticipating natural space threats [2.65].

2.3.1   Does Planetary Protection Apply Equally to Both Earth and Off-Earth Locations?

Our mandate for Planetary Protection (PP) is made clear in Article IX of the 1967 UN Outer Space Treaty. "Parties to the Treaty shall pursue studies of outer space including the Moon and other celestial bodies, and conduct exploration of them so as to avoid their harmful contamination and also adverse changes in the environment of the Earth resulting from the introduction of extraterrestrial matter and, where necessary, shall adopt appropriate measures for this purpose..." [2.85]. This version of the PP principle seems to make it clear: we are morally obligated to protect both Earth and other celestial bodies.

The risk of contamination goes in two directions, forward and backward. The possibility of "forward contamination" alerts us to the risk of disturbing an already existing ecosphere; the introduction of Earth's microbes carried by our spacecraft or equipment could be deleterious to an existing habitable environment. During the SARS-CoV-2 pandemic, NASA issued two interim directives (NIDs) to protect both the Moon and Mars from forward contamination brought by earthlings [2.53]. "Back contamination" would occur if a returning spacecraft brings rocks or soil samples that contain life forms not easily integrated into our terrestrial habitat. When Apollo 11 returned to Earth in 1969 from its Moon landing, it brought Moon Rocks which President Richard Nixon divided into rice grain sized pieces and distributed to each of the 50 states and 135 nations as souvenirs. To this date, no earthling has suffered from back contamination.

Are we earthlings equally concerned about both our own planet and each off-Earth site? Not in practice. In practice, prevention of backward contamination trumps protection against forward contamination. Although forward contamination is a matter of concern, some forward contamination is permissible, opine NASA scientists Catharine Conley and John Rummel. What is not at all permissible is backward contamination. Preventing harmful contamination of the Earth must be of the "highest priority" for all missions [2.73].

Even so, PP in principle includes moral concern for protecting off-Earth bodies from Earth's contamination. "Planetary protection covers explicitly the search for extraterrestrial life and also the potential for Earth life to interfere with future human objectives" [2.17] [2.18]. As we proceed to refine astroethics, we should ask: Might one or another off-Earth biosphere hold intrinsic value? And, if so, might this heighten our responsibility for protecting it from contaminants from our home planet?

2.3.2   Does Off-Earth Life Have Intrinsic Value?

In 2020 a possible biosignature was spotted on Venus. It is not life itself, to be sure. Rather, it is a spectral fingerprint, a light-based signature of phosphine in Venus' harsh sulfuric atmosphere. Even if there is no life on the 900-degree surface, perhaps microbes make their home in the Venusian clouds.

NASA scientists have long given credence to the panspermia hypothesis, namely, that a wandering asteroid entered our solar system during planet formation and seeded both Mars and Earth with life. This means life on Mars and Earth would turn out to be sisters, so to speak. But life on Venus and Earth would be strangers.

On the one hand, if the panspermia hypothesis gets confirmed, we could work with the assumption that the universe underwent a single origin of all life. This would make Martian life like Earth life. On the other hand, if the Venusian hypothesis gets confirmed, then this would mark a 2nd genesis of life. The sulfuric origin of life on Venus would likely differ from the carbon-based life that has evolved on Earth. Would it matter ethically if extraterrestrial life shares our genesis or derives from a second genesis?

Regardless of whether it is due to a shared genesis or a second genesis, astrobioethicists to date have been leaning toward ascribing intrinsic value to off-Earth biospheres. But they have not leaned far enough to tip completely. With the term "intrinsic value," we intend "value that is truly independent of valuing agents" [2.43]. At work is a widespread assumption that intelligent life would warrant intrinsic value but non-intelligent life would not. Is this assumption sound?[6]

---

[6]Intelligence establishes moral status. Unless you are Erik Persson, bioethicist at Lund University in Sweden.

Persson appeals to sentience within the larger category of life. "According to sentientism, one has to be sentient to have moral status whether terrestrial or extraterrestrial and whether biological or nonbiological [such as post-biological].... The most plausible theory for moral standing seems to be sentientism that connects directly to the basic idea behind modern ethics: that ethics is about dealing with situations where one's own actions affect others in a way that matters to them....If we accept sentientism, microbial life and plants do not have moral status, but there are reasons for protecting someone or something other than being a moral object" [2.55]. In my judgment, sentience will not work as a general ethical category, except for vegetarians. Here on Earth we have already committed ourselves to eating meat. Meat-eating requires the death of sentient creatures. We discriminate between pets, which we do not eat, from stock, which we do eat. Vegetarians object to this practice on moral grounds, on the grounds that we have an equal responsibility to all sentient creatures. If we are to export to extraterrestrial realms a categorical respect for all sentient organisms, then for the sake of consistency we would need to adopt vegetarianism back at home. A consistent ethic based upon sentience would require vegetarianism on Earth as well as on all space expeditions.

Sentience will not help for another reason. To date, those contributing to this discussion have drawn on ethical precedents set by environmentalists and eco-ethicists. This ethical posture is oriented holistically toward entire ecosystems, toward protecting entire habitats with their resident living creatures regardless of level of sentience or intelligence. This holistic approach seems intuitively relevant to what we might discover on Mars or a moon orbiting Saturn. Once engaged, we would not discriminate between one species on behalf of another species. Rather, we would assume we are responsible for each entire biosphere with its already established life forms. Entailed in a holistic commitment to an entire ecosystem is an indispensable level of commitment to simple life forms and even to abiotic contributors to this ecosystem.

In sum, we may have to live for a period with a generic respect-for-life's-intrinsic-value principle until we have entered into actual engagement with extraterrestrial life forms. At that point we will rearticulate the quandary and reformulate our responsibility. By no means is this a form of kicking the ethical can down the road. Rather, we are simply marking specific areas where we will need to respond to actual rather than hypothetical situations.

This is our quandary: Does life—even unintelligent life—have intrinsic value? Or, does the value of living organisms depend on the usefulness they have for us? Is worth inherent or instrumental? Do we terrestrial *Homo sapiens* have a responsibility toward extraterrestrial life based upon that life's intrinsic worth or based strictly upon its usefulness to us? [2.64]. Almost no one to date has risen up to defend a brute instrumentalism. The predominant discussion takes place within the intrinsic value option.

Richard Randolph and Christopher McKay believe "that new operational policies for space exploration and astrobiology research must be developed within an ethical framework that values sustaining and expanding the richness and diversity" [2.49]. This applies to entire biospheres, not merely individual organisms. Even so we ask: Just how close to ascribing intrinsic value is this?

Taking a minimalist position, Charles Cockell acknowledges that extraterrestrial microbial life will make some level of demand on us earthlings: "Telorespect or teloempathy merely captures our recognition that extraterrestrial life, including life independently evolved from the biology that we know on Earth, placed demands on our behavior if we think it has intrinsic value" [2.16].

Kelly Smith moves one notch closer toward intrinsic value, distinguishing off-Earth species from what we know on our home planet. With the label "Mariophilia," Smith posits that extraterrestrial "life would be extremely valuable and should be defended against *petty* demands of human beings, but also that human interests can in principle trump those of Martians" [2.79]. With this conditioned appeal to intrinsic value, terrestrial interests still trump extraterrestrial interests.

Octavio Chon-Torres goes beyond Smith. "The proposal that I have presented would include safeguarding the 'rights' of the Martian life to exist, that is, having an intrinsic value. And why not? Every form of life follows Darwinian mechanics and seeks to develop, insofar as it has that 'interest' has a value in itself. A separate question is whether the human being wants to respect it" [2.14].

We get some help from Oxford ecotheologian Celia Deane-Drummond, who would be satisfied with ranking value. "It is possible to hold to the notion of intrinsic value, while also being able to discriminate between different forms of life and non-life in terms of their worth" [2.24]. Or, to say it another way, even if we impute intrinsic value to all living things, within this broad category we may identify some living things to be of greater value or worth.

Even so, the astrobioethicist will ask: How do we decide? Without appealing to instrumental criteria for discriminating between greater or lesser worth, we should look for criteria within the scope of intrinsic value. Regardless of the answer, an astroethics of responsibility will enjoin us earthlings to care for off-Earth microbes and their respective biospheres.

2.3.3   Should Astroethicists Adopt the Precautionary Principle?

Earth's ecologists are used to debating and embracing the precautionary principle. Might astroethicists borrow it? The astroethical principle might look like this: when in doubt, protect off-Earth life in its respective biosphere [2.65].

The so-called Wingspread definition of the precautionary principle was formulated at the 1998 United Nations Conference on Environment and Development: "When an activity raises threats of harm to human health or the environment, precautionary measures should be taken even if some cause and effect relationships are not fully established scientifically" [2.86]. In this context the proponent of the process or product, rather than the public, should bear the burden of proof.

When space scientists and ethicists met at Princeton for a COSPAR workshop in 2010, they embraced a variant formulation: "we define the *precautionary principle* as an axiom which calls for further investigation in cases of uncertainty before interference that is likely to be harmful to Earth and other extraterrestrial bodies, including life, ecosystems, and biotic and abiotic environments" [2.19]. In sum, employment of the precautionary principle for space exploration provides the kind of middle axiom that connects the larger value of life with practical policies that facilitate off-Earth explorations.

2.3.4   Who's Responsible for Space Debris?

According to NASA's count, 22,000 pieces of space junk in the form of defunct human made objects are orbiting Earth. We have turned our upper atmosphere into a trash dump for nonfunctioning space craft, abandoned launch vehicle stages, and fragments of unusable satellites. Do we want to pollute circumterrestrial space just as we have befouled our terrestrial nest? [2.45] [2.77].

The problem with our orbiting landfill is not merely that it is ugly. It is also dangerous. It risks danger to future space flights and future satellites. The Kessler Syndrome, named after NASA scientist Donald J. Kessler, proffers a scenario: debris in low Earth orbit (LEO) may become sufficiently dense that collisions between objects could cause a cascade where each collision would generate smaller particles in increased number; and the increased number of objects would then increase the number of collisions. The cascade would pulverize anything that comes within its region. The LEO region would become impassable for future launch vehicles.

To date, no one has been held financially responsible for space junk. Those who make profits or who otherwise gain from sending this material into space are not required to recycle or dispose of their waste. Space waste accumulates, but nobody is required to pay for cleaning it up. Nations or corporations treat the Greater Earth as their ashtray, as a public trash dump. Follow the money.

If we define Greater Earth as a part of the galactic commons, then we find ourselves already beset with a classic moral problem: those with power and influence utilize common space for their own profit while the population as a whole absorbs the cost of deterioration or degradation of what is publicly shared. If and when Earth's planetary society consolidates its diversity into a single community of moral deliberation, then responsibility will need to be parsed and parceled according to a renewed principle of distributive justice.

The European Space Agency has set up a Space Debris Office to coordinate research activities in all major debris disciplines, including measurements, modeling, protection, and mitigation, and coordinates such activities with the national research efforts of space agencies in Italy, the United Kingdom, France and Germany. Together with ESA, these national agencies form the European Network of Competences on Space Debris. In parallel, the Japan Aerospace Exploration Agency (JAXA) is testing to see if a tethering technique might begin the process of debris-gathering. What we are missing is a planetwide public policy regarding fiscal responsibility on the part of spacefaring parties.

2.3.5   How Should We Govern Satellite Surveillance?

Earth's residents are losing their privacy faster than politicians lose their scruples. The telescopes on board reconnaissance satellites are pointed toward Earth, not toward the stars. Mission tasks include high-resolution photography; measurement and signature intelligence; communications eavesdropping; covert communications; monitoring of nuclear test ban compliance; and detection of missile launches. With the improvements in technology, today's spy satellites have a resolution capacity down to objects as small as ten centimeters. Surveillance satellites also provide us with efficient communications, weather reporting, Google maps, and many more public services.

Satellite spying is international, not just national. The Echelon spy network coordinates satellite snooping by the governments of the United Kingdom, the United States, Canada, Australia, and New Zealand. The Echelon network spies, sorts, decrypts, archives, and processes three million telephone calls transmitted by satellite every minute. The United States government sells pictures taken by satellites; but it keeps certain subjects from public review. Sensitive facilities such as military installations are restricted, as are remote pictures taken over Israel. Similarly, private companies use satellites for remote sensing and sell their pictures.

"Can a State gather information about the natural riches and resources of another sovereign State without having obtained the latter's prior agreement?" asks Arnould. "Is it not up to the remote sensing State to ask for the prior permission of the State [2.81] whose territory is being observed?" [2.3]. This sounds like a reasonable ethical question. Yet, it presupposes the present situation of sovereign nation states, a political system that may have made sense prior to the current thrust toward economic and technological globalization. Satellite surveillance and communication services, right along with other space activities, are playing into an emerging planetary consciousness.

Protecting national boundaries from foreign intelligence or even public transparency may soon be an artifact of history, an era we remember but no longer live in. Perhaps the way forward is to support an ethic of maximal "information without discrimination." Rather than attempt to police information gathered from remote sensing, it would be healthier and easier to prevent such information from deleterious usage.

2.3.6   Should We Weaponize Space?

Should nations weaponize space? Should militaries establish orbital beachheads from which to launch attacks? No. At least according to the 1967 United Nations Outer Space Treaty, which stresses that celestial locations could be used "exclusively for peaceful purposes." The treaty explicitly prohibits the "placing in orbit around the Earth any objects carrying nuclear weapons or any other kinds of weapons of mass destruction."

In 2020, the United States formulated the doctrine to guide its new Space Force.

> "Military space forces are the warfighters who protect, defend and project spacepower. They provide support, security, stability, and strategic effects by employing spacepower in, from, and to the space domain. This necessitates close collaboration and cooperation with the U.S. Government, Allies, and partners and in accordance with domestic and international law" [2.80] [2.70].

Because of the inability of the UN to enforce its rule, regulations of military equipment in space are today the responsibility of unilateral, bilateral and multilateral agreements, not the United Nations. No global community of moral deliberation exists. At least not yet.

"For modern warfare, space has become the ultimate high ground, with the U.S. as the undisputed king of the hill," writes Lee Billings [2.7]. "China and Russia are both developing capabilities to sabotage crucial U.S. military satellites" [2.7]. One can only imagine a skirmish that could lead to Star Wars or, more precisely, Satellite Wars.

2.3.7   Which Should Have Priority: Scientific Research or Making a Profit?

Are we about to witness in space the equivalent of a gold rush? An economic and political frenzy for gaining dominance in space may break loose over the next decades. The telecommunications industry is already accustomed to the cost-effective use of satellites. We are on the brink of an era of space tourism, with the first trips to suborbit and low orbit vacations in the planning stages. Visits to the Moon will most likely follow. Establishing research laboratories on the Moon and Mars are being envisioned as is the mining of asteroids [2.91]. Might we be wise to ready ourselves for an El Dorado type of gold rush to the new extraterrestrial world? If so, should we try to put policies and policing mechanisms in place in advance?

Up until this point we have thought of outer space as a sandbox for Earth's scientists to play in. Governments have found the money to fund modest exploratory adventures; and scientists have organized to conduct experiments which have yielded an abundant harvest of new knowledge about our cosmos. Frequently, scientific goals have been mixed with military goals, because leaders in the military have been willing to share their budgets for scientific purposes.

Scientific experiments do very little damage, if any. Somewhere on the Moon is a golf ball left by visiting astronauts. Landing on Mars or on Titan has not infected or contaminated anybody's ecosystem, as far as we know. NASA decontaminated its first Mars lander, but more recently NASA has saved the money spent for decontamination under the assumption that a little contamination of Mars doesn't matter. The impact on our solar system by scientific activity is benign.

This situation is about to change. The private sector is now ogling space for profit. What about space tourism? Simply flying a few wealthy passengers high enough to experience weightlessness is not likely to provoke anyone's moral ire. But, what about tour busses roaming the surface of the Moon? Busses will leave tire tracks. Perhaps trash. No doubt tourists will want to visit that golf ball as well as historical sites where astronauts first landed. Will the crowds of visitors damage those sites? Are those sites sacred? Protectable? Who will decide and what will be the criteria by which they decide?

The market does not always react the way the marketers predict. Low cost and frequent flights to suborbit heights might actually encourage increased participation by scientists. These scientists will want to do research on the "ignorosphere." The ignorosphere is a level just above balloon traffic but too low for satellites. Scientific researchers might buy tickets with the tourists and then look out the windows [2.81].

2.3.8   Should We Earthlings Terraform Mars?

Should we earthlings terraform Mars? Or, any other planet or moon, for that matter? Let's ask two theologians, one Buddhist and one Christian, and then ask a NASA astrobiologist.

Francisca Cho, Associate Professor of Buddhist Studies at Georgetown University, raises the quandary: Should earthlings terraform Mars? "A Buddhist would apply neither an intrinsic nor instrumental value of life or nature to the question of terraforming Mars. The idea of an intrinsic value would go against the principle of emptiness. Instrumental value, on the other hand, would be problematic because one could not ensure that the instrumental objectives and the proper motivations.... There is no intrinsic worth to nature but neither is there intrinsic worth to human beings.... There is no option between them, so you have to transcend that framework all together" [2.12]. From a Buddhist perspective, neither an appeal to the intrinsic value of life nor an appeal to life's utilitarian value to human beings provides ethical guidance for the terraforming question.

Now, let's ask Christian theologian Cynthia Crysdale. "We need to think of ourselves as living within an ethic of risk, not an ethic of control. I say this in direct reference to the actions we take in terraforming or colonizing or exploring other planets. My caution is to point out that the conditions of possibility that we establish in the hopes of one outcome may at the same time establish conditions under which totally unforeseen schemes of recurrence become established" [2.21]. Dr. Crysdale has wisely invoked the Precautionary Principle based upon her observations about human nature—that is, human sinfulness. No ethical justification could suffice without acknowledgement of who we are as humans, including our human proclivity to mess things up. Nevertheless, anticipating the unforeseen damage we humans are capable of is a principle one must incorporate into any such project, not merely going to Mars.

NASA's Christopher McKay provides ethical justification for his plan to terraform the red planet [2.48]. McKay hypothesizes that Mars is lifeless. At least it is lifeless today. The red planet may have been home to life in the past; but Mars must have lost its atmosphere and its ability to sustain life for reasons yet unknown. Its thin atmosphere is replete with carbon dioxide, but not oxygen. Let us speculate: Suppose we would transplant living organisms from Earth that take in carbon dioxide and expel oxygen into the atmosphere? Then, when enough oxygen suffuses the atmosphere, we could introduce oxygen inhaling organisms that expel greenhouse gases. These greenhouse gases would warm up Mars, and life would thrive. A self-regenerating ecosystem could run on its own. In less than a century, estimates McKay, we could establish a biosphere that would last ten to a hundred million years.

McKay calls this terraforming project "planetary ecosynthesis." Is such an ambitious plan ethically justifiable? Yes. McKay starts with a simple axiom: life is better than non-life. If life is better than non-life, says McKay, then it would be our moral responsibility to sponsor ecosynthesis on that planet. Transferring terrestrial life forms to Mars would be better than leaving Mars lifeless.

Curiously, McKay appeals to both intrinsic value and instrumental or utilitarian value when justifying planetary ecosynthesis. First, the intrinsic argument. Because life has intrinsic value, Mars with life would be ethically of greater worth than a lifeless Mars, even if it is transplanted life. Second, the instrumental argument. Because we on Earth would learn so much from the Mars project about sustaining a biosphere, we could apply what we learn on Mars to sustaining Earth's biosphere in the face of our imminent ecological challenges. "Both utilitarian and intrinsic worth arguments support the notion of planetary ecosynthesis" [2.46].

Should we terraform Mars or any other celestial body within our solar neighborhood? On the one hand, McKay's argument that life is better than non-life provides a sound point of departure. On the other hand, transplanting terrestrial life to an extraterrestrial location looks a great deal like colonizing. As we bring the history of terrestrial colonization to mind, we cannot avoid recalling the imperialism and greed that motivated colonization and the devastating impact of exploitation and genocide on the lands colonized. The Crysdale incorporation of risk based upon what we know from history about human nature gives one pause.

Our pause cannot last too long. The Mars Society is already making plans to colonize the red planet.

2.3.9   Should We Establish Human Settlements on Mars?

Should we earthlings become a transplanetary species? Should we begin establishing human settlements on Mars? [2.44].

Colonize Mars? Yes, says Robert Zubrin, director of the Mars Society, because it's our destiny. "Mars can and should be settled with Earth émigrés" [2.96]. No, cautions NASA consultant Linda Billings, because colonization would exacerbate terrestrial inequality. "It would be unethical to contaminate a potentially habitable planet for further scientific exploration and immoral to transport a tiny, non-representative, subset of humanity—made up of people who could afford to spend hundreds of thousands to millions of dollars on the trip—to live on Mars" [2.8]. Stealing resources from the lower classes to send the wealthier classes to Mars would violate the principles of distributive justice.

Some critics argue that, because we humans have messed up Earth, it would be immoral to do the same to Mars. Adler Planetarium astronomer Grace Wolf-Chase admonishes us to clean up Earth's mess before we mess up another planet. "Considering the possibility of extraterrestrial species motivates us to re-evaluate humanity's history as stewards of Earth, and to examine critically human behaviors before migrating to other worlds" [2.93].

Would colonization be legal? The UN Outer Space Treaty, recall, holds that Mars, like other celestial locations, cannot be subject to national appropriation by claim of sovereignty, by means of use or occupation. Might the USA or China—whichever country lands first—simply stake a claim? "I don't see how Mars could be anything but a land grab driven by homesteading rules," pines Christopher Wanjek [2.91].

Among the many quandaries that the prospect of colonization raises is this: will earthlings living on Mars still be earthlings? Or, will evolution require such a level of adaptation that humans will become posthuman? Polish scientist Konrad Szocik along with his colleagues entertains this quandary and responds by recommending CRISPR gene editing to enhance adaptation. "It is worth keeping in mind that living in different—let's call them unnatural places, which are not a part of the environment of evolutionary adaptedness—locations is not problematic per se, if humans are prepared in an appropriate way to live there" [2.82]. Astroethicists of responsibility must engage the question: Is it our moral right if not obligation to engineer the future of human evolution so that a successor species—a posthuman species—emerges?

Here is my tentative response: if a biosphere exists on Mars, then we should treat it as having intrinsic value. But if Mars is currently lifeless then, despite the interplanetary necessity for genetic engineering, we should take advantage of the opportunity to seed the Red Planet with life for the sake of its future. This becomes the moral warrant for both terraforming and colonization.

What about the mega vision of extending the habitat of *Homo sapiens* to outer space, turning earthlings into a transplanetary species? I applaud such a grand vision. I only add two grumbles, one scientific and the other theological. Scientifically, once Earth's colonists have adapted to a significantly new off-Earth environment, their descendants may no longer be human. We will not be able to say confidently that a single transplanetary species has come into existence. Theologically, we ought not to expect a utopian life to commence on a new planet. We ought not to expect we will create El Dorado or a heaven in the heavens. Earth's colonists to new worlds will bring with them a very ancient yet perduring pattern of living: sin.

2.3.10   How Do We Protect Earth from the Sky?

Even with colonists migrating from the third to the fourth planet, the vast majority of earthlings will remain living on Earth. Earth will continue to be our home for the foreseeable future. The ecological-ethical mandate is clear: if we *Homo sapiens* do not get our act together we'll go extinct before Martian microbes will.

Earth is a dangerous place to live. The heavens threaten. The Sun occasionally launches solar flares, which fry electricity grids by generating intense currents in wires. More rare than solar flares but equally potent are blasts of radiation from a nearby γ-ray (gamma ray) burst. A short-hard γ-ray burst, caused by the violent merger of two black holes or two neutron stars or a combination, provides the most frightening scenario. If one such blast would be directed at Earth from within 200 parsecs away (less than 1% of the distance across the Milky Way), it would zap Earth with enough high-energy photons to wipe out 30% of the atmosphere's protective ozone layer for nearly a decade.

There are more threats coming from Earth's heavens. Comets and asteroids, when large in size, can explode on Earth's surface with the impact of a nuclear bomb. It is widely believed among scientists that sixty-five million years ago an asteroid ten kilometers in diameter hit Earth and triggered the mass extinction of dinosaurs. Can we protect Earth from future asteroid catastrophes? The UN's Science and Technical Subcommittee's Near-Earth Object Working Group and its internal panel, Action Team 14, have been working on the details of an international approach since 2001.

The astroethical response to possible and probable futures is to prepare. These damage scenarios lead us to think ahead. We need to plan for our planet's future, and we need to incorporate such possibilities into our planning. With regard to solar flares, fortunately, there are ways to mitigate the damage should it occur: engineers can protect the grid with fail-safes or by turning off the power in the face of an incoming blast. With regard to a comet or asteroid strike, we will be given advanced notice. A diversion strategy could be effective, perhaps by hitting the object while it is yet far away with a nuclear bomb. We have no way to prevent gamma ray bursts from striking our Earth, but we could provide protective shields in sanctuaries for life forms we wish to restart following the event. These matters belong to our quandary. Just how will we respond?

2.4   Levels of Intelligence in the Milky Way Metropolis

Is it likely that yet-to-become neighbors are already living in our Milky Way metropolis? Yes, indeed. "A conservative estimate," speculates University of Arizona astrobiologist Chris Impey, "might be a billion habitable 'spots'— terrestrial planets in conventionally defined habitable zones, plus moons of giant planet harboring liquid water—in the Milky Way alone. That number must be multiplied by 10^{11} for the number of 'petri dishes' in the observable cosmos" [2.34].

When we meet them, will they be like us? No, says philosopher of biology Michael Ruse. Something like us, perhaps; but not us. "It seems that natural selection can and does produce intelligent beings all the way up to humans. I confess that even if this can happen, I would think selection would more likely produce humanoids—beings like humans but not necessarily identical to us. There might be at least as many Wookies in the universe as there are humans" [2.75].

The first question the ratiocentric astrobiologist will ask about our extraterrestrial contacts will be this: Are they intelligent? If so, just how intelligent? We mean it when we employ a term such as ETI, extraterrestrial intelligence.

In anticipation of contact with alien intelligence in a form that reasonably resembles *Homo sapiens* on Earth, perhaps we should consider engagement with three possibilities: extraterrestrial biotic individuals who are inferior to us (intellectually less intelligent), our peers (equal in intelligence), and superior to us (more highly intelligent) [2.58]. Each of these three categories implies a different set of moral responsibilities [2.58].

Curiously, for decades prior to the advent of the field of astrobiology, astronomers and science fiction writers measured the variety of extraterrestrial beings according to scales of intelligence. The mere existence of astrocognitionists among astrobiologists demonstrates the preoccupation we have with intelligence. "The multidisciplinary field of astrocognition," according to David Dunér, "could be generally defined as "the study of the origin, evolution and distribution of cognition in the Universe," or simply "the study of the thinking Universe" [2.28]. Until recently, coffee conversation among astrobiologists distinguished between stupid microbial life, on the one hand, and intelligent or even super-intelligent aliens, on the other. Elsewhere I have argued that *all life is intelligent, even microbial life; and what we are dealing with are relative levels of intelligence* [2.61].[7]

Regardless of my position on continuity of intelligence, the astroethicist must speculate about possible and probable ETI scenarios. We may make new galactic friends with beings inferior to us in intelligence, equal to us in intelligence, and even superior to us in intelligence. Some might be hostile. Some might be friendly. Others might even be benevolent. Each possible extraterrestrial scenario would shape our terrestrial response and our responsibility. In what follows we will get specific about each scenario.

---

[7]David Dunér expands on my work, adding *intersubjectivity* to my list of traits of intelligence. "A recent attempt to define intelligence in connection to extraterrestrial life and evolution of intelligence is more elaborated. Ted Peters defines intelligence in terms of seven traits: interiority, intentionality, communication, adaptation, problem solving, self-reflection, and judgment [2.61]. Even microbes exhibit the first four traits; humans, along with some other animals, exhibit all of them. Where there's life, there's intelligence, so to speak, according to Peters. Intelligence seems to be a matter of degree rather than of kind. However, to this list of traits, I would add a most critical one: intersubjectivity, which I will explain in more detail in the following. Intersubjectivity, the ability to understand other minds, is an important trait in order to explain intelligence and how an intelligent creature can evolve complex communication, civilization, and technology" [2.28].

2.4.1   What is Our Responsibility Toward Intellectually Inferior ETI?

If intelligence can be measured in terms of higher and lower, what if the extraterrestrials we engage exhibit lower intelligence than Earth's *Homo sapiens*? Might the ethical framework for discerning our responsibility toward intellectually inferior ETI be analogous to our responsibility toward Earth's animals? [2.66].

If we answer affirmatively, then we would find ourselves in a classic dialectic. On the one hand, instrumentalist values are obtained. The human race exploits all other life forms—both plants and animals—for human welfare. Animals provide food, work, clothing, and even company. Animals can be sacrificed in medical research to develop therapies that will benefit only human persons. On the other hand, intrinsic values are obtained. We human beings have a sense of responsibility toward the welfare of animals. We respect them as intelligent beings; and we are concerned about preventing suffering to animals. In some instances, we exert considerable energy and effort to preserve their species from extinction and to insure the health of individual animals. In the case of pets, we love them to a degree that rivals loving our own family. We treat our pets as if they possess intrinsic value. In sum, we have inherited this double relationship to our inferiors already here on Earth.

What about ETI whose intelligence level is similar to that of the animals we have come to know? In terms of our responsibility, I believe we should take the initiative to extend concern for the welfare of such ETI on the model of our current concern for the subjective quality of animal experience. We should do what we are able to protect ETI from suffering and enhance their experience of wellbeing. In short, an astroethics of responsibility suggests that we respect ETI and show them care.

2.4.2   What is Our Responsibility Toward Peer ETI?

In the event that the aliens with whom we make contact and engage in transplanetary community approximate the intelligence level of *Homo sapiens* on Earth, an astroethics of responsibility strongly suggests that we earthlings would treat them as possessing dignity. That is, we would treat our extraterrestrial equals as possessing intrinsic value; and imputing dignity is the principal form in which response and care would become manifest [2.52] [2.57] [2.66].

Might the Golden Rule provide an ethical superstructure? Jesus' version of the Golden Rule is familiar to us all: NRS Matthew 7:12 "In everything do to others as you would have them do to you." Even though philosopher Immanuel Kant found weaknesses in the classical Golden Rule, his categorical imperative universalized it. The formal principle from which all moral duties are derived is this: "I ought never to act except in such a way *that I also will that my maxim should become a universal one*" [2.38]. In short, we should treat peers as equal to ourselves; and we should care for their welfare just as we would care for our own.

Is there any reason to expect that our new friends living on a hypothetical exoplanet will have developed moral standards that correspond to ours? Yes, answers Michael Ruse. After all, extraterrestrial creatures must have evolved and adapted to the same laws of physics operative everywhere in the universe. Their logic and mathematics would be the same. Morality also? Yes, perhaps.

> "Two of the greatest and most widely accepted enunciations of the supreme principle of morality are the Greatest Happiness Principle and the Categorical Imperative. The former specifies that one's actions ought to be such as will maximize happiness [John Stuart Mill]. The latter...entreats one to regard one's fellow humans as ends, and not simply as means to one's own gratification [Kant]. Either or both of these could find their equivalent on our hypothetical planet elsewhere in the universe" [2.74].

Our takeaway is this: if we earthlings ascribe intrinsic value and treat intelligent aliens with dignity, it is reasonable to expect the aliens will understand us and perhaps even respond in kind. Earthling care for ETI might be accepted and, hopefully, reciprocated.

Messaging Extraterrestrial Intelligence International (METI) founder and director, Douglas Vakoch, gives voice to such an astroethical responsibility at the moment of contact. "Relevant responsibilities to address include (1) looking out for the interests of humankind as a whole, (2) being truthful in interstellar messages, and (3) benefiting extraterrestrial civilizations" [2.87] [2.88].

With METI in mind, let's refine the peer ETI category into two subcategories: hostile and peaceful. If hostile, ETI could become a threat to us. Should we hide in hopes that we can avoid detection? Should we mount a defense? A preemptive offense?

Fearing such a threat, Stephen Hawking along with some of METI's own advisors argue that Active SETI should be stopped. We should hide Earth electronically so that hostile aliens cannot find us. Such a rejoinder arises from the fear that aliens will turn out to be just like earthlings: beset with the desire to conquer, subjugate, and pillage.

What about a preemptive offense? One of the problems we have learned from history is that we earthlings can become unnecessarily hostile and pose a threat to our fabricated enemies. We are subject to demagoguery. Even if the ETI civilization in question is peaceful, a terrestrial demagogue could persuade us earthlings that the aliens are hostile; and we earthlings, like a mob, would support military action against them. Theologians are particularly worried about human hostility, regardless of how pacific ETI might be.

Whether hostile or peaceful, peer ETI should be afforded respect in the form of dignity. In the event that peer ETI prove to be neutrally peaceful or even benevolent, then the principles giving expression to Enlightenment values should prevail without challenge: equality, liberty, dignity, and mutuality. And, yes, we should care about their well-being if not flourishing.

2.4.3   What is Our Responsibility Toward Superior ETI or Even Post-Biological Intelligence?

Among the array of possible futures, intellectually superior ETI we encounter could be hostile, peaceful, or even benevolent. In all cases, our inherited Enlightenment values would require that we treat them with dignity. If hostile, we earthlings might find ourselves enslaved to the superior extraterrestrials. We *Homo sapiens* are certainly not ready to develop a slave morality; yet a ratiocentric ethic would mandate protocols for slavelike behavior on our part. We might find ourselves relating to superior ETI just as our pets on Earth relate to us.

What would be our responsibility should intellectually superior aliens turn out to be altruistic, even salvific? This scenario of salvation coming to Earth from another planet has already risen to mythical status in our culture [2.56]. Here's the logic of the ETI myth. Because we on Earth have not yet achieved the level of rationality necessary to see that international war and ecological degradation are inescapably self-destructive, we could learn from ETI more advanced than we. "All technological civilizations that already have passed through their technological adolescence and have avoided their self-destruction...must have developed ethical rules to extend their societal life expectancy," says Guillermo Lemarchand [2.40]. Because they are beyond war, spacelings can help earthlings get beyond war [2.57] [2.66]. If ETI saves Earth from self-destructive habits such as war or even ecocide, might gratitude be a fitting response?

In short, we should treat superior ETIs with dignity, respecting and even caring for their welfare. If they are hostile and enslave us, we should invoke an appropriate slave morality that respects their superior minds and maintains their dignity. If ETI are peaceful toward us and open up avenues of conversation and commerce, then the principles of justice and the striving to maintain peace should be obtained. If out of their superior wisdom and altruistic motives ETI seek to better our life here on Earth, we should accept the gifts they bring and respond with an attitude of gratitude.

Now, what would be our responsibility if superior ETI turn out to be postbiological? This is a reasonable speculation, given how here on Earth we are already imagining a posthuman scenario that leads to merging humanity with technology as the next stage of our human evolution. Transhumanism, also known as Humanity Plus (H+), is calling us forward. Our mental lives in the future may take place within a computer or on the internet. What we have previously known as *Homo sapiens* will be replaced by *Homo cyberneticus*. "As humanism freed us from the chains of superstition, let transhumanism free us from our biological chains" [2.95]. The terms "posthuman" and "postbiological" refer to who we might become if the transhumanists among us achieve their goals.

Might a race of extraterrestrials have already arrived at the posthuman and postbiological stage [2.60]? Templeton Prize winning astrophysicist Martin Rees engages this scenario. "The most likely and durable form of life may be machines whose creators had long ago been usurped or become extinct" [2.72]. In outer space we earthlings may meet our own posthuman future.

To make it more complicated, we can speculate that extraterrestrial postbiological intelligence might not even take an individual form. It might be communal, at least according to Eric Korpela. "I will attempt to dispense with terms such as civilization and species, as such terms presume, to some extent, that ETI will be like us: organized groups of independent biological organisms.... the universe may surprise us. The most common type of 'civilization' might consist of a single electronic intelligence" [2.39].

At this point, a subtle shift away from intelligence toward consciousness requires our attention. Philosopher Susan Schneider, a recent holder of the Baruch S. Blumberg NASA/Library of Congress Chair in Astrobiology, acknowledges that "the most advanced alien civilizations will likely be populated by forms of SAI (Super Artificial Intelligence)" [2.76]. What might this imply for the astroethicist? We observe that we have no moral compunction to allow our laptop computer battery to run down and render it nonfunctional. Our electronic computers may appear intelligent; yet they possess no intrinsic value, no dignity. We earthlings would be justified in treating extraterrestrial intelligence without conscious selfhood within a framework of instrumental value.

Table 2.1 Ethics for inferior, peer, and superior ETI.

	Inferior ETI	Peer ETI hostile	Peer ETI peaceful	Sup ETI hostile	Sup ETI peaceful	Sup ETI salvific
Astro-ethics?	Respect/ Care	Dignity	Dignity	Dignity	Dignity	Dignity

Conscious selfhood adds something well beyond intelligence alone, observes Schneider. "Whether SAI is conscious is key to how we should value postbiological existence…. Consciousness is the philosophical cornerstone here, being a necessary condition on being a self or person, in my view" [2.76]. Unless extraterrestrial intelligence is packaged in a self or a person, we earthlings need not ascribe intrinsic value or treat it with dignity. Before we can determine our astroethical responsibility to an intelligent machine on an exoplanet, we will have to ask it whether it is a self or a person.

A transplanetary community of intelligent persons would warrant an astroethic of responsibility that works out of a vision of a galactic common good (Table 2.1).

2.5 Conclusion

Where have we been? I have proposed an "astroethics of responsibility" founded on a substructure of quandary-responsibility ethics. Atop this foundation, the load-bearing vertical supports included: (1) the moral agent: earthlings as a single planetary community of moral deliberation; (2) the moral norm: the galactic common good; (3) the moral spheres: the solar neighborhood and the Milky Way metropolis; (4) the moral justification: a theological grasp of the common good plus a naturalistic grasp of the Golden Rule. The floor plan designated a conference room for each of thirteen previously formulated ethical issues.

I distinguished two spheres of astrobiological application: the solar neighborhood and the Milky Way metropolis. Within the sphere of the solar neighborhood, ten already articulated quandaries were addressed: (1) planetary protection; (2) intrinsic value of off-Earth biospheres; (3) application of the Precautionary Principle; (4) space debris; (5) satellite surveillance; (6) weaponization of space; (7) scientific versus commercial space exploration; (8) terraforming Mars; (9) colonizing Mars; and (10) anticipating natural space threats [2.66]. Within the sphere of the Milky Way metropolis in which the "galactic commons" becomes the astroethical norm, engagement with intelligent extraterrestrials was analyzed within three categories: (1) ETI less intelligent than Earth's *Homo sapiens*; (2) ETI equal in intelligence; and (3) ETI superior in intelligence in both biological and postbiological forms [2.66].

I have argued that we should nominate earthlings in the form of a "single planetary community of moral deliberation" to the office of moral agent. This earth-born community should then look to the sky and allow our planet's place in the immense and unfathomable universe to affect our shared consciousness. I support the mandate of naturalist Steven Dick regarding the decisive role that cosmic consciousness needs to play in developing an astroethics of responsibility. "A cosmic perspective is surely in order as we expand our views of the space environment, including (and especially) life. Such a view will not happen overnight, but perhaps, humanity will increase its awareness in stages, as we encounter the universe in increasingly intimate ways that will become a basic part of what it means to be human, or post-human" [2.27].

References

[2.1] Aldrin, B., Preface, in: *Icarus' Second Chance: The Basis and Perspectives of Space Ethics*, p. v, Springer, Vienna, 2011.

[2.2] Arnould, J., The Emergence of the Ethics of Space: The Case of the French Space Agency. *Futures*, 37, 245–254, 2005.

[2.3] Arnould, J., *Icarus' Second Chance: The Basics and Perspectives of Space Ethics*, 1st ed, Springer, Vienna and New York, 2011.

[2.4] Arnould, J., Astrotheology, Astroethics, and the New Challenges. *Theol. Sci.*, 16, 4, 380–381, 2018.

[2.5] Arnould, J., Space Exploration: Current Thinking on the Notion of Otherness. *Theol. Sci.*, 16, 1, 54–61, 2018.

[2.6] Bastianel, S., *Morality in Social Life*, Convivium Press, Miami, FL, 2010.

[2.7] Billings, L., Are We on the Cusp of War-in-Space? *Sci. Am.*, 313, 4, 14–18, 2015.

[2.8] Billings, L., A US Space Force? A Very Bad Idea! *Theol. Sci.*, 16, 4, 385–387, 2018.

[2.9] Billings, L., Should Humans Colonize Mars? No! *Theol. Sci.*, 17, 3, 341–346, 2019.

[2.10] Bullock, M.R., Cosmology, in: *Encyclopedia of Science, Technology, and Ethics*, vol. 4, pp. 1, 437–442, Macmillan Gale, New York, 2005.

[2.11] Casebeere, W.D., Scientific Ethics, in: *Encyclopedia of Science, Technology, and Ethics*, vol. 43, pp. 1726–1731, Macmillan Gale, New York, 2005.

[2.12] Cho, F., An Asian Religious Perspective on Exploring the Origin, Extent and Future of Life, in: *Workshop Report: Philosophical, Ethical, and Theological Implications of Astrobiology*, N. R., M. S., C. Bertka (Eds.), pp. 208–218, AAAS, Washington DC, 2007.

[2.13] Chon-Torres, O., Astrobioethics. *Int. J. Astrobiology*, 17, 1, 51–56, 2018.

[2.14] Chon-Torres, O., Moral challenges of going to Mars under the presence of non-intelligent life scenario. *Int. J. Astrobiology*, 19, 1, 49–52, 2020a.

[2.15] Chon-Torres, O., Astrobioethics: A brief discussion from the epistemological, religious and societal dimension. *Int. J. Astrobiology*, 19, 1, 61–67, 2020.

[2.16] Cockell, C.S., The Ethical Status of Microbial Life on Earth and Elsewhere: In Defense of Intrinsic Value, in: *The Ethics of Space Exploration*. J. S. S, a. T. Milligan, (Ed.) pp. 75–92, Heidelberg: Springer 2016.

[2.17] Conley, C.A., Planetary Protection, in: *Handbook of Astrobiology*, pp. 819–834, Taylor and Francis, London, 2019.

[2.18] Conley, C.A., Planetary Protection, in: *Handbook of Astrobiology*, V.M. Kolb (Ed.), pp. 819–834, Taylor and Francis, London, 2019.

[2.19] COSPAR, *Workshop on Ethical Considerations for Protection in Space Exploration*, Princeton University, COSPAR, Paris, 2010.

[2.20] Crawford, I.A., Widening Perspectives: the intellectual and social benefits of astrobiology. *Int. J. Astrobiology*, 17, 1, 57–60, 2018.

[2.21] Crysdale, C.S.W., God, evolution, and astrobiology, in: *Exploring the Origin, Extent, and Future of Life: Philosophical, Ethical, and Theological Perspectives*, C.M. Bertka (Ed.), pp. 230–250, Cambridge University Press, Cambridge, UK, 2009.

[2.22] Curran, C.E., How Does Christian Ethics Use Its Unique and Disctinctive Christian Aspects. *J. Soc Christ. Ethics*, 31, 2, 23–36, 2011.

[2.23] D'Costa, G., Postmodern and Religious Plurality: Is a Common Global Ethic Possible or Desirable?, in: *The Blackwell Companion to Postmodern Theology*, G. Ward (Ed.), pp. 131–143, Blackwell, Oxford, 2001.

[2.24] Deane-Drummond, C.E., The alpha and the omega: reflections on the origin and future of life from the perspective of Christian theology and ethics, in: *Exploring the Origin, Extent, and Future of Life*, C.M. Bertka (Ed.), pp. 100–120, Cambridge University Press, Cambridge, UK, 2009.

[2.25] Denning, K., Hawking Says That Discovering Intelligent Life Elsewhere Would Spark Greater Compassion and Humanity among Us..But Why Wait? *Theol. Sci.*, 15, 2, 142–146, 2017.

[2.26] Dick, S.J., *Astrobiology, Discovery, and Societal Impact*, Cambridge University Press, Cambridge, UK, 2018.

[2.27] Dick, S.J., Humanistic Implications of Discovering Life Beyond Earth, in: *Handbook of Astrobiology*, pp. 741–756, Taylor and Francis, London, 2019.

[2.28] Duner, D., Mind in the Universe: On the Origin, Evolution, and Distribution of Intelligent Life in Space, in: *Handbook of Astrobiology*, V.M. Kolb (Ed.), pp. 701–716, Taylor and Francis, London, 2019.

[2.29] ELCA, *Genetics, Faith, and Responsibility, Evangelical Lutheran Church in America*, Chicago, 2011.

[2.30] European Space Agency, *The Ethics of Space Policy*, UNESCO (COMEST), Paris, 2000.

[2.31] Ferkiss, V.C., *The Future of Technological Civilization*, George Braziller, New York, 1974.

[2.32] Hart, J., Cosmic Commons: Contact and Community. *Theol. Sci.*, 8, 4, 371–392, 2010.

[2.33] Hart, J., *Cosmic Commons: Spirit, Science, and Space*, Cascade Books, Eugene OR, 2013.

[2.34] Impey, C., The First Thousand Exoplanets: Twenty Years of Excitement and Discovery, in: *Astrobiology, History, and Society*, D.A. Vakoch (Ed.), pp. 201–212, Springer, Heidelberg, 2013.

[2.35] Jonas, H., *The Phenomenon of Life: Toward a Philosophical Biology*, Northwestern University Press, Evanston, IL, 1966.

[2.36] Jonas, H., *The Imperative of Responsibility: IN Search of an Ethics for the Technological Age*, University of Chicago Press, Chicago, 1984.

[2.37] Jonsen, A.R., Responsibility, in: *The Westminster Dictionary of Christian Ethics*, J.F. C and J. Macquarrie (Eds.), pp. 545–549, Westminster John Knox, Louisville, KY, 1986.

[2.38] Kant, I., *Groundwork of the Metaphysics of Morals, tr.*, H.J. Paton, Harper, New York, 1948.

[2.39] Korpela, E.J., SETI: Its Goals and Accomplishments, in: *Handbook of Astrobiology*, V.M. Kolb (Ed.), pp. 727–738, Taylor and Francis, London, 2019.

[2.40] Lemarchand, G.A., Speculations on First Contact, in: *When SETI Succeeds: The Impact of High Information Contact*, pp. 153–164, The Foundation for the Future, Bellevue, WA, 2000.

[2.41] Losch, A., What is Life? On Earth and Beyond: Conclusion, in: *What is Life? On Earth and Beyond*, A. Losch (Ed.), pp. 303–311, Cambridge University Press, Cambridge, UK, 2017.

[2.42] Lupisella, M., The search for extraterrestrial life: epistemology, ethics, and worldviews, in: *Exploring the Origin, Extent, and Future of Life: Philosophical, Ethical, and Theological Perspectives*, C. Bertka (Ed.), pp. 190–210, Cambridge University Press, Cambridge, UK, 2009.

[2.43] Lupisella, M., Cosmological Theories of Value: Relationalism and Connectedness as Foundations for Cosmic Creativity, in: *The Ethics of Space Exploration*. J. S. S a. T. Milligan, (Ed.) pp. 75–92, Heidelberg: Springer, 2016.

[2.44] Mariscal, C., Universal Biology Does Not prescribe Planetary Isolationism. *Theol. Sci.*, 15, 2, 150–152, 2017.

[2.45] Marks, P., Clearing the heavens, one piece at a time. *New Sci.*, 209, 2799, 22–23, 2011.

[2.46] McKay, C.P., Planetary ecosynthesis on Mars: restoration ecology and environmental ethics, in: *Exploring the Origin, Extent, and Future of Life:*

Philosophical, Ethical, and Theological Perspectives, C.M. Bertka (Ed.), pp. 250–270, Cambridge University Press, Cambridge, UK, 2009.

[2.47] McKay, C.P., Astrobiology and Society: The Long View, in: *Encountering Life in the Universe*, A.H. S, W. S., C. Impey (Eds.), pp. 158–166, University of Arizona Press, Tucson, AZ, 2013.

[2.48] McKay, C.P., Astroethics and the Terraforming of Mars, in: *Astrotheology: Science and Theology Meet Extraterrestrial Intelligence*, M. H., J.M. M., R.J. R. 2.48, T. Peters (Eds.), pp. 381–390, Cascade Books, Eugene OR, 2018.

[2.49] McKay, C. P. and R O., R., Protecting and Expanding the Richness and Diversity of Life: An Ethic for Astrobiology Research and Space Exploration. *Int. J. Astrobiology*, 13, 1, 28–34, 2014.

[2.50] Mitcham, C., Responsibility: Overview, in: *Encyclopedia of Science, Technology, and Ethics*, vol. 2, C. Mitcham (Ed.), pp. 1609–1616, p. 3, Macmillan Gale, New York, 2005.

[2.51] Moe-Lobeda, C., *Healing a Broken World: Globalization and God*, Fortress Press, Minneapolis, MN, 2002.

[2.52] Narveson, J., Martians and morals: How to treat and alien, in: *Extraterrestrials: Science and Alien Intelligence*, E. Regis (Ed.), pp. 245–266, Cambridge University Press, Cambridge, UK, 1985.

[2.53] NASA, NASA updates Planetary Protection Policies for Robotic and Human Missions to Earth's Moon and Future Human Missions to Mars, Washington DC NASA, 2020, 7/9/2020; HYPERLINK "https://www.nasa.gov/feature/nasa-updates-planetary-protection-policies-for-robotic-and-human-missions-to-earth-s-moon"https://www.nasa.gov/feature/nasa-updates-planetary-protection-policies-for-robotic-and-human-missions-to-earth-s-moon.

[2.54] Niebuhr, H.R., *The Responsible Self*, Harper, New York, 1963.

[2.55] Persson, E., The Moral Status of Extraterrestrial Life. *Astrobiology*, 12pp, 976–985, 2012.

[2.56] Peters, T., Astrotheology and the ETI Myth. *Theol. Sci.*, 7, 1, 3–30, 2009.

[2.57] Peters, T., The implications of the discovery of extra-terrestrial life for religion. *Philos. Trans. R. Soc. A*, 369, 644–655, 2011.

[2.58] Peters, T., Astroethics: Engaging Extraterrestrial Intelligent Life-Forms, in: *Encountering Life in the Universe*, pp. 200–221, University of Arizona Press, Tucson, AZ, 2013.

[2.59] Peters, T., Intelligent Aliens and Astroethics, in: *Space Exploration and ET: Who Goes There?*, pp. 1–20, ATF Press, Adelaide Australia, 2014.

[2.60] Peters, T., Outer Space and Cyber Space: Meeting ETI in the Cloud. *Int. J. Astrobiology*, 17, 4, 282–286, 2016.

[2.61] Peters, T., Where There's Life There's Intelligence, in: *What is Life? On Earth and Beyond*, pp. 236–259, Cambridge University Press, Cambridge, UK, 2017.

[2.62] Peters, T., Toward a Galactic Common Good: Space Exploration Ethics, in: *The Palgrave Handbook of Philosophy and Public Policy*, pp. 827–843, Macmillan Palgrave, New York, 2018a.

[2.63] Peters, T., Got Ethics for the Galaxy? Quandary-Responsibility Ethics for Space Exploration, in: *Science-Religion Dialogue and Its Contemporary Significance*, pp. 9–36, ATC Publishers, Bengaluru, India, 2018b.

[2.64] Peters, T., Does Extraterrestrial Life Have Instrinsic Value? *Int. J. Astrobiology*, 17, 2, 1–7, 2018c.

[2.65] Peters, T., Astroethics and Microbial Life in the Solar Ghetto, in: *Astrotheology: Science and Theology Meet Extraterrestrial Life*, M. H., J.M. M., R.J. R., T. Peters (Eds.), pp. 391–415, Cascade Books, Eugene OR, 2018d.

[2.66] Peters, T., Astroethics and Intelligent Life in the Milky Way Metropolis, in: *Astrotheology: Science and Theology Meet Extraterrestrial Life*, M. H., J.M. M., J. R., T. Peters (Eds.), pp. 416–446, Cascade Books, Eugene OR, 2018e.

[2.67] Peters, T., Astrobiology and Astrotheology in Creative Mutual Interaction, in: *Theology and Science: Discussion about Faith and Facts*, pp. 25–43, World Scientific Publishing, Singapore, 2019.

[2.68] Pryor, A., Intelligence, Non-Intelligence…Let's Call the Whole Thing Off. *Theol. Sci.*, 16, 4, 471–483, 2018.

[2.69] Race, M.S., The implications of discovering extraterrestrial life: different searches, different issues, in: *Exploring the Origin, Extent, and Future of Life: Philosophical, Ethical, and Theological Perspectives*, pp. 210–220, Cambridge University Press, Cambridge, UK, 2009.

[2.70] Race, M.S., A US Space Force? It's Complicated. *Theol. Sci.*, 16, 4, 382–384, 2018.

[2.71] Rathman, K. a., Common Heritage of Mankind Principle, in: *Encyclopedia of Science, Technology, and Ethics, 1st ed.*, vol. 4, pp. 1:361–363, Macmillan Gale, New York, 2005.

[2.72] Rees, M., *Our Final Hour*, Basic Books, New York, 2003.

[2.73] Rummel, J. D. and C. C. A., Planetary protection for human exploration of Mars. *Acta Astronaut.*, 66, 5-6, 792–797, 2010.

[2.74] Ruse, M., Is rape wrong on Andromeda? An introduction to extraterrestrial evolution, science, and morality, in: *Extraterrestrials: Science and Alien Intelligence*, E. Regis (Ed.), pp. 43–78, Cambridge University Press, Cambridge, UK, 1985.

[2.75] Ruse, M., Charles Darwin and the Plurality of Worlds: Are We Alone?, in: *Handbook of Astrobiology*, V.M. Kolb (Ed.), pp. 125–136, Taylor and Francis, London, 2019.

[2.76] Schneider, S., Superintelligent AI and the Postbiological Cosmos Approach, in: *What is Life? On Earth and Beyond*, A. Losch (Ed.), pp. 178–197, Cambridge University Press, Cambridge, UK, 2017.

[2.77] Science, E., Researchers Eye Tethers for Space Debris. *Science*, 343, 6175, 1062, 2014.

[2.78] Smith, K.C., Manifest complexity: A foundational ethic for astrobiology. *Space Policy*, 30, 209–214, 2014.

[2.79] Smith, K.C., The Curious Case of the Martian Microbes: Mariomania, Intrinsic Value and the Prime Directive, in: *The Ethics of Space Exploration*. J. S. S a. T. Milligan, (Ed.) pp. 75-92, Heidelberg: Springer, 2016.

[2.80] Smith, M., U.S. Space Force's First Doctrine Signals Expansion, Space Policy Online, Washington DC, 2020, (10 August); HYPERLINK "https://spacepolicyonline.com/news/u-s-space-forces-first-space-doctrine-signals-expansive-plans/"https://spacepolicyonline.com/news/u-s-space-forces-first-space-doctrine-signals-expansive-plans/.

[2.81] Stern, S.A., The Low-Cost Ticket to Space. *Sci. Am.*, 308, 4, 69–73, 2013.

[2.82] Szocik, K., W, T., R, M.B., C, C., Ethical issues of human enhancements for space missions to Mars and Beyond. *Sci. Direct (Elsevier)*, 115, 102489, 1–14, 2020.

[2.83] Tarter, J.C., Contact: Who Will Speak for Earth and Should They?, in: *Encountering Life in the Universe*, A.H. S., W. S., C. Impey (Eds.), pp. 178–199, University of Arizona Press, Tucson, AZ, 2013.

[2.84] Tyson, N.d., Holy Wars: An Astrophysicist Ponders the God Question, in: *Science and Religion: Are They Compatible?*, P. Kurtz (Ed.), pp. 73–82, Prometheus Books, Buffalo, NY, 2003.

[2.85] UN, *Treaty on Principles Governing the Activities of States in the Exploration and Use of Outer Space, Including the Moon and Other Celestial Bodies*, United Nations, New York, 1967.

[2.86] UN, *Wingspread Statement on the Precautionary Principle*, United Nations, 1998, http://www.gdrc.org/u-gov/precaution-3.html.

[2.87] Vakoch, D.A., Responsibility, Capability, and Active SETI: Policy, Law, Ethics, and Communication with Extraterrestrial Intelligence. *Acta Astronaut.*, United Nations, 68, 512–519, 2011.

[2.88] Vakoch, D.A., Symposium: Should We Heed Stephen Hawking's Warning? *Theol. Sci.*, 15, 2, 134–138, 2017.

[2.89] VI, P.P., *Pastoral Constitution on the Church in the Modern World: Gaudium et Spes*, Vatican, Vatican City, 1965.

[2.90] Voegelin, E., *5 Volumes: Order and History; The Ecumenic Age*, pp. 1956–1987, Louisiana State University Press, Baton Rouge, LA, 1987.

[2.91] Wanjek, C., *Spacefarers: How Humans Will Settle the Moon, Mars, and Beyond*, Harvard University Press, Cambridge, MA, 2020.

[2.92] Wilkinson, D.A., Why Should Theology Take SETI Seriously? *Theol. Sci.*, 16, 4, 427–438, 2018.

[2.93] Wolf-Chase, G.A., New Worlds, New Civilizations? From Science Fiction to Science Fact. *Theol. Sci.*, 16, 4, 415–426, 2018.

[2.94] S.P.J. XXIII, *Pacem in Terris*, Vatican, Vatican City, 1963.

[2.95] Young, S., *Designer Evolution: A Transhumanist Manifesto*, Prometheus Books, Buffalo, NY, 2006.

[2.96] Zubrin, R., Why We Earthllings Should Colonize Mars. *Theol. Sci.*, 17, 3, 305–316, 2019.

3

Moral Philosophy for a Second Genesis

Julian Chela-Flores[1,2]

[1]*The Abdus Salam International Centre for Theoretical Physics, Trieste, Italy*
[2]*Institute of Advanced Studies (IDEA), Caracas, Bolivarian Republic of Venezuela*

Abstract

Firstly, we discuss possibilities for constraining the evolution of life in the universe. Secondly, we consider the possibility of searching for a second genesis and, finally, we review some moral implications that are implied in astrobiology. In an extra-terrestrial environment a human level of intelligence may be favored when the combined effect of natural selection and convergent evolution are taken together. This is independent of the particular details of the phylogenetic tree that may lead to the non-human organism.

We discuss a moral question that ethicists consider to embrace all human beings everywhere on Earth, independent of creed, gender and ethnic origin. But argue that ethical arguments may be universalizable, extended to non-human species on Earth and to life elsewhere in the universe, whether microbial or multicellular.

In astrobiology, being eventually and inevitably confronted with the inexorable possibility of discovering a second genesis, either in our cosmic neighborhood, or on bodies around stars in our galaxy, we are induced to assume moral principles that have been forced upon us, either by religious creed, or by empirical experiences.

An example is given to illustrate how the role of ethics naturally arises with advanced instrumentation to be used during our previous search proposals for the exploration of the Jovian system. We conclude that a new (second) morality in astrobioethics is unnecessary. This point of view has been reinforced in our work, especially with extrapolations of what philosophers have learned on Earth about our ethical treatment of non-human species—animals.

Keywords: Moral philosophy, a second genesis, astrobiology, evolutionary convergence, universalizable ethical arguments

Email: chelaf@ictp.it

Octavio Alfonso Chon Torres, Ted Peters, Joseph Seckbach and Richard Gordon (eds.) Astrobiology: Science, Ethics, and Public Policy, (57–78) © 2021 Scrivener Publishing LLC

3.1   Moral Philosophy on Earth and Elsewhere

3.1.1   The Origin of Ethics and Its Universal Relevance

Astrobioethics is a new, exciting, hybrid, cultural sector of science and philosophy. It is a timely new approach worth discussing in detail, in spite of its outstanding difficulties, including those intrinsically of moral philosophy [3.45]. There is a vast work in philosophy that goes back to ancient times. Aristotle's contribution, *Nicomachean Ethics* [3.1], in some sense is not useful from the present point of view, as he accepted inequality with no objection to slavery. Secondly, we may recall Spinoza's *Ethics* [3.58] in which the philosopher makes the relevant general point that taking larger matters into account, human life, including evil and suffering, is an infinitesimal part of life in the universe [3.50].

However, we are convinced that discovering extraterrestrial life would induce a fruitful dialogue between ethics and other sectors of culture that are not always willing to approach one another. For these and other reasons, we feel that exploring the problem of the unanswered ethical questions related to the distribution of life in the cosmos, remain an important enquiry.

The origin of ethics is one of two possibilities [3.66]: Either ethical precepts are independent of human experience, or alternatively they are human inventions. Moral philosophy, or ethics, is not necessarily connected within the Western tradition with any religion, either Jewish or Christian, or even with those of other non-Western traditions. Different approaches to the origin of ethics is not a debate between science and religion, but rather a debate between two lines of thought:

(i)   Transcendentalists
This line of thinking (cf. Glossary), which included Hegel's philosophy, is the last all-embracing Western philosophical system that searches for the significance and comprehension of the whole of reality. According to Hegel, nature is the manifestation of the Divinity turned into matter; everything that takes place in the universe is nothing but the transformation of matter into spirit. Knowledge consists in the discovery and comprehension of this process. With Hegel, the intelligibility and meaning of the universe are within reach of human reason.

(ii)   Empiricists
This line of thinking (cf. Glossary) goes back to Rationalism, which is the doctrine that attempted to construct an

entirely deductive philosophy in the manner of geometry, with the intention of demonstrating that reality was intelligible and meaningful. Subsequently, empiricism considered that the only source of knowledge was the sensorial experience. Hence, everything that was beyond the experience of the senses would not be a subject of knowledge, but instead would be in the domain of belief, closer to the approach of transcendentalism.

3.1.2   Why Should We Act Morally?

We are concerned with the question of extending an area of philosophy, ethics, into our actions during the exploration of the cosmos; more specifically, not just ethics, but what we have learned from life on Earth—bioethics—our cosmic environment, into a new subfield of philosophy, astrobioethics, that forces upon us a fundamental question to which there is still no consensus:

"Why should we act morally when faced now and in the near future with options that were unknown during the Age of Enlightenment?"

Some of these questions are planetary protection, terraforming and care, respect, with possible non-humans, including microorganisms inhabiting the various Solar System oceans that are already known to us. These concepts are to be taken into account during our further steps in planetary exploration.

To address this enquiry, we should not only consider the question:

"Why should we act morally?"

which has some bearing on astrobioethics; but also

"How are we going to treat non-human life?"

When we are concerned with the eventual discovery of non-human life in the universe, we should persevere with these difficult philosophical issues, whose answers still have no consensus. One early point of reference comes from Socrates. Socrates (with Thrasymachus) discuss why one ought to be moral [3.47]. Their main concern is justice rather than ethics, even though their thoughts are relevant, even if their translation into astrobioethics concerns may not be acceptable: Thrasymachus' essential argument is that individuals should act unjustly because acting unjustly confers greater benefits to an individual than acting justly. Additional support followed throughout the history of philosophy, but some recent insights come from ethical studies in the area of finance, outstanding amongst which we find those of the Queen Mary College philosopher, Kara Tan Bhala [3.4].

In our discussion, we extend ethics to non-human species, not in a terrestrial context that has been mainly concerned with animals, especially animal rights, but rather our discussion will be extended in the context of Astrobiology to eventual non-human extraterrestrial species. Such arguments are constrained by reason and by meta-ethics (cf. Glossary) [3.56].

3.1.3   Is a New Morality Needed for Astrobiological Explorations?

We may wonder whether there are reasons for suggesting that a new morality would be needed in successful astrobiological explorations. There are two different cases that need to be discussed: Firstly, the more urgent present question is the exploration of the Solar System. In this case, discussed below from preliminary plausible suggestions that we have been familiar with in our earlier research, we may have to choose the type of instrumentation that our technological capabilities may offer, as the most efficient way to approach microbial, or multicellular life in the ocean worlds, or on terrestrial planets.

Secondly, a less urgent case is the exploration of other solar systems. In the very distant future, well beyond the present stage, we may have to choose the type of instrumentation that future technological capability may offer to probe microbial, or non-human life on exoplanets, or their exomoons.

Ethical principles are continually changing according to the evolution of science and technology (cf. Glossary). It is useful to reconsider the philosophical consequences of the biogeocentric point of view, a deep-rooted assumption that has been in the way of progress in astrobiology [3.12]. Indeed, biogeocentrism in one form or another has been advocated in the past by both evolutionists [3.40], as well as by molecular biologists [3.43]. Mayr went as far as assigning extraterrestrial life an "improbability of astronomical dimensions." On the other hand, Monod believes that: "The present structure of the biosphere certainly does not exclude the possibility that the decisive event occurred only once." The biogeocentric position has also been maintained from the point of view of paleobiology [3.18]: If indeed we are alone and unique, a possibility, however implausible, that cannot yet be refuted, then we have special responsibilities.

Almost a century and a half separate our insights on biogeocentrism and its deep connection with the dialogue of faith and reason from the beginning of philosophical thinking. In fact, Sir Charles Lyell was concerned with the position of man amongst all living organisms. It seems that the underlying difficulty in the dialogue of faith and reason is still related to inserting Darwinian evolution into our culture.

Darwin was prudent enough to avoid ideological issues. He even avoided philosophical issues. Bertrand Russell in his book, *Religion and Science,* inevitably raised these issues: "From evolution, so far as our present knowledge shows, no ultimately optimistic philosophy can be validly inferred" [3.52]. Charles Darwin himself concentrated on the narrow, but transcendental problem of establishing the theory of evolution of life on Earth, but prudently postponed the wider issue of the position of man in the universe, whether our life is a lonely isolated case or one of multiple phenomena elsewhere. Clearly, if astrobiology research concludes that we are not alone in the universe, there are some unavoidable ethical, philosophical and theological consequences for which we still have no answer. For instance, outstanding amongst them are those found in theologies of different traditions, which have made substantial progress in questions that are shared with the new science of astrobiology [3.53] [3.54]. They have also been considered by Christian thinkers of different traditions: firstly, Martini and Coyne [3.38] and, secondly, Ernan McMullin [3.42]. These authors expose the wide range of implications, in which astrobiology has influenced the cultural sector, including astrotheology. These questions have made remarkable contributions to the humanities.

3.2 Identifying the Lack of Ethical Substance in Science Communication

3.2.1 Understanding the Boundaries of Knowledge

Philosophy is an important component of knowledge. There is a clear formulation of the boundaries between the main areas of Western culture. They were highlighted by Bertrand Russell in his major work on popularization of philosophy [3.51], which we have often cited in the past ([3.11], p. vii, [3.13], p. vi):

"Philosophy, as I shall understand the word, is something intermediate between theology and science. Like theology, it consists of speculations on matters as to which definite knowledge has, so far, been unascertainable; but like science it appeals to human reason rather than authority, whether that of tradition or that of revelation...But between theology and science there is a No Man's Land, exposed to attack from both sides; this No Man's Land is philosophy."

The transparent and illuminating Russellian insertion of science within its humanistic neighbors, becomes especially relevant when we focus our attention on the frontiers of astrobiology and astrobioethics. To begin with, consider, for instance, when we have to choose between good and evil: We enter the area of ethics. Indeed, for an average reader, who recognizes the frontiers of culture, it should be evident that ever since science and philosophy were born in ancient Greece, they never stopped flourishing. They have invaded our questions of natural history, not excluding enquiries that are deeply related to astrobiology.

We should place the sciences of astrobioethics and astrobiology in an ethical context that is easily readable and accessible to both scientists and lay readers. But amongst these potential readers, we have also taken special care to present the dialogue that should be sustained between believers and non-believers, who are not familiar with the technical language of science. Assuming that the frontiers of science are a delusion has led to a frequent misinterpretation of Darwinism, which is partly our fault as scientists [3.14].

Due to the necessary specialization that modern science imposes on us to do research of the highest level, we do not allow enough time in our tight working schedules for our moral obligation to popularize science and make it comprehensible for the general public. But to share science with society should be a major objective of scientists, due to the enormous costs involved in supporting our research. We owe it to the society supporting us to provide a clear and honest indication of our achievements and, especially, our limitations.

We should not be too harsh when they identify lack of ethical substance in science communication. After all, such statements are not harmful, once we realize they are meaningless, beyond the personal opinions of their authors. A fractionated culture either hides the natural horizons of science, or simply is exposed to scientific specialization that is compulsory, in order to achieve results in the highly specialized fields. Evidently, as believers, who are also scientists, we are a minor fraction of the world population. In our privileged case we are able to confidently evaluate science communication that ignores the frontiers of science. The other fraction of believers, who are not scientists, are at a disadvantage.

Sadly, they are the great majority [3.46]. The third group, the remaining population of agnostics (cf. Glossary), is not particularly affected, due to their lack of concern for any religion. We hope that in the future a complete education will be the general case for both science communicators, as well as for their readers. It will not have escaped the attention of our readers that culture, not only includes philosophy and theology, but that it incorporates many other areas, such as classical studies, history, languages, literature, music, theatre and dance; not forgetting the visual arts—painting, sculpture, as well as the arts that require multimedia [3.5]. There are specific cases where ethics enters in the instrumentation of solar system exploration, as we have shown above.

3.2.2 Implications of the Limits and Horizons of Science

To illustrate the lack of ethical substance in astrobiology communication we recall the important question of a presumed conflict between astrobiology and religion [3.13]. Indeed, the absence of a conflict is evident when we accept the limits and horizons of science [3.37].

Unnecessary difficulties arise when the horizons of science are ignored, as it has happened with the views, not of agnostics, but rather of atheists, who have ignored the frontiers of science. In the literature they are referred to as representing a "New Atheism" [3.22] [3.33] [3.34]. These views have led to much confusion in the general public that is not necessarily fluent with the terminology of science. Within the realm of theology, there seems to be a contradiction with what science has maintained, a misconception that has deep roots in Western philosophy.

It is remarkable that even at the time of Plato's contemporaries, well before modern science emerged, there was an apparent conflict with the rationalization of the origin of the world in terms independent of a Creator. Saint Augustine of Hippo's philosophy is Platonic, as he was familiar not only with Plato's philosophy, but had read accepted physical knowledge, logic and ethics of his contemporaries. For well over a millennium and a half, it has been evident from *The City of God* [3.2], that when there appeared to be a clear conflict between what Augustine took to be the historical account of the Bible and an ascertained fact of rational science, the philosopher of Hippo understood the Bible allegorically. That profound thinker did not accept that all the contents were of an allegorical nature. These views on biblical interpretation provide a stimulating early example of the dialogue of faith and reason, which is glossed over by the New Atheists.

Fortunately, there is a growing body of literature [3.3] [3.20] [3.39] of high-level science communication efforts by well-qualified writers. These authors have provided ethical accounts that do not ignore the evident horizons of science that have been outlined since the time of Galileo Galilei in the 17th Century.

Besides, the role of chance and necessity can be misinterpreted when it is taken outside the frontiers of science [3.48]. The significance of chance (historical contingency) is not a synonym of "blind." Rev. Sir John Polkinghorne emphasizes that bringing in the word chance with its interpretation as a synonym of blind is a metaphysical claim that should not be presented, as if there were no frontiers between science and the humanities. In other words, the claim should not be presented as if it were a scientific conclusion.

3.3   Going from Astrobiology to Astrobioethics: A Big Step for Science and Humanism

3.3.1   The Pathway from Ethics to Bioethics and to Astrobioethics

The need to consider astrobioethics continued to emerge gradually, well before the early establishment of astrobiology as a new science at the turn of this century when moral questions appeared in relation to the possibility of learning about non-human species beyond terrestrial animals [3.28]. The bioethicist [3.55] has put an emphasis on "practical" ethics, especially considering the question of equating the lives of humans and non-humans. (Peter Singer had in mind non-human animals.) This point of view renders it possible to make a smooth transition from orthodox bioethics to astrobioethics, in the new context of moral obligations, when we study astrobiology, as we will discuss in this work. With the birth of space exploration, the search for non-human species acquires novel ethical considerations.

3.3.2   The Question of the Role of Ethics in Astrobiology

Some of the deepest questions that humans have raised are not confined within the boundaries of science. Instead, philosophers, including ethicists, and theologians have approached such questions with their own expertise of competence [3.10] [3.13]. One such example is provided by the question of the role of ethics in astrobiology, whether of transcendental or empirical origin. This question should encourage us to provide appropriate answers in the areas of philosophy, science, or theology. It also may be argued that the task of a scientist should be independent of that of the other areas of human culture. On the other hand, it is surely useful to be aware that this view of the role of science that is "divorced" from both philosophy and natural theology can also be seen from the point of view of close relationship [3.60]: Because science and religion are evolving, and are similar in their search for truth, convergence of these independent searches for truth may occur in the future.

All aspects of astrobiology involve, not only the frontiers of science, but are also relevant in a wider cultural context in which ethics in intercultural communication is vital, and often sadly neglected, creating unnecessary confusion. But it is rewarding to notice that the question of ethics in the wider context of astrobiology has been given its proper significant position in a cultural context—astrobioethics—a branch of astrobiology that studies the moral implications related to the search of non-human life beyond the Earth [3.15] [3.16] [3.17]. This new area of philosophy should particularly be treated with care in science communication, as we shall attempt to illustrate later on.

3.4   Would There Be New Ethical Principles if There Were a Second Genesis?

3.4.1   Inevitability of the Emergence of a Particular Biosignature

In an extraterrestrial environment the evolutionary steps that led to humans would probably never repeat themselves. However, the possibility remains that a human level of consciousness may be favored when the combined effect of natural selection and convergent evolution are considered. This is independent of the particular details of the phylogenetic tree that may lead to a non-human organism: Conway-Morris suggests that the role of contingency in evolution has little bearing on the emergence of a particular biological property [3.18].

What we are interested in is not the origin, destiny, or fate of a particular lineage, but the likelihood of the emergence of a particular property, say consciousness. In these cases, it is of primary importance to have clearly defined ethical concepts. Convergence suggests that the tape of life, to use Gould's metaphor (cf. Glossary), can be run as many times as we like and, in principle, intelligence will surely emerge [3.8] [3.18].

We may return to convergent evolution, a phenomenon that has been recognized by students of evolution for a long time [3.62]. As an illustration, an example may be found in the phylum of mollusks: The shells of both the camaenid snail from the Philippines and the helminthoglyptid snail from Central America, resemble the members of European helcid snails. In spite of having quite different internal anatomies, these distant species (they are grouped in different Families) have grown to resemble each other outwardly over generations in response to their environment. In spite of considerable anatomical diversity, mollusks from these distant families have tended to resemble each other in a particular biological property, namely, their external calcareous shell.

A second example is found with swallows—Passeriformes—a group that is often confused with swifts—Apodiformes—but they are unrelated to them. Members of these two orders differ widely in anatomy. Their similarities are the result of convergent evolution on different stocks that have become adapted to the same life styles in ecosystems that are similar to both species.

Many other examples have been provided by Conway-Morris. Together they provide us with a convincing argument that the results of evolution can be predicted to a certain extent. It is not to be understood as if humans will emerge elsewhere in the cosmos, but rather that brains and consciousness are bound to emerge in any case. From ethical relativism (cf. Glossary), accepting the possibility of a second genesis, morality in our own living community (humanity and non-human terrestrial species) may not be ethical elsewhere—if we are confronted with a second genesis.

3.4.2 Universalizable Ethical Criteria

Only a basic moral principle, such as equality amongst living beings, can allow us to defend a form of equality that embraces all living beings everywhere on Earth, independent of creed, gender, and ethnic origin. However, equality embracing all living beings is not necessarily constrained to humans alone. Universalizable ethical criteria can be considered to also go beyond humanity. Singer proposes that it should be extended to genera beyond our own species, indeed to non-human animals [3.55]. Basic moral principles are hereby assumed to rest on the principle of equal consideration of interests. But Singer returns to a crucial question that can be easily understood: "Can they suffer?"

In view of our astrobiological perspective, namely, endeavoring to understand the distribution of life in the universe, the basic moral principle of equality should be extended beyond equality for animals. We maintain that given the possibility of discovering a second genesis, the moral principle of equality should be envisaged in a truly universal context.

3.5   Astrobioethics is Subject to Constraints on Chance

The fact of whether Astrobioethics is subject to constraints on chance is a preliminary question to be answered, for it is an aspect for deciding whether we should expect an alternative morality. Indeed, it is instructive to appreciate the implications of several constraints on chance that apply in the wider context of astrobiology [3.11]. Analogous arguments also apply to constraints due to convergent evolution. These constraints are relevant to the question of whether life exists elsewhere, since such constraints have been discovered by careful analysis of terrestrial evolution.

Hence, analogous constraints will arise when we come into contact with the process of life in our exploration of the Solar System and beyond. Various examples of constraints on chance have been enumerated by the Belgian Nobel Laureate Christian de Duve [3.24] [3.26].

3.5.1   Not All Genes Are Equally Significant Targets for Evolution

The genes involved in significant evolutionary steps are few in number; they are the so-called regulatory genes. In these cases, mutations may be deleterious and are consequently not fixed. Besides, not every genetic change retained by natural selection is equally decisive. Such changes may lead more to increasing biodiversity, rather than contributing to a significant change in the course of evolution.

3.5.2   Evolutionary Changes Are Constrained

Given an evolutionary change that has been retained by natural selection, future changes are severely constrained; for example, once a multicellular body plan has been introduced, future changes are not totally random, as the viability of the organisms narrows down the possibilities. For instance, once the body plan of mammals has been adopted, mutations such as those that are observed in *Drosophila*, which exchange major parts of their body, are excluded. Such fruit fly mutations are impossible in the more advanced mammalian body plan.

Implicit in Darwin's work, chance is represented by the randomness of mutations in the genetic patrimony and their necessary filtering by natural selection. Astrobiology, which encompasses evolution in the cosmos, forces us to accept that randomness is built into the fabric of the living process. Yet, contingency represented by the large number of possibilities for evolutionary pathways is limited by a series of constraints. Natural selection seeks solutions for the adaptation of evolving organisms to a relatively limited number of possible environments. From astrochemistry we know that the elements used by the macromolecules of life are ubiquitous in the cosmos.

To sum up, a finite number of environments force upon natural selection a limited number of options for the evolution of organisms, so when we are wondering whether we will encounter microbial or non-human life, we can prepare ourselves to a certain extent to formulate in anticipation a suite of ethical principles for future astronauts to bear in mind.

We expect that convergent evolution will occur repeatedly, wherever life arises. In view of the assumed universality of biology [3.21] [3.23], it makes sense to search for the analogues of the attributes with their corresponding constraints that we have learned to recognize on Earth, such as intelligence and sensitivity to pain.

These remarks clearly militate in favor of the appropriateness of most searches for a second genesis, as currently undertaken by the major space agencies, including planetary protection in our cosmic neighborhood, as well as by the eventual information exchange that may be open to future generations thanks to the SETI project (cf. Glossary) [3.7].

One concluding thought aimed at inviting a broad cross section of scientists and technologists, the dialogue with philosophers has been held mainly by theoretical physicists and mathematicians [3.25], probably because of a common meeting ground in abstraction. The resulting cosmological picture consists of a comprehensive view of the physical world, leaving out the cosmic phenomenon of life, in all aspects—biochemical, molecular biological, evolutionary. All these specializations and specialists should be inserted into a comprehensive approach to the universe.

3.6   How Are We Going to Treat Non-Human Life Away from the Earth?

3.6.1   Can Ethical Behavior Be Extended into a Cosmic Context

To make sense of a rational approach to ethics with respect to non-human species, we should first clarify whether the cosmos is a well-determined ordered system, and hence intelligible to all observers. If so, then our ideas of ethical behavior can be extended into a cosmic context with no motivation for viewing the origin and evolution of the universe and life itself, as being absurd or pointless, as in an opposite to intelligible—existentialist—viewpoint (cf. Glossary).

Strong support has been given to intelligibility: Darwin developed an explanation of the origin of humans based on natural selection and the underlying random mutations. Clearly, this approach made it unnecessary, within the scope of the scientific method, to appeal to metaphysical concepts that cannot be rejected by experiments such as Divine Action. The development of a mathematical formulation of randomness and its increasing application to a wide range of scientific disciplines, such as physics and biology, allowed the substitution of the concept of necessity—determinism—for the concept of chance—probability—in order to explain both the existence of phenomena, as well as their type of structure.

3.6.2   Instrumentation for the Search of Life

Certain general ideas are clear, even before our technology will allow us to reach the possibility of developing instrumentation for the search of life on exoplanets. For instance, convergence in evolution is widespread and can be expected when we consider the possibility of being faced with a second genesis. This possibility is a suggestive term that was introduced by Christopher McKay to denote the presence of independent lines of evolution—the emergence of life in places other than on Earth (cf. Glossary)—that will inevitably present us with significant challenges arising from the ever-increasing pace of the exploration of the Solar System and, more generally, from the rapid progress of the space sciences. There is compelling motivation for a careful discussion of the frontier of science and the humanities. Amongst the priorities of the new science of astrobiology, we should consider the search for other lines of biological evolution elsewhere in the universe This search is likely to be relevant, not only to astrobiology, but also philosophy (ethics).

3.7   Ethical Principles in Early Proposals for the Search for Non-Human Life in the Solar System

3.7.1   Ethical Considerations in Previous Research in the Solar System

The foundational ethical principles for planetary protection in space exploration, whether for the extension of our knowledge of planetary science, or for the search for life in our local cosmic environment, has been extensively discussed in the literature. In particular, it has been noted that we need to examine with care the exploration of the Solar System taking into consideration our own environment or other planetary bodies [3.49]. However, we will not review this topic, as there are other reviews in the literature. Instead, we will restrict ourselves to different ethical considerations that have been closer to our previous research experience to illustrate aspects of the ethical concepts raised in this chapter, in which the universality of concepts discussed above apply.

The possibility of exploring Europa's habitability was raised very early. Soon after, the Galileo mission was proposed to examine the Jovian system with novel instrumentation—a submersible called a "hydrobot" coupled to an ice-breaker called a "cryobot" [3.35] [3.61]. For valid reasons, sometime later, the hydrobot-cryobot proposal was rightly criticized and subsequently abandoned [3.32], even though a suggested improved proposal was later defended [3.67].

The relevant question—a possibility that was not considered—was whether the cryobot was to be driven with nuclear energy in order to penetrate the Europan icy surface more efficiently. The question of considering nuclear-powered cryobots had arisen in the context of solar system exploration, including the Europan icy surface penetration with our previously suggested pair of cryobot-hydrobots. The levels of fission power and related harmful radiation are normally small for humans.

A study was carried out in order to investigate whether a surface fission power system was applicable for landed missions. In the course of their study, it became clear that the application of such a power system was possible for a wide range of missions for solar system exploration [3.63], especially for Mars [3.29]. Their study was concerned with fission electrical power of the order of a few kilowatts (kW), abbreviated as kWe. This level is needed for payloads that are feasible, since the alternative of radioisotope power systems (RPS) becomes unmanageable, being too massive and expensive.

3.7.2   Instrumentation That Might Harm Exo-Microorganisms

Even though the radiation levels are tolerable for humans, the possibility remains that once we envisage non-human species elsewhere in the Solar System, we are *a priori* ignorant of the harm we could make to other forms of life, even microorganisms. Profound ethical arguments arise due to the use of innovative instrumentation that may not consider harming potential encounters with microorganisms:

"What is wrong with killing with radiation a putative living community that is being searched?"

We believed, and still do, that living microorganisms are likely to dwell near hydrothermal vents on the Europan ocean floor. The need and relevance of searching for life on Europa had already been extensively discussed at that time, soon after the arrival of the Galileo probe in the Jovian System [3.6]. However, we should take due care when going beyond the early proposal of a cryobot technology housed inside the metal casing with banks of electrical heaters, which cause the probe to descend vertically through the ice driven by gravity [3.35].

Special care is needed if going beyond harmless gravity into harmful fission power to help cryobot penetration. Unlike the original cryobot-hydrobot, an appropriate technology which in principle poses no ethical difficulties for the exploration of the Solar System planets and satellites is the penetrator [3.64] [3.65], which was furthered developed by the UK Penetrator Consortium for lunar surficial research [3.57]. These instruments consist of small projectiles that can be delivered at high velocity to reach just beneath the surface of planets or their satellites for probing samples of surficial chemical elements, amongst other investigations for a review and additional references, cf. [3.8], p. 162). If budgetary constraints force a choice between penetrators and landers, some advantages of the penetrator approach are evident: the low mass of these instruments combined with their agility in deployment, make them worthy complements to orbiter missions launched without landers.

Penetrators have a long history of feasible technological development by several space agencies. Penetration of the surfaces of the icy worlds of the Solar System to investigate the possible presence of biosignatures by means of penetrators was discussed in our earlier paper [3.31] and in others [3.63]. More recent options for surficial exploration with penetrators have been suggested [3.36] [3.59].

3.8   Conclusion

In view of our consideration of ethical questions that are part of our philosophical patrimony, as well as with the options we may encounter in the near future in space exploration, we may conclude that a new (second) morality in astrobioethics is unnecessary. This point of view has been reinforced in our work, especially with extrapolations of what philosophers have learned on Earth about our ethical treatment of non-human species—animals.

Glossary

Agnosticism: The doctrine according to which humans cannot know the existence of whatever lies beyond natural phenomena on which we could perform experimental tests. The term has also been applied in a slightly different context: skepticism in religious matters. Progress of science since the Renaissance has supported agnosticism to a certain extent. In spite of this, science is not in a position to reject faith which lies beyond its own horizons.

Astrotheology: This is an aspect of humanism that discusses the space sciences and standard topics of theology for the purpose of constructing a comprehensive and meaningful understanding of our human situation within the universe. Within the growing field of Science and Religion, astrotheology, as well as astrobioethics, are also concerned with questions raised by astrobiology [3.44].

Biogeocentrism: A term that reflects a tendency observed in some contemporary scientists, philosophers and theologians, according to which life is only likely to have occurred on Earth.

Convergent evolution: In biology and biochemistry, features that become more, rather than less, similar through independent evolution will be called "convergent" [3.9] [3.19]. Convergence in biology is associated with similarity of function, as in the evolution of wings in birds and bats. At the biochemical level, convergent evolution is manifest in whole proteins: serine proteases have evolved independently in bacteria (e.g., subtilisin) and vertebrates (e.g., trypsin): in each, the same set of three amino acids form the active site [3.27].

Empiricists: Believers that ethics arises from the human mind.

Ethical relativism: Maintains that ethics is relative to the moral standards of a given society and may differ in different cultures.

Ethics (moral philosophy): A branch of philosophy closely associated with anthropology, economics, politics, and sociology. As in all philosophical pursuits, it investigates the basic concepts underlying a human activity. It is concerned with what is morally good and bad with a given activity, right and wrong. Ethics is divided into metaethics, normative ethics, and applied ethics.

Existentialism: This philosophical thesis considered human life as a contingent phenomenon. According to these philosophers, life in the universe is not intelligible (it is rather pointless and absurd). Existentialism is an attempt to live logically in a universe that is ultimately absurd. They further addressed the isolation of man in what was considered to be an alien universe.

Gould's metaphor: Stephen Jay Gould raised the question that if one could imagine rewinding the origin and evolution of life back to its initial starting point, the same results would follow when such a "tape" ran forward again [3.30].

Meta-ethics: This concept means that we are not directly concerned with arguing about ethics not covering internal topics such as the question of why killing is wrong.

New Atheism: New atheism describes a recent radical position promoted by a minority of atheists, in which there is emphasis on the absence of cultural frontiers between science, philosophy and theology.

Second Genesis: This is a suggestive term introduced by Christopher McKay to denote the presence of independent lines of evolution—the emergence of life in places other than on Earth [3.11] [3.41].

SETI project: Frank Drake, one of the most important pioneers in astrobiology, suggested searching for extraterrestrial intelligence with the SETI (search for extraterrestrial intelligence) Project. His main objective was to test the hypothesis that other intelligent forms of life may have evolved elsewhere on planets, or moons around stars besides our own.

Transcendentalists: Believers that moral guidelines are independent of the human mind.

References

[3.1]   Aristotle, *Nichomachean Ethics*, p. 229, Dover, New York, 1998.

[3.2]   Augustine, St., *Concerning the City of God against the Pagans*, cf., Introduction by J.J. O'Meara, p. xxii, Penguin Classics, London, 417-427 AD.

[3.3]   Ayala, F.J., The Evolution of Life: An overview, in: *Evolutionary and Molecular Biology: Scientific Perspectives on Divine Action*, R.J. Russell, W.R. Stoeger, F.J. Ayala, (Eds.), pp. 21–57, Vatican Observatory & Center for Theology and Natural Sciences, Vatican City State/Berkeley, CA, 1998.

[3.4]   Bhala, K.T., The Philosophical Foundations of Financial Ethics, in: *Research Handbook on Law and Ethics in Finance and Banking*, Edward Elgar Publishing, Cheltenham, UK, 2019.

[3.5]   Bod, R., *A New History of the Humanities*, p. 384, Oxford University Press, New York, USA, 2014.

[3.6]   Chela-Flores, J., The Phenomenon of the Eukaryotic Cell, in: *Evolutionary and Molecular Biology: Scientific Perspectives on Divine Action*, R.J. Russell, W.R. Stoeger, F.J. Ayala (Eds.), pp. 79–99, Vatican Observatory and the Center for Theology and the Natural Sciences, Vatican City State/ Berkeley, California, 1998.

[3.7]   Chela-Flores, J., Testing the Drake Equation in the solar system, in: *A New Era in Astronomy*, G.A. Lemarchand and K. Meech (Eds.), vol. 213, pp. 402–410, Astronomical Society of the Pacific Conference Series, San Francisco, California, USA, 2000.

[3.8]   Chela-Flores, J., *The New Science of Astrobiology from Genesis of the Living Cell to Evolution of Intelligent Behavior in the Universe*, p. 279, Kluwer Academic Publishers, Dordrecht, The Netherlands, 2001.

[3.9]   Chela-Flores, J., Testing evolutionary convergence on Europa. *Int. J. Astrobiology*, 2, 307–12, 2003.

[3.10] Chela-Flores, J., Fitness of the cosmos for the origin and evolution of life: from biochemical fine-tuning to the Anthropic Principle, in: *Fitness of the Cosmos for Life: Biochemistry and Fine-tuning*, J.D. Barrow, S.C. Morris, S.J. Freeland, C.L. Harper (Eds.), pp. 151–166, Cambridge University Press, Cambridge, Great Britain, 2008.

[3.11] Chela-Flores, J., *A Second Genesis: Stepping Stones Towards the Intelligibility of Nature*, p. 248, World Scientific Publishers, Singapore, 2009.

[3.12] Chela-Flores, J., *The Science of Astrobiology A Personal Point of View on Learning to Read the Book of Life*, Second Edition. Book series: Cellular Origin, Life in Extreme Habitats and Astrobiology, p. 250, Springer: Dordrecht, The Netherlands, 2011, http://www.ictp.it/~chelaf/ss220.html.

[3.13] Chela-Flores, J., *Astrobiology and Humanism: Conversations on the frontiers of science, philosophy and theology*, Cambridge Scholars Publishing, Newcastle upon Tyne, United Kingdom, 2019, http://www.ictp.it/~chelaf/ss220.html.

[3.14] Chela-Flores, J. and Seckbach, J., Divine Action and Evolution by Natural Selection, in: *Divine Action and Natural Selection: Science, Faith and Evolution*, J. Seckbach and R. Gordon (Eds.), pp. 1034–1048, WSP, Singapore, 2008.

[3.15] Chon-Torres, O.A., Astrobioethics. *Int. J. Astrobiology*, 17, 51–56, 2018.

[3.16] Chon-Torres, O.A., Disciplinary nature of astrobiology and astrobioethic's epistemic foundations. *Int. J. Astrobiology*, 1–8, 2018, https://www.semanticscholar.org/paper/Disciplinary-nature-of-astrobiology-and-epistemic-Chon-Torres/4a19ed476f0e71138c3d377f17cb2f6befdd9421.

[3.17] Cockell, C., Essay on extraterrestrial liberty. *J. Br. Interplanet. Soc*, 61, 255–275, 2008.

[3.18] Conway-Morris, S., *The Crucible of Creation. The Burgess Shale and the Rise of Animals*, pp. 9–14, 222–223, Oxford University Press, New York, 1998.

[3.19] Conway Morris, S., *Life's Solution: Inevitable Humans in a Lonely Universe*, Cambridge University Press, Cambridge, 2003.

[3.20] Cornwell, J., *Darwin's Angel: A Seraphic Response to "The God Delusion"*, Profile Books, London, 2007.

[3.21] Dawkins, R., Universal Darwinism, in: *Evolution form Molecules to Men*, D.S. Bendall (Ed.), pp. 403– 425, Cambridge University Press, London, 1983.

[3.22] Dawkins, R., *The God Delusion*, Bantam Books, New York, USA, 2006.

[3.23] Dawkins, R., *Science in the Soul Selected writings of a passionate rationalist Richard Dawkins*, G. Somerscale (Ed.), pp. 119–150, Random House, New York, 2017.

[3.24] de Duve, C., *Vital Dust. Life as a Cosmic Imperative, Basic Books*, pp. 296–297, A Division of Harper-Collins Publishers, New York, 1995a.

[3.25] de Duve, C., *Vital dust: Life as a cosmic imperative*, p. 362, Basic Books, New York, 1995b.

[3.26] de Duve, C., *Life Evolving Molecules Mind and Meaning*, New York, Oxford University Press, 2002.

[3.27] Doolittle, R.F., Convergent evolution: the need to be explicit. *Trends Biochem. Sci.*, 19, 15–18, 1994.

[3.28] Ekers, R., Cullers, K., Billingham, J., Scheffer, L. (Eds.), *SETI 2020: A Roadmap for the Search for Extraterrestrial Intelligence*, p. 549, SETI Press, Mountain View, California, 2002.

[3.29] Elliott, J.O., Lipinski, R.J., Poston, D.I., Mission Concept for a Nuclear Reactor-Powered Mars Cryobot Lander, in: *AIP Conference Proceedings*, vol. 654, p. 353, 2003, https://doi.org/10.1063/1.1541314.

[3.30] Gould, S.J., *Wonderful Life*, W. W. Norton and Co, New York, 1989.

[3.31] Gowen, R.A., Smith, A., Fortes, A.D., Barber, S., Brown, P., Church, P., Collinson, G., Coates, A.J., Collins, G., Crawford, I.A., Dehant, V., Chela-Flores, J., Griffiths, A.D., Grindrod, P.M., Gurvits, L.I., Hagermann, A., Hussmann, H., Jaumann, R., Jones, A.P., Joy., A., Sephton, K.H., Karatekin, O., Miljkovic, K., Palomba, E., Pike, W.T., Prieto-Ballesteros, O., Raulin, F., Sephton, M.A., Sheridan, M.S., Sims, M., Storrie-Lombardi, M.C., Ambrosi, R., Fielding, J., Fraser, G., Gao, Y., Jones, G.H., Karl, W.J., Macagnano, A., Mukherjee, A., Muller, J.P., Phipps, A., Pullan, D., Richter, L., Sohl, F., Snape, J., Sykes, J., Wells, N., Penetrators for *in situ* sub-surface investigations of Europa. *Adv. Space Res.*, 48, 725–742, 2011.

[3.32] Greenberg, R., *Europa – The Ocean Moon Search for An Alien Biosphere*, Springer, New York, 2005.

[3.33] Harris, S., *The End of Faith: Religion, Terror, and the Future of Reason*, W. W. Norton & Company, London, 2004.

[3.34] Hitchens, C., *God Is Not Great: The Case Against Religion*, p. 320, Atlantic Books, London, 2007.

[3.35] Horvath, J., Carsey, F., Cutts, J., Jones, Johnson, J., Landry, E., B., Lane, L., Lynch, G., Chela-Flores, J., Jeng, T.-W., Bradley, A., Searching for ice and ocean biogenic activity on Europa and Earth, in Instruments, Methods and Missions for Investigation of Extraterrestrial Microorganisms, *Proc. SPIE*, vol. 3111, pp. 490–500, 1997, http://www.ictp.it/~chelaf/searching_for_ice.html.

[3.36] Luo, H., Li, Y., Liu, G., Yu, C., Chen, S., *Buffering Performance of High-Speed Impact Space Penetrator with Foam-Filled Thin-Walled Structure*, Hindawi, Shock and Vibration, London, UK, 2019, https://doi.org/10.1155/2019/7981837.

[3.37] Martini, C.M. and Chela-Flores, J., Dialogo, in: *Carlo Maria Martini Orizzonti e limiti della scienza Decima Cattedra di non credenti*, E. Sindoni and C. Sinigaglia (Eds.), pp. 65–68, Raffaello Cortina Editore, Milan, 1999.

[3.38] Martini, C.M. and Coyne, G., Dialogo, in: *Carlo Maria Martini Orizzonti e limiti della scienza Decima Cattedra di non credenti*, E. Sindoni and C. Sinigaglia (Eds.), In the series "Scienze e Idee", directed by G. Giorello, pp. 37–40, Raffaello Cortina Editore, Milan, 1999.

[3.39] McGrath, A.E., *Science & Religion: A New Introduction*, Second Edition, p. 264, Wiley-Blackwell, Hoboken, New Jersey, U.S.A., 2010.

[3.40] Mayr, E., *The search for extraterrestrial intelligence in Extraterrestrials. Where are they?*, 2nd ed., B. Zuckerman and M.H. Hart (Eds.), pp. 152–156, Cambridge University Press, London, 1995.

[3.41] McKay, C.P., The search for a second genesis in the solar system, in: *The First Steps of Life in the Universe*, J. Chela-Flores, T. Owen, F. Raulin (Eds.), pp. 269–277, Kluwer Academic Publishers, Dordrecht, The Netherlands, 2001.

[3.42] McMullin, E., Tuning fine-tuning, in: *Fitness of the Cosmos for Life: Biochemistry and Fine-Tuning*, J. Barrow, S. Conway Morris, S. Freeland, Jr Harper (Eds.), pp. 70–94, Cambridge Astrobiology. Cambridge University Press, Cambridge, UK, 2007.

[3.43] Monod, J., *Chance and Necessity An Essay on the Natural Philosophy of Modern Biology*, p. 136, Collins, London, 1972.

[3.44] Peters, T., *Astrotheology*, Oxford Research Encyclopedia of Religion, Oxford, United Kingdom, 2019a.

[3.45] Peters, T., Astrobiolgy and Astrotheology in Creative Mutual Interaction, in: *Theology and Science: From Genesis to Astrobiology*, J. Seckbach and R. Gordon (Eds.), pp. 25–43, World Scientific Press, Singapore, 2019b.

[3.46] The Future of World Religions: Population Growth Projections, 2010-2050, Pew Research Center, Washington, D.C., USA, 2015. www.pewforum.org/2015/04/02/religious-projections-2010-2050/.

[3.47] Plato, *The Republic*, 1987, Penguin Classics, London, Great Britain, 360 BC.

[3.48] Polkinghorne, J.C., *Scientists as Theologians*, pp. 46–47, SPCK Publishing, London, Great Britain, 1996, Chapter 4: Cosmic Scope, especially.

[3.49] Rummel, J.D., Race, M.S., Horneck, G., The Princeton Workshop Participants, Ethical considerations for planetary protection in space exploration: a workshop. *Astrobiology*, 12, 1017–1023, 2012.

[3.50] Russell, B., *History of Western Philosophy and its Connection with Political and Social Circumstances from the Earliest Times to the Present Day*, p. 562, Routledge, London, 1991a.

[3.51] Russell, B., *History of Western Philosophy and its Connection with Political and Social Circumstances from the Earliest Times to the Present Day*, p. 13, Routledge, London, 1991b.

[3.52] Russell, B., *Religion and Science*, pp. 49–81, Oxford University Press, New York, 1997.

[3.53] Russell, R.J., Life in the Universe: Philosophical and Theological Issues, in: *The First Steps of Life in the Universe*, J. Chela-Flores, T. Owen, F. Raulin (Eds.), pp. 365– 374, Kluwer Academic Publishers, Dordrecht, The Netherlands, 2001.

[3.54] Seckbach, J. and Gordon, R. (Eds.), *Theology and Science: From Genesis to Astrobiology*, p. 417, World Scientific Press, Singapore, 2019.

[3.55] Singer, P., *Practical Ethics*, Second edition, pp. 55–62, Cambridge University Press, United Kingdom, 1993.

[3.56] Singer, P. (Ed.), *Ethics*, p. 415, Oxford University Press, United Kingdom, 1998.

[3.57] Smith, A., Crawford, I.A., Gowen, R.A., Ambrosi, R., Anand, M., Banerdt, B., Bannister, N., Bowles, N., Braithwaite, C., Brown, P., Chela-Flores, J., Cholinser, T., Church, P., Coates, A.J., Colaprete, T., Collins, G., Collinson, G., Cook, T., Elphic, R., Fraser, G., Gao, Y., Gibson, E., Glotch, T., Grande, M., Griffiths, A., Grygorczuk, J., Gudipati, M., Hagermann, A., Heldmann, J., Hood, L.L., Jones, A.P., Joy, K., Khavroshkin, O.B., Klingelhoefer, G., Knapmeyer, M., Kramer, G., Lawrence, D., Marczewski, W., McKenna-Lawlor, S., Miljkovic, K., Narendranath, S., Palomba, E., Phipps, A., Pike, W.T., Pullan, D., Rask, J., Richard, D.T., Seweryn, K., Sheridan, S., Sims, M., Sweeting, M., Swindle, T., Talboys, D., Taylor, L., Teanby, N., Tong, V., Ulamec, S., Wawrzaszek, R., Wieczorek, M., Wilson, L., Wright, I., Lunar Net —A proposal in response to an ESA M3 call in 2010 for a medium sized mission. *Exp. Astronomy*, 33, 2, 587–644, 2012.

[3.58] Spinoza, B.de, *Ethics and Treatise on the Correction of the Intellect*, p. 306, Everyman, London, 1993.

[3.59] Ahrens, C.J., Paige, D.A., Eubanks, T.M., Blase, W.P., Mesick, K.E., Zimmerman, W., Petro, N., Hayne, P.O., Price, S., Small penetrator instrument concept for the advancement of lunar surface science. *Planet. Sci.*, 2, 38, 2021.

[3.60] Townes, C.H., *Making Waves*, p. 166, American Institute of Physics, Woodbury, NY, 1995.

[3.61] Trowell, S., Wild, J., Horvath, J., Jones, J., Johnson, E., Cutts, J., Through the Europan Ice: Advanced Lander Mission Options, in: *The Europa Ocean Conference*, San Juan Capistrano Research Institute, San Juan Capistrano, CA, p. 76, 1996.

[3.62] Tucker Abbott, R., *Compendium of land shells*, pp. 7–8, American Malacologists, Melbourne, Florida, USA, 1989.

[3.63] Ulamec, S., Biele, J., Funke, O., Engelhardt, M., Access to glacial and sub-glacial environments in the solar system by melting probe technology. *Rev. Environ. Sci. Biotechnol.*, 6, 71–94, 2007.

[3.64] Weiss, P., *System Study and Design of a Multi-Probe Mission for Planetary In-Situ Analysis*. Ph.D. Thesis, p. 178, The Hong Kong Polytechnic University, Hong Kong, 2010.

[3.65] Weiss, P., Yung, K.L., Ng., T.C., Komle, N., Kargl., G., Kaufmann, E., Study of a melting drill head for the exploration of subsurface planetary ice layers. *Planet. Space Sci.*, 56, 1280–1292, 2008.

[3.66] Wilson, E.O., *Consilience The Unity of knowledge*, Alfred A. Knopf, New York, 1998, Chapter 11 Religion and Ethics.

[3.67] Zimmerman, W., Bryant, S., Zitzelberger, Nesmith, B., A radioisotope powered cryobot for penetrating the Europan ice shell. *AIP Conf. Proc.*, vol. 552, pp. 707–715, 2001.

4

Who Goes There? When Astrobiology Challenges Humans

Jacques Arnould

Centre National d'Etudes Spatiales, Paris, France

Abstract

"Who goes there?" For millennia, human beings have wondered about the existence and nature of extraterrestrial beings. For a long time, this question was reserved for the fields of philosophy and theology. Now, with the birth of the space age, this issue has become more pressing and concrete. While the hypothesis of extraterrestrial life continues to question the limits of theological systems to recognize those of scientific knowledge, it also leads to an expanded application of the principles of precaution and responsibility beyond terrestrial borders. To seek to offer an extraterrestrial answer to the question "Who goes there?" is also to accept knowing more about human nature.

Keywords: Copernican principle, extraterrestrial life, precautionary principle, principle of responsibility, theology, heresy

4.1 Introduction

For millennia, human beings have questioned the existence and nature of extraterrestrial beings, whether they worship, wait for them or fear them. With the birth of the space age, this questioning took on a more scientific dimension: we are now talking about astrobiology. In addition, extraterrestrial life remains a hypothesis that refers humans to the knowledge of the limits of their knowledge, to the extent of the consequences of their

---

Email: jacques.arnould@cnes.fr
Jacques Arnould: http://jacques-arnould.com/

Octavio Alfonso Chon Torres, Ted Peters, Joseph Seckbach and Richard Gordon (eds.) Astrobiology: Science, Ethics, and Public Policy, (79–94) © 2021 Scrivener Publishing LLC

actions, and to the awareness of the responsibility of their choices. It's up to them to answer the question, "Who's going there?"

4.2   The Copernican Revolution

Bernard Le Bouyer de Fontenelle, perpetual secretary of the Paris Academy of Sciences, once chose "to know how this world we inhabit is made, if there are other similar worlds, and which are inhabited too"; in 1686 he published his famous *Entretiens sur la pluralité des mondes* (Conversations on the plurality of worlds). In this book, the scientist decided to convince a learned marquise that the Moon is inhabited:

> "Suppose that there was never any trade between Paris and Saint-Denis, and that a bourgeois of Paris, who never left his city, is on the towers of Notre-Dame, and Saint-Denis[1] from afar; he will be asked if he believes that Saint-Denis is inhabited like Paris. He will boldly say no; for, he will say, I see the inhabitants of Paris, but those of Saint-Denis I do not see them, we have never heard of them. [...] Our Saint-Denis is the Moon, and each of us is this bourgeois of Paris, who never left his city."

Fontenelle wrote these lines eighty years after Galileo's first astronomical observations of the Moon, yet his demonstration remains accurate: to what extent can we judge things, their existence and appearance, when we are too far away from them and are therefore forced to make assumptions based on the incomplete information we have? From the towers of Notre-Dame or through a telescope, our understanding of reality remains partial and limited. Fontenelle himself acknowledges this:

> "We want to judge everything, and we are always in a bad position. We want to judge ourselves but we are too close; we want to judge others but we are too far away."

We are therefore most often obliged to develop speculation; to try to see beyond our eyes and telescopes, we are forced to formulate laws, to set principles.

We owe Copernicus one of the first and most famous principles of modern astronomy: the *Copernican Principle*, named after the Polish canon and scholar, that postulates that there is no privileged point of view in the

---

[1]"Saint-Denis" is a town in the suburbs of Paris; "Notre-Dame" is the cathedral of Paris that burned down in April 2019.

universe, especially none that is related to the human observers that we are. This principle means that cosmological models established by terrestrial astronomers must be able to be validated from observations made from anywhere in the universe. The Copernican principle has three consequences: the Earth is not the only one to lose the privileged status of center; the Copernican principle rejects any temptation, any attempt to give a center to the universe; universe, according to the consecrated formula, is "a sphere whose center is everywhere and circumference nowhere."

To this principle are attached two others: the *Cosmological Principle* that the universe is globally homogeneous and isotropic; and the *Principle of Mediocrity* that our solar system is a mundane system. These three principles constitute what historians of science and philosophers call the "Copernican revolution." Accordingly, the Earth and its inhabitants are not the center of the cosmos and therefore are not, according to previous philosophical or religious convictions, the summit of the divine work of creation (on the grounds that humans had been created in the image and according to the likeness of God) or, on the contrary, the cloaca of the cosmos, a place of imperfection and death. The Copernican revolution and early modern astronomers promoted the Earth to the same rank as the stars.

4.3   Religious Reactions to the Copernican Revolution

With this conclusion, the earthlings quickly imagined themselves as companions of the stars and explorers of the universe: "Let us create ships and sails adapted to the celestial ether," Johannes Kepler wrote as early as 1610 to his colleague Galileo, "and there will be plenty of people unafraid of the empty spaces." Humanity would conquer its place in the universe. However, the fate of those who had occupied this "celestial ether" had yet to be resolved.

For, in this revolution of cosmic dimensions, the promotion of some led to the degradation of others, to the point of threatening their existence: one day, Yuri Gagarin, returning from the first flight of an earthling in space, will claim not to have met God there; or rather Nikita Khrushchev will lend him that word. While waiting for the time of the first space journeys, theologians and philosophers were forced to speculate on the status of celestial beings, since the boundaries that separated them from the earthlings could be crossed by the earthlings, and even permanently deleted.

As soon as the Copernican revolution was triggered, theologians began to discuss among themselves the status of the possible inhabitants of other worlds: belittled to the rank of human creatures, had they also "eaten the apple" of an alien Garden of Eden, contracting a variant of original sin? Was it necessary to advance the frightening assumption that Christ would be forced to visit each of the planets populated by alien sinners, to suffer passion and to rise again every time to save them?

These religious perspectives, without reaching the level of heresy of pluralistic thought defended by Giordano Bruno at the end of the 16th century, nevertheless brought Valladolid's controversy back to the level of a mere moral point. Admitting that Indians and "all other peoples who may later be discovered by Christians" must be considered "true human beings" required a bull and a letter from Pope Paul III, and then a trial that ended in 1551; but the question of the status of the Indians and the existence of their souls did not involve the very foundation of the Christian tradition, the Christological dogma.

This was not the case with the possible existence of intelligent creatures on planets other than Earth: the leaders of the churches of the time, in Rome, Geneva or Wittenberg, were horrified. Fortunately, Bruno's pyre was the last to be lit by the Inquisition on Campo dei Fiori in Rome, and Galileo's trial was the last one against a modern-day scientist. The plurality of worlds has still sparked many disputes between astronomers, philosophers and theologians, as in 19th-century England, and contemporary advances in astrobiology have often revived them. However, the theological dimension of the plurality of worlds is no longer used today for the slightest scientific argumentation, but only for theological debate in order to illuminate the elements of the Christian tradition that seem most obvious as the most basic.

It is often necessary to go to the extreme limit to measure the importance, the difficulty, the "thickness" of a question. Vercors, in his book *Les animaux dénaturés* (which he adapted for the theatre under the title *Zoo ou l'Assassin philanthrope*), imagines the discovery by explorers of a people of half-human half-ape beings, who could be representatives of the missing link in the evolution of the human species. Faced with the impossibility of reaching a scientific consensus about these beings belonging to the human species, one of the heroes of the novel ends up killing the offspring of a female, artificially fertilized with human sperm: to judge this act, the court is compelled to decide whether or not the victim belonged to the human race. The question of the redemption of extraterrestrials and, more generally, of the status to be granted to them, is not far removed from the staging of Vercors: if it is hypothetical, if it is speculative, it nevertheless illuminates the limits of our knowledge as well as those of our beliefs.

C. S. Lewis, the English author better known for *The Chronicles of Narnia* than for his apologetic commitment and his work defending the Christian faith entitled *Mere Christianity*, wrote in the 1950s: "I think a

Christian can admit himself to be happy if his faith never encounters more formidable difficulties than these conjectural ghosts." By these strange "conjectural ghosts," Lewis referred precisely to hypothetical extraterrestrial beings and probably was not wrong: the earthly reality (the condition of our existences, our businesses and our knowledge, our hopes and our fears as human beings) seems sufficient to question, sometimes unceremoniously, our religious traditions and our ideological systems.

4.4   Astrobiology and Speculation

Certainly, "we want to judge everything, and we are always in the wrong light," but it is up to our nature to ask ourselves what is the case with the inhabitants of Saint-Denis as well as the Moon. Speculation belongs to our human nature as much as our imagination. We simply have to pay attention to the reduction we make when we subject reality, in any form, to the dimensions of our own capacities for knowledge; we must recognize that, far from being limited to extraterrestrial territories, this exploration always sends us an image of ourselves. All speculation is the work of a Narcissus.

Fortunately, one of the attractions of science, one might write, is to "turn and rest our eyes on an object that is not ourselves at last." The discipline that today calls itself exobiology or astrobiology seems to correspond to such a perspective since it has taken on the object of study or, more precisely, researched what is known as extraterrestrial life. This attraction seems all the more pronounced because this discipline, juvenile in its denomination, methods and results, has roots which feed on the philosophical reflections of Democritus and Aristotle as well, with the modern era, on the science fiction litterary works. At least, astrobiology obliges to define life.

At first glance, the scientific interest in extraterrestrial life lies in the spirit that animates modern science, especially the pretense to observe phenomena for themselves, as they would happen apart from ourselves. This claim to a form of objectivity lacks neither foundation, good will, nor even success: what Sigmund Freud calls the "destruction of the narcissistic illusion" has been illustrated and relied upon by three scientific, astronomical, biological and psychoanalytic revolutions, following the work of Copernicus, Darwin and Freud. Therefore, discovering beings beyond the borders of our Earth or, even worse, being visited by them would constitute a fourth and perhaps ultimate revolution. But should we not apply to all human pretensions the formula carved out by Arthur C. Clarke about the believing attitude? "A faith that cannot survive the collision with the truth deserves little regret." Does knowledge that cannot survive the collision with the truth deserve more regrets?

4.5   Heretics

"We say, pronounce, judge and declare that you, Brother Giordano Bruno of the Order of Preachers, are a heretic, unrepentant, tenacious and obstinate." A few days later, on 17 February 1600, the sentence handed down against the most famous rebel of his time led him to a pyre at Campo dei Fiori in Rome. "I think he will have gone to tell, in these other worlds that he had imagined, how the Romans used to treat the ungodly and the blasphemers," said Caspar Schoppe, a witness to the ordeal.

In the eyes of his Roman judges, such as representatives of churches born of the Reformation who, in Geneva and Wittenberg, had also declared him heretical before excommunicating him, Bruno did not deserve to be condemned solely on the grounds of having defended the plurality of worlds; his file contained other theologically and philosophically heavier exhibits. But the idea that in more or less evolved living forms, "life" and, why not, "soul" could exist elsewhere than on Earth, this idea inevitably led to think (and, for these men, to believe) out of the box, outside the framework accepted by the religious tradition to which they belonged and of which they considered themselves the guarantors of orthodoxy. Such an idea had to be fought and its author severely punished by exclusion or even death.

In reality, these men feared neither the strange nor the supernatural; perhaps they were even more accustom to it than we are today: the strange had for them a consistency which today can make us smile, astonish us, perhaps frighten us; the strange belonged to the framework of their cultures, their religious convictions, their dogmas. But this was not the case of the extraterrestrial who at first aroused a real and deep fear because it threatened the very foundation of their religion, of their faith: the extraterrestrial seemed to put God himself in peril or, rather, the image they had built. Is it the same today?

On the occasion of a somewhat exceptional discovery in the field of exoplanets or astrobiology, journalists rarely fail to question the director of the Vatican observatory: could the discovery of extraterrestrial life not lead to the ruin of Christianity? The scientific priest always responds with reassuring words and sometimes goes so far as to evoke the possible baptism of these beings discovered in an extraterrestrial New World. For her part, Jill Tarter, considered the "pope" of the search for extraterrestrial intelligence (SETI), is convinced that the discovery and assured existence of an extraterrestrial intelligence would signal the end of the monotheistic religions that have merely maintained regimes of war and destruction on Earth. They would be, Tarter continues, necessarily competing with extra- terrestrial religious and philosophical systems; probably wiser than ours, she is persuaded.

Both misunderstand the "power" of the extraterrestrial: the ecclesias- tical underestimates it, the American astronomer overestimates it. The Christian religion remained too anthropocentric, despite the efforts of men like Pierre Teilhard de Chardin, not to suffer from the discovery of other beings who could also be created in the image of God; but humans are too naturally religious, too naturally inclined to ritualize their existence, to refer it to powers other than their own, so as not to rush to replace one religion in difficulty with another.

In any case, the extraterrestrial, if it were to exist, would not leave any human being indifferent, any philosophical system untouched, any scien- tific paradigm unscathed. The Giordano Bruno case teaches us in its own way: "by nature" the extraterrestrial is a vehicle, a surrogate of heresy.

No one can claim the title of heretic: it is not enough to profess or to support an opinion or a doctrine that claims to be contrary to conventional wisdom; a competent authority still needs to hear a case, conduct a trial, pronounce a judgment. However, the etymology of the term introduces an additional nuance to this institutional approach: heretics exercise their right to implement their ability to choose one idea, one conviction, one opinion over another, in this case rather than the more official one.

The heretic is a being with free thought. It is for this reason that he frightens the guarantors of order: whether he likes it or not, his life, his thoughts, his teachings, his mores call into question all forms of author- itarianism and dogmatism, in other words any exaggerated exercise of authority, everything a narrow-minded teaching of dogma. A heretic like Giordano Bruno does not preach relativism but the link between ideas, beings, things; it calls into question borders and walls of any kind. A her- etic like Bruno does not rule out dogmas, but refuses to turn them into systems that paralyze the act of thinking. On the contrary, dogma must serve as a marker to leave the havens of intellectual and spiritual peace, to face the darkness of ignorance, to explore the untouched areas on the maps of our sciences, our techniques, our beliefs. Heresy and exploration have more than one trait in common.

4.6   The Many Worlds Hypothesis

Certainly, the hypothesis of the plurality of worlds, of the existence of extraterrestrial life still belongs (and no one can say for how much longer) to the fascinating world of "conjectural ghosts"; but it has already shown its ability to inspire and to fuel revolutions and it will undoubtedly still be able to do so in the future. It would therefore be wrong and unwise to underestimate the "powers" of this heresy: even treated by fire, it has left traces, sometimes even cracks, in apparently solid buildings. Just think of the existence, within the divine creation, of other beings created by God in his image, the controversy over the hypothetical fossilized bacteria discovered in the Martian meteorite ALH84001, or even the affair of the canals of Mars and the radio hoax made in 1938 by Orson Welles. Without praising it, it seems fair to support the pressure and the heretical tension that represents the extraterrestrial, even if, apart from periods richer in discoveries or hypothesis, simple precautionary measures are sufficient to keep the open questions and possible debates, even if they already appear as desecration.

4.7   Desecration of Planets Beyond Earth

"Life on Mars!": the title is spread out on the cover of a French popular science magazine, with the illustration of an American Mars mission. Enough to attract the eye, attention and curiosity of a reader who is a fan of astronomical sciences... until he reads, in small print, the subtitle of the catchphrase: "It was the man who brought it." The writing of the magazine thus marks a double success: that of addressing the question of extraterrestrial life, whose media interest does not suffer any decline, and that of highlighting a phenomenon directly related to the exploration of space by man, the possible and even probable human contamination of the planets it has undertaken to explore.

For nearly sixty years, ever since we began to lay probes and then humans on the surface of worlds other than our Earth (the Moon, Mars, the comet Churi), since we brought back to Earth these ships and their occupants (as during the Apollo missions) or space materials (as the Japanese Hayabusa missions carried out), we participate, with a "success" and an "efficiency" that remains to be measured, in (hypothetical) phenomena of panspermia. We have become extraterrestrials from other space lands.

Today, specialists readily distinguish between *sensu stricto* contamination (i.e., the deposit of materials and structures, products and sometimes beings in places that are foreign to them) and proliferation (reproduction and dissemination of all that has the power). Therefore, while it is possible that Mars exploration missions deposited organisms of Earth's origin on the surface of the Red Planet, obviously of very small size and in a dormant state, they were most likely not able to reproduce and proliferate under the conditions on the surface or even in the first centimeters of soil on Mars: as we know them today, they are indeed unfavorable to the proliferation of organisms from Earth. However, how can we be sure that they do not become so in the future? Without pouring into science fiction, it would therefore be irresponsible of us not to question the possible consequences of our projects, of our space exploration programs, since they are not limited to the observation of the sky using instruments installed on Earth.

The Committee on Space Research (COSPAR) was established in 1958 on the occasion of the International Geophysical Year; it is now mandated by the United Nations to develop the recommendations that form the basis of what is now known as planetary protection. The policy of planetary protection introduced and defended by COSPAR is therefore clear here: it is a question of protecting both scientific research (in this case, the possibility of finding extraterrestrial life forms) and the Earth itself; in other words, the first objective is not to protect extraterrestrial environments.

Let us begin by clarifying the position of the researchers, as expressed in this policy. A first observation is necessary: caution is required. The existence of life in the solar system outside of Earth is declared unlikely, which means in practice that the measures developed and implemented by COSPAR are a precautionary measure, not a prevention. Prevention would have been an issue if the risks had been considered proven, their existence known and demonstrated, their occurrence and their consequences reasonably estimated; this is not the case. In other words, the risk of contaminating a planet during a scientific expedition remains hypothetical: it escapes any possibility of an estimate, an assessment of its probability as well as its conditions and effects. And the discovery of exoplanets, the active study of areas of habitability fundamentally do not change this impossibility: it only increases the value of a probability that remains unknown to us today. Astrobiology is a precautionary imperative.

4.8   The Precautionary Principle

I know how the idea of precaution and especially the precautionary principle do not necessarily have "good press" in our societies, as a result of sometimes untimely or at least diverted use. This is a pity if we refer to its original statement, the Rio Declaration on Environment and Development, adopted in June 1992:

> "In order to protect the environment, the precautionary approach shall be widely applied by States according to their capabilities. Where there are threats of serious or irreversible damage, lack of full scientific certainty shall not be used as a reason for postponing cost-effective measures to prevent environmental degradation."

The "lack of full scientific certainty": is this not precisely the main characteristic of astrobiological research and even its main basis? Astrobiological research would not be necessary if we had already discovered life outside of Earth, if we knew all the properties and, consequently, all the dangers for us and our planet. We must never forget, as obvious as it may seem that everything about the extraterrestrial belongs to the *terrae incognitae*, the unknown lands that have fascinated and continue to fascinate our humanity.

Thus, formulated and based on the lack of knowledge, or even ignorance, the precautionary principle, far from promoting inaction, is to the contrary an invitation to encourage action: "In doubt, do not abstain, but act as if it were proven." And this invitation is necessary in a world where the scientific approach, based on this doubt, this ignorance, makes us dream simultaneously and, when combined with the development of techniques, easily suggests, "dreaming that all that is possible is desirable because it is achievable" (Jean-Jacques Salomon). The precautionary principle is one of the pillars of an ethic that is not reduced to making judgments or to granting authorizations, but is understood as the questioning of the obvious and the reasonable management of possibilities.

Why shouldn't the precautionary principle, so interpreted and implemented, apply beyond the Earth, to these planets that pique our curiosity and now have the audacity to be put within reach of our investigations, our knowledge? And it is indeed in a similar spirit of precaution that COSPAR has gradually established rules of planetary protection that depend on the nature of the space mission and the celestial body targeted. The risk of contamination and, consequently, the need for protection are considered to be increasing from a simple flyby of the planetary body (the risk then consists of a navigational error that would cause the space probe to crash) when landing a human crew on the planet's soil and its exploration, going through an orbiting of the planet. Similarly, this risk increases with the degree of habitability of the celestial body by organisms from the Earth. It should be added that these rules and procedures cannot be fixed: they must

evolve with the knowledge acquired both on the physical and possibly biological conditions of the planets explored and on the multiple forms of life
on Earth and their ability to adapt to environments considered extreme,
at least for us humans. Taking the measure of time, the evolution of our
knowledge and that of the environments we approach, is a way of combin-
ing precaution and responsibility.

We must not forget Hans Jonas' teaching about the principle of respon-
sibility (in German, *Vorsorgeprinzip*). Inspired by Emmanuel Kant, the
German philosopher who had to emigrate to Palestine and then to the
United States, a categorical imperative is formulated in four sentences:

> "Act in such a way that the effects of your action are compatible with
> the permanence of an authentically human life on Earth";

> "Act so that the effects of your action are not destructive to the future
> possibility of such a life";

> Do not compromise the conditions for the indefinite survival of
> humanity on Earth"; and

> "Include in your choice the future integrity of man as the secondary
> object of your will."

Of course, in its writing and in Jonas's thinking, this principle of responsibility concerns above all the human species and its earthly habitat; but nothing prevents it from extending to other planets and applying to astrobiology
the spirit it encourages: to voluntarily include, deliberately, the concern for
humanity, its survival and its integrity, in any form of decision and action.
Is this concern marked by excessive or even pretentious anthropocentrism?
This is worth discussing and above all being confronted with the political
and economic issues that are now unique to human enterprises in space.

At a time when a race to the Moon or to Mars is happening and when
NewSpace entrepreneurs are talking about lunar tourism and the exploitation
of asteroids' mineral resources, we may indeed wonder about the possible
effectiveness of the spatial law and the principles on which it was developed
fifty years ago and, above all, on their application at a time and in a context
in which the search for profit and the realization "at all costs" of personal
ambitions and dreams seem to prevail. In such a context, what responsibility can we honestly and reasonably assume with regard to planets that are
still unknown or imperfectly known? Should a planet that may supports life
forms or more simply precursors be protected from biological intrusion of
terrestrial origin? What right and to what extent can we change nature, disseminate living forms where they are not previously found, voluntarily and

artificially create aliens from Earth, aliens that are not totally from them? Would there be limits to human activities depending on whether they are part of a natural process or, to the contrary, seem to transgress it?

To undertake a directed panspermia does not necessarily mean blindly obeying a kind of law of determinism of nature. Undertaking a panspermia program could simply be a way of participating in the seemingly general course of life: after Carl Sagan or Francis Crick, many researchers in astronomy and astrobiology defend the idea, which could facilitate the conquest of new habitats by earthly life forms. In their view, the question of the danger of encountering other forms of life must certainly be taken seriously, but weighted by taking into account their level of development: on the other hand, if primitive forms are indeed at risk of disappearing, the possibility of defending against the spread or even the intrusion of terrestrial forms would be allowed. Finally, several scientists argue that the use of panspermia could offer the possibility of creating a veritable "Noah's Ark," responsible for ensuring the survival of species threatened by a dramatic evolution of the Earth's ecosystem and by predicted evolution of the solar system.

We are being urged to implement the objectives of the expansion of humanity and its conscience, as its presence would mean much more than disobeying the rules established by COSPAR, which have no prospects but the protection of scientific work itself and the protection of the Earth. What then would be the principles of precaution and responsibility? Would they be forgotten, scorned? Nothing is more certain, especially if we introduce an idea underlying many space operations, even if it is rarely expressed, that of desecration.

To desecrate, in other words, to violate the integrity of a place, an object, a person declared as sacred, consecrated. The expression is fraught with meaning, rich in tension, even at a time when relativism is often claimed or denounced; it conveys the ideas of separation and prohibition, defilement and purification. Therefore, declaring or invoking the sanctity of life or, simply, of the living, could be an argument capable of jeopardizing the discourse of the "colonizers of space," of those who want to transform Red Mars into Green or Blue Mars; but would it be enough to prevent it from actually being achieved (or at least starting with pollution or seeding)? I am not sure that it is necessary to promote a conflict situation, I hear opposition between the promoters and the critics of the panspermia, voluntary or involuntary, between the "Greens" (the supporters of a transformation of Mars into a "green" planet, habitable and inhabited) and the "Reds" (the proponents of a sanctuarization of Mars in case even simple, primitive life forms are discovered there). I think it would be better to resort to the idea of desecration and the notion of sacredness.

4.9   The Sacred Beyond Earth

The sacred, explains Roger Caillois in *L'homme et le sacré* (*Man and the Sacred*), demands organizing the transgression of the forbidden, the crossing of borders; without this possibility, the sacred is in danger of disappearing, unless it becomes impossible to establish, to restore. In fact, the ritual apparatus that accompanies the multiple expressions of the sacred present in all human cultures, serves as much to preserve the integrity of the declared sacred realities as to foresee its desecration: it must not be the consequence of a thoughtless, unreasonable act, but on the contrary, is thoughtful and community-minded. We find ourselves in the conditions described by Vercors in *Les animaux dénaturés*: no laws, no rules, no recommendations have been written to tell us the way forward, the border not to cross, as long as it is the living. And the fact that we confront the extraterrestrial does not change anything: the extraterrestrial is not the bearer of any special precautions, the guarantor of no special protection. Here too (as in most of the challenges of contemporary knowledge and techniques that deal with matter, life or consciousness), we must learn to resist the engineer's drive, not to do, nor to undertake for the simple reason that we claim to be capable of it or possess the ad hoc knowledge. The "We Can Do It" has other limitations than that of acquired knowledge, of available power; it must be here as of the law enacted by David Hume which forbade the logical transition from being (*is*) to the duty to be (*ought*): anything that is possible is not desirable for the sole reason that it is achievable. It is imperative that we apply the principles of precaution and responsibility. We must not stop defining what is sacred. We must keep the reasonable and conscious decision to desecrate it. In short, we must always choose to be human.

4.10   Who Goes There?

"Who goes there?": the first and final question to which we humans, collectively and individually, have sought and will seek to give an answer. In the past, it opened the horizon for us and invited, sometimes forced, us to travel the world as if we were trying to multiply and vary the opportunities to ask it. By gaining a foothold outside the Earth's atmosphere, discovering the excess of space, we have increased their numbers to exorbitant proportions, taking the risk of never being able to answer this question. And, in fact, this is the situation in which astrobiology and we find ourselves today: faced with the hypothesis of extraterrestrial life forms, we are reduced to juggling probabilities, to scaffold hypotheses, to compare conjecture;

sometimes to take precautions without knowing if they really deserve to be taken. And nothing I said can put an end to this situation, except to express a belief, a conviction that has no real reason to prevail over others.

So, what is the way forward for astrobiology? Born with the exploration of space, it could suffer from the evolution towards exploitation and colonization activities that predict, for example, the promoters and actors of NewSpace: the precautions taken by researchers, under the control of COSPAR and as part of its recommendations, will be ineffective if the interest and gains prove to be up to the current forecasts of these prophets. The NewSpace entrepreneurs will only use the knowledge acquired and produced by astrobiology to manage for their own benefit the possible presence of living forms on the celestial bodies that they have decided to conquer or to limit dangers associated with the importation of extraterrestrial products to Earth.

Astrobiology may also see its activities decline or even disappear due to a lack of funding: if public interest in this area were to wane or if the lack of concrete results were to continue, governments could reduce budgets for this area of science; no one in the astrobiological community is unaware that in the late 1990s NASA interrupted its financial commitment to the search for extraterrestrial intelligence. At that time, the argument was financial: why pay researchers to listen to the waves that travel through space in an attempt to spot possible extraterrestrial messages? What if tomorrow, other politicians felt that the lack of safe and concrete results provided by astrobiology causes them to stop funding?

But can astrobiology really fail? In reality, this outcome seems difficult to imagine: as long as astrobiology does not provide evidence of the existence of life forms, even extraterrestrial intelligences, it can continue its quest since the unlimited character, perhaps infinite, of the universe makes unacceptable the conclusion that there is certainly no other form of life than that occupying the Earth. And even if astrobiologists "miss" life forms so unpredictable that they would be unable to detect them, it would not be appropriate to speak of failure, but only of temporary lack of success.

4.11　Conclusion: The Astrobiological Apocalypse

Two outcomes remain to be considered that correspond to the two meanings usually given to the term "apocalypse." The most common meaning is "end of the world," "end of time"; it owes nothing to the etymology of the word, but to a common use that associates it with epidemics, cataclysms, and disasters. The collision of a good-sized celestial body with our planet could bring about the end of the world or, more precisely, the end of our world: the question of extraterrestrial life would not be solved, but those who ask it would simply and definitively be gone.

The other meaning of apocalypse, that used by the authors of the Bible, falls under another register, that of revelation: to possess the ultimate answer to the question of extraterrestrial life would indeed be a form of revelation. I am not thinking about the discovery of living forms that would complement, confirm or disprove the knowledge we have acquired about the living, its possible definition or its earthly history; I am thinking about acquiring knowledge that would be as much about others (extraterrestrials and earthly) as it is about ourselves. We would know who "they" are and, as a result and under the mirror effect already mentioned, who we are; then we would have perhaps reached the end of the story. Everything would then be known and thus accomplished. This revelation would indeed mean the end of the world, without catastrophe or, rather, without cataclysm. For, according to Greek etymology, a catastrophe merely marked the end of an act in the theatre, a chapter in one book and the beginning of another; and the discovery of other forms of life, other than on our planet, would probably look like a similar catastrophe. Would we have completed the reading of the "Book of Life"? I am absolutely not sure, and even I do not believe it; rather, the universe would only begin a new chapter in its history, with or without us.

Further Readings

Arnould, J., Icarus' Second Chance. *The Basis and Perspectives of Space Ethics*, Springer, New York, NY, 2011.

Arnould, J., *Impossible Horizon: The Essence of Space Exploration*, ATF Press, Adelaide, Australia, 2017.

Arnould, J., *Ethics Manual for the Space Odyssey*, ATF Press, Adelaide, Australia, 2020.

Milligan, T., *Nobody Owns the Moon: The Ethics of Space Exploration*, McFarland, Jefferson, NC, 2016.

Schwartz, J.S.J. and Milligan, T. (Eds.), *The Ethics of Space Exploration*, Springer, New York, NY, 2016

5

Social and Ethical Currents in Astrobiological Debates

Kelly C. Smith

Department of Philosophy & Religion, Clemson University, Clemson, South Carolina, USA

Abstract

As the newest scientific discipline, astrobiology is mostly a collection of unanswered questions, especially when it comes to the many social and ethical issues it raises. Enormous progress has been made in recent years, but it remains the case that discussions in astrobiology often involve a degree of speculation many (especially those trained in more mature disciplines) find problematic. In this chapter, I identify some of the factors that play an important (though largely unrecognized) role in astrobiological debates, and explore in a preliminary way the challenge posed by open empirical questions, different time frames, and implicit conceptual commitments. Explicitly recognizing these factors can help bind our analyses and avoid some of the unnecessary and counterproductive confusion that so often characterizes these debates by enriching our understanding of how different scholars approach social and ethical issues in astrobiology.

Keywords: Astrobiology, ethical, moral, conceptual, implicit bias

5.1 Introductory Musings

Social and ethical debates are increasingly common in astrobiology [5.7] [5.9] [5.15] [5.20] [5.21] [5.27] [5.28]. In the broad terms of NASA's roadmap [5.6], astrobiology unavoidably deals with questions far beyond traditional biology—indeed, far beyond science. In particular, its focus on "the future of life on Earth and beyond" forces astrobiologists to consider all

Email: kcs@clemson.edu

Octavio Alfonso Chon Torres, Ted Peters, Joseph Seckbach and Richard Gordon (eds.) *Astrobiology: Science, Ethics, and Public Policy*, (95–112) © 2021 Scrivener Publishing LLC

sorts of issues that are not normally part of a scientist's job. While humanities scholars and social scientists are beginning to enter the fray,[1] the literature is still dominated by scientists and engineers whose training does
not prepare them well for this kind of work. That doesn't mean scientists
aren't allowed to have opinions, of course, much less that their opinions are
necessarily wrongheaded or problematic. While philosophers, theologians,
and social scientists have no corner on social and ethical insight, they are
much better trained to anticipate the many pitfalls that can plague this
kind of analysis. In particular, they are probably less prone (though hardly
immune) to the influence of unconscious bias in such contexts.

One important difference between human and artificial intelligence is
that humans engage in selective attention [5.11]. Computers pay attention to every scrap of information they are fed, relying on brute processing power to make sense of it, whereas humans filter the information they
receive, focusing on some things at the expense of others.[2] Thus, missing
details and ignoring things are not bugs of human cognition, but features—
inextricable aspects of a clever evolutionary strategy. Unfortunately, like all
evolutionary strategies, it's not perfect. Because we filter, and because this
is done largely below the level of conscious awareness, we are subject to
various types of what is usually called implicit bias. The key, it is thought, to
defusing such biases is to explicitly identify them, bringing them out in the
open where the critical reasoning facilities of the neocortex can function as
a conscious corrective [5.3] [5.8].

Normally, when we think of implicit bias, we have in mind very general
emotional or attitudinal perspectives rather than complex rational concepts.[3] Thus, we might learn via an implicit association test that we have a
bias toward associating those of another race with negative as opposed to
positive descriptors. This sort of bias is not really the product of reason—
indeed, that's the problem. However, complex rational thoughts can also be
subject to implicit bias (or something much like it). For example, while the
claim that most change in evolutionary history is the result of adaptation

---

[1]The Society for Social and Conceptual Issues in Astrobiology (SSoCIA), established in
2016, is dedicated to building just such a community and has grown to include over 100
members from a wide array of disciplines.

[2]Indeed, this is essential as it's been plausibly claimed that checking the logical consistency
of all the beliefs we hold is simply not possible for our brains to accomplish in a lifetime,
even if we had nothing else to occupy our time.

[3]At least, this is how *experts* view them. Part of the difficulty with fighting such biases in
society more broadly is that the average person thinks of them as rational claims (e.g.,
"I hate differently colored people") and, when they can't find anything like this in their conscious inventory of beliefs, conclude that they are bias free.

may first be formulated and assessed in an explicitly rational process, our brains can eventually decide that it no longer warrants continuing critical review and move it to a place where it still influences our thinking, but without conscious oversight—and so is born another adaptationist. The problem with a perspective like adaptationism is not that it can't be defended rationally—it certainly can be and often is—when critical questions are raised. But in the absence of such questioning, it tends to exert an outsized, and unconscious, influence on our thinking. As Stephen J. Gould once observed of the adaptationist influence on its adherents:

> "...you generally do not support your favored phenomenon by declaring rivals impossible in theory. Rather, you acknowledge the rival but circumscribe its domain of action so narrowly that it cannot have any importance in the affairs of nature. Then, you often congratulate yourself for being such an undogmatic and ecumenical chap" [5.13].

The goal of this chapter is to begin the process of identifying and thinking through some of these unconscious conceptual commitments that seem to be in play beneath the surface of recent debates about social and ethical issues in astrobiology. This is far too complex a task for a single chapter or any single scholar. My goal is thus merely to point out that the issue exists, suggest some initial lines of approach, and (hopefully) inspire others to expand on my highly imperfect schema. To that end, I am attempting to fill the role, not of an interlocutor in these debates (though I am also that), but more an ethnographer observing the dynamics from a position far above the fray. I will thus try to avoid discussing details (both empirical and theoretical) of these debates except briefly and by way of illustration. It's my hope that this kind of analysis can enrich our discussions by broadening our horizons, making the contours of the hidden conceptual landscape visible and thus subject to discussion and debate.

5.2   Uncertainty Opens the Door

One kind of implicit assumption that gets in the way of convergence is empirical in nature. Astrobiology often involves open empirical questions that we will not be able to answer for a long time, but that are clearly crucial to the shape of the discourse. Rather than attempt a comprehensive list of such questions, I will instead content myself with a few examples and observations on the general nature of such questions.

First, it's important to state that, while empirical questions are crucial, they are not everything. Those with a very heavy pragmatic orientation (e.g., scientists and engineers) often grossly overstate their criticality—for example, many biologists seem to take the position that there is no point in even beginning to discuss astrobiology until we have at least one specimen to study.[4] This attitude is especially problematic when applied to social and ethical issues, since it's often possible to get a sense of what should happen without knowing precisely how it might happen—witness the Ethical, Legal and Social Issues (ELSI) arm of the Human Genome Project, which helped us think through potential advances in genetic technology and plan for the many ethical problems they would present.

On the other hand, when discussing something without a clear empirical basis, it's very easy for the discussion to veer off into uncontrolled speculation. This isn't entirely a bad thing, to be sure, as speculation has its place. However, it's important when we are speculating to be clear about this fact and maintain a loose grip on any "conclusions" we draw, which is not always the case. For example, proponents of aggressive attempts to message aliens (METI) often state rather blithely that any aliens capable of receiving our transmissions would surely be benign and thus nothing to worry about. While this may be true, it may also not be true and this needs to be explicitly recognized and factored into the debates [5.26] [5.33].

By way of example, I now briefly discuss two common areas where empirical uncertainty allows interlocutors to insert their own hopes and fears into astrobiological (and other) debates. We are all guilty of this to one extent or another, of course, but when our underlying assumptions are not stated explicitly, they can confuse the debate and allow controversial claims to escape direct critical scrutiny. For example:

1. It's an open empirical question whether it is possible anytime in the near future to establish an economic system that works better than capitalism. To be sure, there are some promising developments which provide reason for hope, and this is clearly something to hope for, since the problems with capitalism are many. But we have been attempting to develop such systems now for at least 100 years without a breakthrough, so there is also reason to be skeptical. Of course, given enough time, anything is possible and thus any notion of monetary exchange will no doubt seem quaint

---

[4]Carol Cleland in particular has made a reputation claiming that we simply are not in a position to discuss the nature of life because we lack sufficient data, comparing those who make the attempt to alchemists trying to make sense of materials without an understanding of their fundamental basis [5.4] [5.25].

should we one day construct a Dyson sphere around the sun and use the limitless power thus captured to run factories that fulfill every material desire for free. But that does not mean we should limit astrobiological prospects now on the assumption that this is an imminent possibility—for example, by arguing that we must delay planning for offworld settlements until we can do so without the taint of capitalism, as some authors imply [5.15].

2. Whether the ecological crisis we currently face will soon become irreversible is just not clear, though this "fact" is sometimes used to argue for a focus on Earth as opposed to space. To be sure, there are good reasons to be very, very concerned about what we are doing to our planet, and thus complacency is simply not an option for any reasonable, well-informed, person. But there are potential solutions that have been proposed [5.1] and one of these might well be successful—perhaps even within the next 50 years. Even if none of the currently envisioned solutions pans out, and global warming continues unchecked for the next 100 years, it's obvious that this is truly an existential threat to humanity (as opposed to "merely" an unprecedented challenge). It is thus not clear yet whether the existence of this crisis means we should not pursue proposed projects in space—especially since some of these projects might well alleviate the strain on Earth's ecosphere.

My point here is not to take a position on these issues, but merely to note that we have to be very careful of allowing our preferred answers to open empirical questions that determine our present actions. At the very least, if such claims play a critical role in our debates about ethical issues in astrobiology, we should state them up front and invite critical analysis. Finally, it is more than a bit selective to use these speculations to argue against astrobiological projects specifically, as opposed to other developments, since these sorts of extremely broad speculative assertions pose problems for a great number of projects humanity is considering that have nothing to do with astrobiology. Thus, if we choose to assume that capitalism is just a passing phase or the world will soon become uninhabitable, then we should radically rearrange all of our priorities, not just those in astrobiology.

The solution is to carefully identify the implicit assumptions that guide our analyses, stating them explicitly at the outset so everyone knows the exact conceptual context that motivates our arguments. This is not really different from what we do when we discuss other types of models in science—a given model can be very useful for a certain purpose, but it must always be clear precisely what it assumes so we know what its limitations are.

5.3   Time Frames

There are so many empirical uncertainties in astrobiology that discussions of its social and ethical issues often take on a speculative air. That isn't necessarily a problem, but it can result in a kind of speculative feedback where more and more implicit conceptual frames are imported without sufficient critical analysis.[5]

By far, the most important kind of implicit framing we see in such discussions has to do with the timescales we have in mind. Timescales are critical, since they place boundaries on the kinds of problems that are appropriate to address in the first place, as well as the technological, social, and economic resources we may have to address them. For example, there has been a fair amount of ink spilled over the question of whether it would be ethical to terraform Mars [5.2] [5.32], but there are multiple unknowns that influence the debate in extremely important ways. For example:

1. Most scholars agree that whether there is life on Mars is the single most important factor in any discussion of how we should approach missions to our red neighbor. But we do not yet know if there is life on Mars—indeed, since proving a negative is not logically possible, one could argue that we will never know for certain that Mars is lifeless. On the other hand, we are likely to have much better information within the next 30 years, so we might not have to wait too long for a reasonable scientific consensus to emerge. Any discussion of planetary protection policies for Mars, therefore, *must* specify whether it applies to current missions or those 30 years or more down the road, on pain of confusing ambiguity.[6]

2. Terraforming is a purely hypothetical technology at present—while we think it's probably possible, we really have only the vaguest notion of how it might be accomplished.

---

[5]The fact that many of these discussions are undertaken by non-scientists may well exacerbate this problem.
[6]Note that 30 years is within the "short" timescale I discuss below, which shows that sometimes additional precision is required.

Moreover, the proposals that have been put forward thus far could not possibly even begin for at least 50 years (and likely far longer). Finally, barring some truly transformative breakthrough, no practical results of terraforming (e.g., settlements without extensive life support) would be seen for hundreds (or thousands) of years.

This does not mean, as some conclude, that discussing the ethics of terraforming now is a pointless exercise, just that it is necessarily highly speculative and the "conclusions" we draw must therefore be seen as highly qualified, as should be the case with any prediction about a distant future we can only dimly perceive.

The lesson to be learned is this: Those involved in ethical debates in astrobiology should explicitly indicate the timescale they have in mind before launching a prolonged analysis (a habit that is currently quite rare). To that end, I propose a very simple framework that can be employed profitably since we do not always need to be precise to avoid the lion's share of confusion:

1. *Short timescales* apply to anything we can reasonably expect to be possible within the careers of people now working in the field. To be generous and allow for round numbers, let's say this encompasses the next 50 years. We can predict, with reasonable confidence, the basic kinds of technology and important social and political systems which will exist at that point (though even here we will doubtless be wrong in at least a few important ways). 50 years ago, for example, rocketry was a mature technology and we were on the cusp of our first manned mission to another world. The shape of geopolitics was roughly the same as it is now, though still defined largely by the cold war. Examples of debates that fall into a short timescale would be things like robotic missions to bodies within our own solar system, manned missions to the moon and Mars, and at least the initial phases of METI and nanosat probes to other systems.

2. *Medium timescales* apply to things we have good reason to believe might be possible within the lifetime of the next generation of researchers. Again, for simplicity's sake, let's assume this extends 100 years into the future. Obviously, it is much harder to predict the contours of future capacities this far out, but some of the broad contours can be identified.

One hundred years ago, evolutionary biology was an established science, we knew about nucleic acids, had learned a great deal about cell biology, and were just beginning to understand viruses. On the other hand, the mysteries of microelectronics were completely unknown and the computer technologies this made possible were the stuff of science fiction. Geopolitics were similar in many ways to what we see today, though communism had just begun to leave the salons of Europe to make a real impact on the world, while critical components of the modern world order like the United Nations and massive multinational free trade networks were dreams yet to be realized. A medium timescale would encompass debates such as self-sustaining settlements on bodies within our solar system, commercially viable asteroid mining, and the sort of infrastructure in space that could make offworld missions several orders of magnitude more efficient.

3. *Long timescales* extend out to 200 years. At this point, it becomes extremely difficult to predict what will be available and thus the error bars on any "predictions" are dauntingly large. In particular, we are almost certain to completely miss multiple major breakthroughs that will fundamentally alter the very nature of future decisions, since their precursors are not even envisioned yet. Such long-term predictions will inevitably appear to our descendants as a strange mix of pre-science and naivete. After all, 200 years ago, we thought we inhabited a static physical universe and were still trying to figure out the speed of light. Though we knew of the existence of microbes, we had little clue as to how they functioned or what role they played in our world—in particular, the idea that they might cause disease was highly speculative. Jules Verne had yet to muse about the possibility of traveling to the Moon (in a ship fired from a giant cannon) and the jury was still very much out on the success of the American experiment in democracy. Even though we have to be careful really believing our speculations here, it's also important to keep in mind that they are not useless. If nothing else, there are moments when a speculative visionary identifies something that, while currently impossible, captures our imagination in such a way that people work to make it possible. For example, evolutionary thinking (broadly conceived)

was certainly "in the air" 200 years ago. Charles Darwin was deeply influenced by his grandfather, Erasmus, who believed in something like evolution, though it would be 30 years before Darwin was able to put forward a viable theory explaining how it was possible (and then another 100 years before the mechanism of inheritance was identified).

4. *Fictional timescales* are anything more than 200 years in the future. At this point, we really have no idea what kind of factors might be in play. Consider that a "mere" 350 years ago, the very idea of science as we know it today was novel and we were just beginning to discover the hidden macro and micro world around us with crude telescopes and microscopes. Moreover, since the rate of scientific progress increases as our knowledge base grows, predictions on long timescales will become less and less reliable over time. Thus, it's hard to say that anything is truly out of bounds, given such a timescale: new nation states on Mars and in the asteroid belt? Why not? Faster than light travel? Maybe. Humans transitioning to a non-corporal form? Possibly. Fictional speculation can still be useful, but mostly as a means to expand the boundaries of our imaginations rather than as a practical guide for our present actions—excellent for entertainment and philosophical discussion, but not much else. Thus, anyone who seeks to further serious work on social and ethical issues in astrobiology should probably avoid these timescales altogether on pain of being dismissed as an unrealistic dreamer.

5.4   Conceptual Frames

Next, I wish to identify a few conceptual continua where implicit positions are often taken that impact our thinking in crucial ways. In theory, a schema of these conceptual dichotomies can illuminate some of the large-scale features of the conceptual landscape on which these debates play out, allowing for deeper appreciation of the most fundamental commitments at work behind the scenes. There is certainly at least some value in identifying where each of us stands on these continua, if for no other reason than to help avoid the common sins of talking past one another and attributing base motives to opponents. At the very least, such an exercise can bring into sharper focus precisely what beliefs are in tension and allow for more fruitful discussion. I do this with trepidation, however, for several reasons.

For one thing, conceptual orientations like this are hard to describe precisely in a way that's also fair to all points of view. For another, they overlap and interact in complex ways. Lastly, it's clear that simplified general classification systems are often abused and can stifle rather than facilitate debate by pigeonholing opponents in superficial ways (witness the innumerable juvenile takes on the highly suspect Myers-Briggs personality scale).

To some extent, whether this project is appealing depends on whether one is a lumper or a splitter—a distinction Darwin posited as the most fundamental difference in biology. Splitters focus on each subtle difference between one individual or population and another, while lumpers emphasize the common elements underlying such variation. Splitters accuse lumpers of imprecision and lumpers accuse splitters of detail worship, but the truth is that neither approach is best in an unqualified sense; they can each make valuable contributions to our understanding, provided that we apply the right technique to the question at hand and always remain mindful of its limits. This project is all about lumping, since my goal is to elucidate the grand contours of the intellectual landscape, not wade into the messy details.

Finally, I want to be clear that the continua I suggest here are in no way definitive. Indeed, they are not even realistic in an important sense, as I intentionally present them as caricatures to make a point. It is certainly true that few careful thinkers would adopt any of these positions without qualification, but that does not mean they don't capture a kernel of truth that can help reveal important motivations for the positions we take. With these caveats, I now examine three of the more obvious conceptual divides.

5.4.1   Error Avoiders vs. Optimizers

Error avoiders tend to focus on what can go wrong. As a consequence, they feel we should only pursue astrobiological projects that we know will improve the prospects for all of humanity—it's not morally permissible to take risks with humanity in general or impose significant costs on some humans to benefit the collective. On the other hand, astrobiological optimizers tend to focus on what can go right. As a consequence, they argue we should pursue any project with a high probability of producing good results for humanity in general—it's permissible to take some risk of a suboptimal outcome or impose costs on some humans if doing so is likely to benefit the collective.

Consider the debate over whether and how to establish settlements on a body like the Moon, where indigenous life is not a factor. Error avoiders often argue that they are a bad idea, at least at present, since an initial settlement will be a hellish place. Not only will early settlers face a high risk of injury or death, but settlements will probably at least recreate, and likely exacerbate, existing social ills such as the undue influence of capital, unrepresentative government, and discrimination against settlers who don't embody our stereotypical notions of "the right stuff" [5.5] [5.23]. Optimizers don't disagree that these may be problems (though they may well disagree on their likelihood or severity), but feel these are risks worth taking in pursuit of a second home for humanity, with all the benefits that promises in the longer term. The most explicit discussions of these issues occur in popular culture—consider the manifest destiny themes so common in science fiction or Gil Scott-Heron's beat poem, "Whitey on the Moon" [5.22].

5.4.2   Ecologicals vs. Anthropocentrists

Ecologicals argue that, since humans are ultimately just one of many forms of life, we should only pursue projects that benefit (or at least don't harm) life in general, even if this approach results in some opportunity cost for humanity. Humans matter, to be sure, but ultimately not more than the many other inhabitants of our shared ecosystem. Anthropocentrists tend to work from the opposite starting point: that we should pursue projects which have the best chance of furthering human interests, even if these come at a cost to other life—ecological considerations are relevant, but mostly only indirectly through their influence on humans.

Again, let's consider the question of whether we should establish offworld settlements, but this time on a world like Mars that may harbor indigenous life. Ecologicals often argue that, given our abysmal track record of protecting the environment of Earth, it would be immoral to inflict ourselves on another world until and unless we establish that we have fundamentally altered our evil ways [5.16] [5.20]. The fact that such a settlement promises benefits for humans does not justify the risk of destroying precious ecosystems elsewhere. Again, anthropocentrists don't entirely disagree and will typically grant that ecological impact is an important factor to consider, though for very different reasons. It's quite clear that other lifeforms benefit humans, whether because we rely on them for our own survival or because they offer opportunities for important scientific discovery. However, when push comes to shove, a sufficiently enticing prospect like a second home for humanity is worth some damage to other life. This particular dichotomy is more explicit than most, given Carl Sagan's famous claim that, if Mars is inhabited, even if merely by microbes, we must "leave Mars to the Martians" [5.19].

5.4.3   Communalists vs. Commercialists

Communalists stress the many evils of unrestrained capitalism and thus argue that we need to do all we can to prevent base commercial motives from tainting our astrobiological projects. While there may be nothing wrong with self-interest and profit in some spheres, our objective in space must be pure—to benefit all mankind. Commercialists counter that, while there is nothing wrong with governments pursuing such high-minded ideals should they choose to do so, they are rarely practical, as market-based mechanisms are the best available means to efficiently harness human ingenuity and tackle complex tasks. So, if we are serious about achieving very difficult things in space (and they are all difficult) within a reasonable timeframe, we must allow commercial enterprise to play a prominent role.

The most obvious example of this dichotomy in action is the debate over private property in space (and any associated rights) [5.14] [5.31]. The Outer Space and Moon treaties explicitly express communalist ideals from the very first article:

> Article I: The exploration and use of outer space, including the moon
> and other celestial bodies, shall be carried out for the benefit and in
> the interests of all countries, irrespective of their degree of economic
> or scientific development, and shall be the province of all mankind
> [5.17].

There is zero provision of a possible role for commercial enterprise here—and indeed, a clear indication that such would simply not be appropriate. Communalists feel these treaties set exactly the right tone and should not be altered—objects in space should always be managed for the collective good. Commercialists counter that this position is an impractical luxury that passed muster only because, in the initial stages of the space age, commercialization of space simply wasn't possible. Now that projects such as asteroid mining are nearing feasibility, we must reassess and treat objects in space like any other resource—as things to be exploited in many different ways by many different actors. Indeed, they often make the point that such a change in attitude is inevitable, pointing to recent attempts by the United States and Luxembourg to attract space mining business with friendly alternations to their legal systems [5.10] [5.30].

No doubt there are many other broad conceptual differences that could be identified—I offer only a few obvious examples to make a point. I certainly do not mean to suggest that one's position on any of these, admittedly somewhat artificial, dichotomies is the only (or even the most important) factor determining most scholars' views. Quite clearly, to take one example, empirical claims about what is possible, likely, or feasible must be carefully assessed and there is ample ground for sincere disagreements even among those of like mind on these and other issues. My point is simply that the debates currently coming to the fore in astrobiology are influenced crucially by these much, much broader commitments and that perhaps some of our energy should be invested in exploring them directly.

5.5   Complications, Connections, and CYA

In some sense, this chapter is attempting the impossible—a simple answer to an extremely complicated question that nevertheless offers useful illumination. It is therefore trivially easy to critique in a number of ways, so now I will list and briefly discuss some of the more obvious criticisms in an attempt to defend the attempt.

First, many will observe that caricatures are by definition unfair, so it's quite strange for me to openly admit that the conceptual frames I put forward are just that. I certainly grant this point in a general way, but note that people (even highly trained scholars) often do think along these lines (whether they admit it, even to themselves). This is especially true when it comes to ethics. For example, there is a real tendency for people to formulate ethical principles in ways that do not admit pragmatic compromise. This can generate a beautiful moral high ground with little relevance to real-world ethical problems. For example, if one uses the powerful machinery of intrinsic value to defend an extremely high moral value for Martian microbes, one immediately encounters the paradox of inclusion: if all things we are considering are intrinsically valuable, and this value cannot be compromised, then we have no means to resolve the complex trade-offs ethical dilemmas inevitably pose.[7] If we insist on reasoning from this moral pinnacle, we are forced to become subtly hypocritical—in effect allowing that, while we say in theory that all these entities are morally equal, in practice we hold that some entities are more equal than others [5.24].

Second, any philosopher worth her salt can point to an immensely complex conceptual background I completely ignored. Again, I agree in a sense, but I am not really speaking to the tiny fraction of those engaged in these debates who understand the history of philosophy. Most within the astrobiology community are only dimly aware of this rich tradition, in which

---

[7]If, on the other hand, intrinsic value comes in degrees, which is possible in theory, then appealing to it does not settle any practical disputes cleanly.

case talking about it in detail doesn't illuminate how most of those actually engaged in these disputes actually think about these issues (as opposed to, arguably at least, how they should think about them). Therefore, at least in one important sense, deep philosophical analysis is a red herring.[8]

For example, there are clearly echoes within my schema of the ancient battle between the ethical theories of utilitarianism and deontology. But even when these terms are actually used in astrobiology, they are typically employed in unsophisticated ways that can do more to obscure than enlighten. Thus, many a seeming ecological is actually a closeted anthropomorphist—if your defense of the value of microbes is based on their scientific potential or their essential role in the ecosystem that humans require for survival, then you are an anthropocentrist, whether you realize it or not. Words do matter, so we should strive to be precise when possible. On the other hand, it's fortunate that very different perspectives can sometimes converge (if unintentionally) in their actual recommendations: for instance, even the most ardent anthropocentrists and ecologicals often agree that a robust precautionary principle should be applied (at least initially) to investigations of any potentially life-bearing body (though for radically different reasons).

It's worth nothing in passing that my use of the "anthropocentric" is another example of an intentional simplification which can cause confusion. Strictly speaking, an anthropocentrist believes that humans have high moral value simply because they are human, which is not a position thoughtful people would ever defend. Instead, the vast majority of scholars who adopt a strong defense of human interests probably do so because they believe humans exemplify other attributes (e.g., reason, the capacity for self-reflection and/or complex culture, etc.)—it's these attributes, not the mere fact of humanity, that account for humans' special, moral value. This is an important distinction, especially in astrobiology where it is not a given that these capacities are uniquely exemplified by a particular lifeform on a rather ordinary planet in a stupendously vast universe.

Third, even if we only consider the very limited schema I have provided, there is much more that could be said. For example, timescales and the other conceptual frames obviously interact. Thus, as a general rule, the

---

[8]It could be argued that my training as a scientist has tainted my thinking here, leading me away from a "properly philosophical" perspective. But there is a growing body of empirical work in ethical reasoning that suggests philosophers would do well to follow suit if they wish to have any real-world impact [5.12] [5.18].

shorter one's timescales, the more likely one is to adopt the perspectives of error avoidance, commercialism, and perhaps ecologism. This can be because people choose to adopt a shorter timescale for practical reasons, as when a scientist chooses to focus on projects she might be able to actually undertake in her lifetime. But it can also result from specific empirical commitments—a large part of the reason Robert Zubrin and his wild-eyed Zubrinistas have pushed so aggressively for an immediate attempt to settle Mars is their belief that all the critical technologies are either available or can be developed quickly [5.34]. And again, there is nothing wrong with such focus as long as it's explicitly stated and isn't allowed to go too far.[9]

Finally, there are a number of interactions between these conceptual frames and other factors that I don't explore. For example, there are clearly some interesting correlations between one's cultural upbringing and the conceptual framing they tend to adopt. For one, academics with an interest in ethics are self-selected to be more idealistic than the norm, thus worrying much less about empirical details and skewing towards longer timescales, communalism, and ecologism. For another, Europeans are much more likely than their American counterparts to adopt a highly skeptical attitude toward commercialism (as well as a view of interdisciplinarity whose boundaries include non-scientific disciplines). Finally, there are likely several Western biases in these debates that are made possible by the fact that the literature is at present almost entirely dominated by Western scholars [5.29]. Carefully thinking through all these connections is far beyond the scope of this chapter however, since it's all too easy for a non-expert to paint with an overly broad brush.

5.6   A Concluding Thought

There is quite clearly a lot more that could be said here, so to the extent my effort succeeds at all, it merely scratches a large and complex conceptual topology, hopefully inviting further discussion and debate. On the other hand, there is real value in reflecting on one's position even within such a simplified and imperfect conceptual schema as I outline here. The essential trick is to recognize biases in oneself and others without reifying them. If we can view this as an opportunity to broaden our conceptual horizons rather than stereotype and pigeonhole our opponents, it can be a powerful

---

[9]Too much practical focus can lead the incautious to conclude that *any* discussion involving longer timescales is somehow inappropriate "pie in the sky" thinking.

tool—in the same way that explicitly identifying other implicit biases is the first step in defusing their power to influence our reasoning in inappropriate ways. For myself, knowing that I tend to be a medium to long timescale, optimizing anthropocentric commercialist makes me more aware of (and more open to) the views I know I tend to underemphasize in my analyses. It's in that spirit that I offer this conceptual schema, which I hope will spark much further debate.

References

[5.1] Aouf, R.S., *Five Geoengineering Solutions to Fight Climate Change*, 2018, https://www.dezeen.com/2018/10/18/five-geoengineering-solutions-climate-change-un-ipcc-technology/.

[5.2] Astrobiology Magazine, *Should we terraform?*, 2004, https://www.astrobio.net/mars/should-we-terraform/.

[5.3] Bodenhausen, G. and Lichtenstein, M., Social stereotypes and information-processing strategies: The impact of task complexity. *J. Pers. Soc. Psychol.*, 52, 871–880, 1987.

[5.4] Cleland, C. and Chyba, C., Defining Life. *Orig. Life Evol. Biosph.*, 32, 387–93, 2002.

[5.5] Cockell, C.S., *Extra-terrestrial Liberty: an enquiry into the nature and causes of tyrannical government beyond the Earth*, Shoving Leopard, London, 2013.

[5.6] Des Marais, D.J., Nuth, J.A., Allamanola, L.J., Boss, A.P., Farmer, J.D., Hoehler, T.M., Spormann, A.M., The NASA Astrobiolology Roadmap. *Astrobiology*, 8, 4, 715–30, 2008.

[5.7] Dick, S.J., *Astrobiology, Discovery, and Societal Impact*, Cambridge University Press, Cambridge, 2020.

[5.8] Djikic, M., Langer, E., Stapleton, S., Reducing stereotyping through mindfulness: Effects on automatic stereotype-activated behaviors. *J. Adult Dev.*, 15, 106–111, 2008.

[5.9] Duner, D., Parthemore, J., Holmberg, G. (Eds.), *The History and Philosophy of Astrobiology: Perspectives in extraterrestrial life and the human mind*, Cambridge Scholars, Cambridge, 2013.

[5.10] Foust, J., *Luxembourg expands its space resources vision*, 2019, https://spacenews.com/luxembourg-expands-its-space-resources-vision/.

[5.11] Fu, D., Weber, C., Yang, G., Kerzel, M., Nan, M., Barrois, P., Wermter, S., What can computational models learn from human selective attention? *Front. Integr. Neurosci.*, 14, 10, 2020.

[5.12] Fumagalli, M. and Priori, A., Functional and clinical neuroanatomy of morality. *Brain*, 135, 7, 2006–21, 2012.

[5.13] Gould, S.J. and Lewontin, R.C., The spandrels of San Marco and the Panglossian paradigm: a critique of the adaptationist programme. *Proc. R. Soc. B*, 205, 1161, 1979.

[5.14] Leahy, B., *Space Access: the private investment vs. public funding debate*, 2006, https://www.space.com/2401-space-access-private-investment-public-funding-debate.html.

[5.15] Milligan, T., *Nobody Owns the Moon: The ethics of space exploration*, MacFarland, Jefferson, 2014.

[5.16] Morton, A., *Should we colonize other planets?*, Wiley, New York, 2018.

[5.17] Outer Space Treaty, 1967. https://2009-2017.state.gov/t/isn/5181.htm.

[5.18] Pascual, L., Rodrigues, P., Gallardo-Pujoli, D., How does morality work in the brain? *Front. Integr. Neurosci.*, 7, 65, 2013.

[5.19] Sagan, C., *Cosmos*, Ballentine Books, New York, 1985.

[5.20] Schwartz, J.S.S., *The Value of Science in Space Exploration*, Oxford University Press, London, 2020.

[5.21] Schwartz, J.S.S. and Milligan, T. (Eds.), *The Ethics of Space Exploration*, Springer, New York, 2016.

[5.22] Scott-Heron, G., *Whitey on the Moon*, 1970, https://www.youtube.com/watch?v=goh2x_G0ct4.

[5.23] Shammas, V.L., One Giant Leap for Capitaliskind: private enterprise in outer space. *Palgrave Commun.*, 5, 10, 2019, https://doi.org/10.1057/s41599-019-0218-9.

[5.24] Smith, K.C., The curious case of the martian microbes: Mariomania, intrinsic value, and the prime directive, *The Ethics of Space Exploration*, S.J. Schwartz and T. Milligan (Eds.), Springer, New York, 2016a.

[5.25] Smith, K.C., Life is Hard: Couintering definitional pessimism concerning rhe definition of life. *Int. J. Astrobiology*, 15, 4, 277–89, 2016b.

[5.26] Smith, K.C., MET or Regretti: Ethics, risk, and alien contact, *Social and Conceptual Issues in Astrobiology*, K.C. Smith and C. Mariscal (Eds.), Oxford University Press, London, 2020.

[5.27] Smith, K.C. and Mariscal, C. (Eds.), *Social and Conceptual Issues in Astrobiology*, Oxford University Press, London, 2020.

[5.28] Smith, K.C., Abney, K., Anderson, G., Billings, L., Devito, C., Green, B.P., Wells-Jensen, S., The Great Colonization Debate. *Futures*, 110, 4–14, 2019.

[5.29] Traphagan, J., Religion, science, and space exploration from a non-western perspective. *Religions*, 11, 8, 397, 2020.

[5.30] Wall, M., *New space mining law is 'history in the making'*, 2015, https://www.space.com/31177-space-mining-commercial-spaceflight-congress.html.

[5.31] Whitfield-Jones, P., *One smalle step for property rights in outer space?*, 2020, https://www.mayerbrown.com/en/perspectives-events/publications/2020/05/one-small-step-for-property-rights-in-outer-space.

[5.32] York, P., *The Ethics of Terraforming*, 2002, https://philosophynow.org/issues/38/The_Ethics_of_Terraforming.

[5.33] Zaitsev, A.L., *Rationale for METI*, 2011, https://arxiv.org/abs/1105.0910.

[5.34] Zubrin, R., *The Case for Mars*, Free Press, New York, 2011.

6

The Ethics of Biocontamination

Tony Milligan

Cosmological Visionaries Project, Department of Theology and Religious Studies, King's College London, London, UK

Abstract

The search for biosignatures and for life itself draws upon a strong background intuition that life matters in a deep way. Not just human life or sentient life, but life as such. Accordingly, biocontamination is a matter which concerns ethics as well as the protection of opportunities for science. However, this idea that life has value is a difficult intuition to tease out and understand, and even harder to justify. It is difficult to make sense of in the light of "the parallel case," i.e., our largely instrumental treatment of microbial life here on Earth. In what follows, some tentative moves will be made in order to help us to understand the widespread intuition about life's importance. These moves will then help us to understand what is problematic about crashing tardigrades onto the lunar surface, and our ethical responsibilities in relation to any microbial life found *in situ* elsewhere in the Solar System. The approach does not call upon rights theory, or focus upon individual microbes as the bearers of such rights. Attention is shaped by a more aggregate focus, and the concept of intrinsic value is qualified in order to draw out the relational dimensions of human valuing, and the pragmatic dimensions of an inclusion of the standing of microbial life within a social ethic.

Keywords: Ethics, microbial life, intrinsic value, planetary environmental protection

Email: Anthony.milligan@kcl.ac.uk
Tony Milligan: https://www.cosmovis.uk, https://kcl.academia.edu/TonyMilligan

Octavio Alfonso Chon Torres, Ted Peters, Joseph Seckbach and Richard Gordon (eds.) Astrobiology: Science, Ethics, and Public Policy, (113–134) © 2021 Scrivener Publishing LLC

6.1   The Beresheet Tardigrades

In 2019, the Israeli lander Beresheet brought life to the Moon. The lander crashed, and parts of a disk imprinted with tardigrades in a desiccated/cryptobiotic state are assumed to have survived the incident. The tardigrades need rehydration to do anything, but they are *in situ*, and are presumed to be alive. Seeding the Moon with life was not the primary goal of the mission, or any sort of explicitly stated goal of the mission. Rather, the aim was to achieve a soft lunar landing, brought about by private and state sector collaboration, with much of the credit going to the private sector. More specifically, to the American venture capitalist Nova Spivack's Arch Mission Foundation which does have a goal of sending life elsewhere in order to "back up" planet Earth. However, there seems to have been no intention to scatter living organisms in an uncontainable way. "For the first 24 hours we were just in shock," Spivack stated afterwards. "We sort of expected that it would be successful. We knew there were risks but we didn't think the risks were that significant" [6.12].

In terms of the ways in which we approach space, and the extension of life's presence, I will assume that the approach was not quite good enough. The fact that the tardigrades were included was a surprise to people involved in the mission from the state side, with the manifest intentionally or unintentionally ambiguous. It is unlikely that their inclusion would have been approved by the Israeli Space Agency (ISA), had the contents of the disk been made clearer. This applies even though the legal requirements for planetary protection, which apply to Mars and the other planets, are not generally considered to apply in quite the same way to the Moon given the definitive absence of indigenous lunar life [6.13]. I will remain officially neutral on the question of whether or not the action was actually illegal under international law, although my suspicion is that international law may simply be indeterminate on the question. And while legal responsibility would ultimately fall upon the Israeli state, as it was the launch state, I have no interest in linking this issue of biological contamination in space into some broader and hostile political narrative. Instead, my interest is in the ethical claim that it was wrong to take the tardigrades to the Moon *in this way* and that the wrongness is nothing to do with the launch state. It would have been just as wrong if the agency involved had been NASA, ESA, Roscosmos, JAXA, or China's CNSA.

But what exactly was wrong about the action is less clear. Not because there may have been nothing wrong with it, but because there are multiple candidates for its wrongness. We might, for example, claim that

"Any transfer of life from Earth to the Moon should be avoided," or that "Such transfer should be avoided until we have better control mechanisms," that "LIFE should only be transferred as a result of some sort of international consensus, in less haphazard ways, and with clear and defensible goals in mind." I am sympathetic to the last two claims rather than the first, which strikes me as unnecessarily restrictive. To say this plays upon a widely shared intuition that transferring life of any sort elsewhere is something that matters, even if the life in question is in the form of microanimals, like tardigrades, or even in the more rudimentary form of unicellular microorganisms.

Our best evidence is that this intuition *is* widely shared, at least across the scientific community and the space community more generally. The discovery of unusual concentrations of phosphene in the atmosphere of Venus in 2017 as a possible biosignature, became something of an event when it found its way into the international press during the COVID-19 epidemic, as a moment of great hope. Yet, nobody imagined that we were on the brink of discovering giant trees or the warring kingdoms of an Edgar Rice Burroughs' novel. This was Venus, not Burroughs' "Amtor." Life, if present, would be rudimentary. Yet, somehow still extremely important. Discovery would be one of the great events of the 21st century…if it occurred. We might also appeal to the hopes surrounding the Allan Hills 84001 meteorite, and the claims made in 1996 that it contained microscopic fossil evidence for life. A claim later rejected by the broader scientific community, but one which accelerated a sense that we might be close to discovery, and accelerated the expansion of astrobiology as a multidisciplinary scientific research field.

What we can draw from these incidents is the breadth and depth of the intuition that discovery would matter a great deal, and not just because of its scientific potential but also as a pivotal event for humanity as a whole. Partly, it could be explained as a massive advance in astrobiology and our understanding of life. But such a boost is not actually entailed, as anything like a matter of logical necessity, or of some looser conception of necessity. The kind of rudimentary life that we might discover could well reflect our local conditions within the Solar System, and be much the same as microbial life as we already know it. Or, it could be different, but in a way that falls within a range of extremophiles whose boundaries are of a familiar sort. We might even imagine finding something similar here, on the ocean floor or in a cave somewhere. This would give us something interesting and important, but on its own not anything radically new in terms of biology, and it might not clarify our understanding of what life everywhere *must* be like. Discovery may instantly open up new vistas of microbiology, but it may instead open up lines of enquiry which are strongly continuous with those we already engage in.

Whatever scientific pathways it opens up, discovery would be a momentous event in our shared human history. Yet, it is difficult to say more than this, and tempting to treat it as a "just so" story, to say in answer to the question of why we have this intuition and this attitude towards life, "we just do." What follows will try to push our understanding of the intuition a little further. Not too far, because it involves matters of depth, and progress in our understanding of them is often slow or else does not happen at all. The innovative moves will be distributed throughout the text, with a heavy concentration toward the end of the chapter, where something will be said about planetary environmental protection and the ethical side of the avoidance of contamination. The preliminaries and mid-section will also involve some familiar moves, in an attempt to tease out our conflicting intuitions about microbial life, and the idea advanced by people such as myself [6.20] and Charles Cockell [6.5] [6.6] that such life might sometimes have ethical standing in its own right. In a sense, it is easier to understand why something that matters *in its own right* should also matter to us. Yet the idea here, of something mattering in its own right should not be mistaken for a claim that organisms of this sort have "rights." Which is one way to think about having ethical standing, but not the only way of doing so. In the case of microbial organisms, it may not be our best option.

In a terminology familiar from other areas of ethics (e.g., environmental ethics and animal ethics), it may help to talk about life as something that matters, as something with "intrinsic value." Although, this concept of "intrinsic value" is often associated with arbitrary bestowals of importance, and it can be a little misleading in terms of what it tracks. By this, I mean that there are many examples of ethicists picking out some or other property as the locus of intrinsic value, and then using it to drive their entire position, even at the expense of implausible claims about the equality of all value bearers, from insects to humans and other primates [6.7]. In what follows, I will try to avoid this arbitrariness, and will comment upon why "intrinsic value" is such an awkward terminology. Nonetheless, my point is that what sits behind our sense of the importance of discovery is not just aspirations for a certain kind of important scientific progress, but a deep sense if life's own importance. A sense that even in its most rudimentary instances, life *can* matter, given the right context and circumstances. And this is one of the reasons why we have a reasonable sense of caution about how we spread it elsewhere. The next section will try to tease out possible grounds for such an attitude towards life, as something that can matter in its own right, in spite of a conflicting intuition that many of us also have about the relative unimportance, or strictly instrumental importance, of certain kinds of life.

6.2 Our Conflicting Intuitions

It is difficult to make sense of how humans can reasonably hold two apparently conflicting attitudes about microbial life at the same time: (1) our widely shared view that finding rudimentary life elsewhere would be extremely important, and that such life *ought* to be protected, for more than scientific reasons; coupled with (2) our mundane treatment of microbial life here on Earth in a strictly instrumental way. Microbial life is indispensable to our lives. Indeed, as Donna Haraway [6.10] points out, we are outnumbered in our own bodies, with more bacteria than cells carrying our DNA. Yet, we also depend upon oxygen to survive, and we are not about to claim that it has some sort of intrinsic value. Routinely, and often intentionally, we destroy microbial life, and we have no guilty conscience about doing so. We destroy it as something which has no importance in its own right. Yet, the life that we would find wonderful in the first of these attitudes may be much the same, in structural terms, as the life we destroy in routine ways. Let us call this "the parallel case."

Of course, we might set aside all such deliberations by appealing to the basic unreliability of ethical intuitions and the attitudes they underpin. In which case there is no problem to be explained. Our intuitions about the importance of life elsewhere, its discovery and protection, would then say something about our psychology, rather than about the genuine importance of any of these things. However, this is a move that comes at a high price. The actual practice of ethics, and especially of attempting to think in some orderly way about ethics, requires us to draw upon our shared ethical intuitions. We may then discard them in the light of further deliberation, or we may come to regard some intuitions as more important than others in the light of various theories about ethics, or in the light of various thought experiments, or in the light of described experience. We may pursue what John Rawls called a "reflective equilibrium" between theories and intuitions, or between intuitions, principles and a multiplicity of theories. Or we may adopt some similar approach which acknowledges that intuitions themselves have an important functional role in ethics. They shape background requirements and ways of speaking across ethical differences between peoples, times and cultures. They may not be where we finish when engaging in ethical enquiry, but they are often a useful starting point or part of the mix, in our best deliberations about ethics. I will take it, then, that a "one fell swoop" approach of simply dismissing our intuitions about the value of life will generate too many other problems. Our intuitions are something to be explained and examined, not dismissed.

What may make the combination of the two attitudes above harder to understand as a rational, justifiable combination, is a way of addressing ethical questions about the non-human by appealing to a concept of "value" in the sense of "intrinsic value." A concept I will use here, but with a cautionary warning. The concept can make us think about inherent properties as the locus of ethical importance, as something we might separate from everything else and still regard as having the same significance. Exactly what sort of entities are taken to have such value has, of course, shifted over time. So, for example, there was a time when many philosophers believed that only humans had such value, because the inherent property which was the locus of value was rational autonomy [6.14] or, more traditionally, the possession of an immortal soul [6.2]. With changes in our attitudes towards non-human animals [6.26] [6.30], the relevant property shifted to sentience. Something less demanding than rational autonomy. (Which some animals may actually have, but others probably do not.) Yet the structure of justification remained the same: something inherent was sufficient grounds, on its own, for attributions of value [6.4]. Environmental ethics then led to a further readjustment, allowing us to attribute value to places and things without requiring them to be sentient. While we may try to think like a mountain, in the manner proposed by Aldo Leopold [6.15], mountains do not actually think. They are not sentient. It is only in science fiction [6.1] that we might consider Olympus Mons as inherently valuable because of an imagined sentience.

Environmental ethics has traditionally been split in multiple ways. Some versions say that we should care for the non-sentient, including non-sentient life, for strictly instrumental reasons [6.23]. Other versions attribute value to such life and also to various sorts of non-life (mountains, and rivers thought of as distinct from the life they support). Within discussions of how to extend environmental ethics to space, Erik Persson [6.24] takes something closer to the former view, while I take the latter view. But this is not a division between warm-hearted theorists and theorists of a different sort. Advocates of animal rights also tend to withhold value attributions from microbes, yet they could hardly be regarded as lacking is a sense of warmth or care for the non-human. In many cases, those who hold different positions in the debate may urge exactly the same practical forms of protection protocols for microbes, albeit with marginally differing justifications. Is this, then, a difference that makes no difference? I suspect not, for two reasons. First, an intrinsic value approach does help to make our familiar attitude towards discovery more intelligible. Second, it is far from obvious that we can tell any plausible, non-arbitrary, tale about what makes any ethical claims true while placing rudimentary life forms outside of the cluster of things whose value is endorsed. This is more of a methodological and metaethical point, and it takes the discussion off in a slightly more abstract direction. However, the small number of abstract metaethical moves which are made below will be kept to a minimum.

Among the approaches which do allow that microbial life can be of intrinsic value, some take the "intrinsic" part of the idea in a fairly literal sense. However, rather than rationality, or sentience being the special property which is of value, it is "structure," or "structured complexity" which is taken to be important. And there is something to this idea. Structures which are unique contribute to diversity, and they may make individual things irreplaceable. Applications of the idea to the context of space have important strengths [6.31]. Although, they can also go off in unusual directions, e.g., when it is claimed that limiting entropy is itself somehow an ethical goal, and perhaps something that a cosmos-level or cosmocentric ethic could be based around [6.33]. However, moves of this sort can be a little like participating in a game in which we are asked, "Pick a property...any property," and ethical considerability is then associated with whatever property we happen to select: self-consciousness, sentience, structure, and so on.

The most plausible versions of this approach are ones which suggest that structure is at least *part* of the story of why we value things. They may also appeal to a concept of "integrity" which is arguably more intuitively plausible than talk about "rights" in the case of non-sentient life or things [6.16]. If, for example, I ask an old question once put by Chris McKay [6.19], "Does Mars have rights?" there is something a little odd about the question. But if I ask "Does Mars have integrity?" we can more readily understand what is at stake. And what is at stake will be partly a matter of the unique Martian structures of places such as Olympus Mons and the Valles. Structure, then, is part of the story of value if we appeal to the more plausible ethical concept of integrity, but it is not the whole story. After all, everything has structure, yet we cannot value everything in the way that we may value living things and special places.

Also, if we appeal *only* to structure, we will be back where we started, dealing with the parallel case. As indicated, the awkward truth is that any microbial life that we find elsewhere, and value, may well turn out to be structurally similar to microbial life here on Earth, i.e., the kind of life that we routinely kill in large numbers as part of our daily lives. Life which we also kill intentionally when using sanitizing gel, or when we clean kitchen worktops and bathrooms. Yet in such cases, it seems absurd to suggest that we are doing anything wrong. There have, admittedly, been voices making precisely this suggestion, but they have tended to draw upon a special sort of religiously-inflected ethic about life. Here, I am thinking of the Jain tradition, which has been in something of a long-standing care conflict with Buddhism, and geared to showing a broader range of care than that extended by Buddhists to all sentient beings. And, up to a point, I am thinking also of Albert Schweitzer, who drew upon Jain ideas in the 1920s, and combined them with a Christian-inflected reading of respect from Immanuel Kant to produce a sort of ur-theory of life's value, later drawn upon environmentalists in the 1960s and 1970s [6.29]. Schweitzer's view was that all life had value. Indeed, all life had *equal* value, but the pragmatics of being human required us to set this aside in cases where it was simply impractical. We are flawed beings and cannot value all that we ought to, and in the way that we ought to.

Structure, if thought of as just another move in the property identifying game, may fare no better than the rest (rational agency, sentience, and so on). What we may need to break out of the game is a more relational understanding of what is at stake, rather than one which treats value talk as if it involved the search for some mysterious inner thing of wonder, a kind of invisible inner gold, undetectable by our best science. Appeals to integrity already tend to do this, either deliberately or by accident. They shift towards more of a "systems" view of things, locating individual things in a broader supporting and enabling context. For example, Holmes Rolston III [6.27] is notorious for making this move of an appeal to "integrity" of other planets in his early attempt to theorize environmental ethics in space: unique structures bear the marks of unique history. The value of whatever has value is not then to be thought of exclusively in terms of its immediately present structure, but in terms of the combination of structure, the unique past that it is a signature for, and its contribution to diversity [6.18]. Such an approach has traction when it comes to the parallel case. It allows us to make sense of our differing attitudes towards what may be structurally similar life forms as more than an issue of pragmatics, more of a sense that we need to kill microbes routinely *here*, but do not need to kill them *there*. Rather, microbes *here* and *there* have different histories, and it is (at least in part) as a result of these differing histories, and our overall differing relation to them, that we are justified in treating them differently.

To back up this more relational shift in our thinking about value and about practices of valuing, we may consider our most striking experience of what it is to value something other than ourselves, i.e., love. Consider an old thought experiment from Derek Parfit: if you love someone who steps into a replicator and is disassembled, should you love the person who steps out at the other location? They will be a physiologically exact duplicate. Parfit believes that we should. Some of us think that cases of this sort are more problematic [6.9] [6.21] [6.22]. What matters, from the standpoint of loving another, is not structure alone, but unique causal history. For example, I want to go home to the same Suzanne that I met at the end of my teens and sat out with under the stars, not a physiologically or structurally exact copy. Just as I want to see the original paintings by Titian, and not a brilliant duplicate. (And yes, I know that exact duplication may only be possible in theory, but that is enough for such thought experiments to run. Conceivability does not always entail possibility of a stronger sort.)

But what this drives us towards is not simply a one-dimensional account of the role of relations in attributions of value, one which connects structure and history and does little else. Rather, it drives us towards acceptance that the whole practice of talking about value is really a way of capturing something about our human relations to each other, and to things that certain agents will tend to regard as important *in their own right*. Another way to make the point is to say that "value talk" can usefully be seen as a sort of convenient shorthand which helps us to recognize what might more formally be referred to as "reasons for action and response." This approach to value talk is very different from treating it as a way of picking out hidden inner gold. Instead, there are reasons for acting and responding (e.g., for feeling compassion and regret) in particular ways, that suitable agents will be able to appreciate. For convenience, I will speak of "reasons for action" and leave the reader to fill in various other sorts of associated reasons for feeling in particular ways, and for response more generally. Additionally, the reasons that I have in mind are not instrumental, or at least they are only ever partly instrumental. They acknowledge things as having a significance in their own right, in ways which can be independent of our interest in them.

In line with this approach, if we can say that all human agents with a reasonably unimpaired rationality and appropriate socialization would affirm something, or recognize something, then from an ethical point of view, *who could ask for anything more?* If only a crazy person, or someone suffering from a certain kind of mental disorder would act in a particular way, this will give us everything that we need to know in order to use a language of value and of ethical truths in stable ways, secure in the expectation that future humans will share much the same attitudes, unless something goes badly wrong.

Of course, ethical disputes often involve disagreement among rational agents, rather than general agreement with only irrational or impaired agents dissenting. And in this respect, they are like disputes about politics and economics. But both of these are cases in which disagreement on its own is not taken to imply the absence of truth. Such disagreement, instead, indicates the difficulty of finding the truth, and particularly so in a context where beliefs are heavily skewed (on all sides) by interests. Similarly, we can navigate our way through ethical disagreements in much the same way by considering the best way to look at matters, among multiple rivals. Sometimes there may be no such best way of looking at matters, but we will often accept that some ways are better than others. And this will narrow the range of our legitimate options. My point here is the simple and familiar one that value talk then presupposes valuers. Value talk makes little sense without them. Yet, it is less misleading for such talk to draw upon metaphors of vision, and of seeing, than it is for such talk to draw upon metaphors of projection. Such talk concerns recognition and response, recognition that various aspects of how the world stands are particularly salient to our deliberations, ways of feeling, and actions. And having the relevant kinds of responsiveness is part of what it is to be human.

Strictly, in terms of metaethical theory, this draws upon what is known as a "response-dependent" approach [6.17] to what makes at least some ethical claims true. Certain kinds of suitable observers are disposed to respond in particular ways, with sensitivity to what others might miss, and certain kinds of things are disposed to produce the relevant response in agents of this suitably sensitive sort. The dispositions are not all on the one side, with the human agent. The dispositional properties are both *here* and *there*. In terms of metaethical theory, this may well be "the best game in town" unless we want to say that ethics is constructed or projected. But it is also an approach which blocks off arbitrary restrictions of value to rational beings, or sentient beings. Instead, it keeps such attributions aligned closely with the broader range of things that suitable and well-placed agents are actually disposed to value.

As an exemplar, and one which draws us back directly to the valuing of living things, we may think of two variants of a familiar story. One draws from Loren Eiseley's short story, "The Star Thrower" (1969), in which a narrator encounters a young boy confronted with a mass of starfish washed up on the beach. He holds one away from the mud while the narrator asks, "Do you collect?" The boy throws the starfish back into the water and replies, "Only like this...And only for the living." The narrator later compares our human predicament to that of stars cast onto an infinite beach by an unknown hurler of suns [6.8]. The story is about small acts of compassion, but the compassion in question presupposes that there is something worth saving and not just some manner of private distress. In the second variant, which owes something to Ronald Hepburn's essay on "Wonder" (1980) [6.11], the boy on the beach finds out that he can get a couple of cents for every starfish that he gathers but does not return to the sea. And so, he gathers a massive number, exchanging their lives for pennies. The thought here is that there is something lacking in his encounter with the creatures. He sees them only as exchangeable goods. A sense of wonder is missing, or at least compromised.

Comparing the two cases, it will take a good deal of work to convince most of us that there is nothing to choose between them. Or that the first is a merely sentimental reaction, and the second a more realistic reaction. The clash here is certainly, in part, a clash between two ways of seeing. And perhaps one way of seeing is a lot "nicer" than the other. But my point is not just that we side with the child who tries to rescue the starfish. Rather, my point is that the child who sees only the financial potential of the echinoderms really is *missing something* about them. That, in some sense, he sees less of what there is to be seen, while the child who throws back the starfish sees more. He has a fuller grasp of the reality. Moreover, this idea does not depend upon his possession of a theory of mind and consciousness. He may turn out to have no view at all about whether starfish or urchins think. If he were to come to believe that they are only reactive, and not conscious in any way which is analogous to our consciousness, we need not imagine that he would then behave any differently. They might still be seen by him as living creatures, and the lives of even one or two might still be seen as something worth saving.

6.3   The Intelligibility of Microbial Value

The above story about value and valuing begins to allow us to make sense of why we think that the presence or absence of even non-sentient life can be a significant matter. And here, we need not mistake the beginning of a fuller account for more than it is. Even so, it helps us to make sense of why the discovery of life or of genuine biosignatures in the atmosphere of Venus would be momentous. And why we might think that randomly populating the Moon with micro-animals is not the kind of thing we should take lightly, as an insignificant accident. In neither case do we need to think of the rudimentary forms of life in question as the possible precursors of more complex life, or beings such as ourselves. Yet, given the choice between a universe in which there is no life, or a universe in which there is only microbial life, with no prospect of evolving into beings of some other sort, most of us would choose the latter. This, again, is a thought experiment, or "only" a thought experiment, rather than a demonstration of a more rigorous sort. Yet it can show us something about our attitude towards life, or it can help us to articulate, more clearly, what that attitude is.

But if we do go down this road, and accept that we may well have reasons for action in relation to living things which are not just about our own interests, or about anything other than the living things themselves, where does it get us? I will take it that several significant problems still remain. And one of these concerns intelligibility. In the case of microbial life, which feels nothing at all, can we even make sense of the avoidance of biocontamination as a matter of "protection" in any familiar sense? Can we, for example, think of ourselves as engaged in an attempt to ensure that no *harm* is done? While it may be difficult to make sense of talk about microbes having rights, we may still have to defend the difficult idea that it makes sense to talk about microbes being harmed, when there is no way for them to experience the harm. The example of the starfish on the beach may perhaps suggest that experience is not really what matters here, but we may still have more to say about what is at stake.

To do so, I will make a softening up move, an argumentative move which begins to erode a familiar intuition in favor of something a little more qualified. It involves consideration of our own case. Ordinarily, when we are harmed, we imagine some physical impact that we are all too well aware of. Or perhaps we might think of psychological trauma. In any case, the paradigm instances of being harmed do seem to involve awareness. What we do not feel and do not know does not harm us. But are such cases of harm really the only ones in which harm occurs? There do seem to be other cases of the sort which have figured in accounts of posthumous harm [6.25], and cases where harm occurs but there is no associated experience.

For an example of the former, consider the possibility that the folios of Shakespeare's plays might have been destroyed in the Great Fire of London, i.e., in an event which occurred after his death. In the imagined world where this occurs, Shakespeare's impact upon the world is much less than it has been in our own world. He is forgotten, or reduced to a footnote. Someone who might figure only in dissertations on obscure playwrights of the Elizabethan era. Perhaps, we might find it difficult to say that Shakespeare himself has been harmed, because he would not be around to be harmed, but perhaps the excellence of his life has been harmed. Or, we might use any one of several alternative descriptions, leaving little doubt that *something* had been harmed, and the something in question is closely connected to the playwright. In this case, the harm would not be contingent upon his awareness of the harm. We need not imagine that he hovers around in the air somewhere. Also, this is not just a peculiar example, but a thought experiment which draws out our familiar intuition that we have duties and responsibilities in relation to the dead, and that these do not reduce to our ways of caring for those who are alive.

Pulling matters even closer to the mundane. Are we really so confident that "What we do not know cannot harm us"? Let us suppose that you are unfaithful to your partner. I am not talking here about open relationships, but straightforward infidelity under conditions where there is a reasonable and unforced expectation of exclusivity and loyalty. The norms of intimacy which are internal to the relationship are violated. Let us also suppose that you are extremely careful, and that there really is no way that your partner can find out. It seems odd to say that you have done something wrong, but that there is no victim of the wrongdoing. That nobody, or nobody's life, has been in any way damaged. That its excellence has not been in any way compromised. Surely, it is better to have a life in which one's partner is faithful, rather than unfaithful, *irrespective of whether or not any infidelity is ever disclosed*? Similar imaginary cases can be run in order to drive home the point. Rapists who ensure that their victims are never aware of having been raped have nonetheless harmed them in terrible ways. These are cases which erode our rather hedonistic sense that what we do not know, and are not aware of, cannot harm us; that harm is akin to pain and benefits akin to pleasure. Rather, in terms which matter here in a direct way, *harm does not always depend upon awareness*.

Of course, we can say that these are cases which involve beings who have been aware at some point in time, even if they are not aware of the harm in question. And that may seem to be a significant consideration to some of us. I think it may be irrelevant, but accept that others might see things differently, and neither of us may be making a blunder of any easily demonstrated sort. So, let us treat this only as a softening up argument, and nothing more. What really gets us over the line to the idea that non-sentient life, such as microbes, can be harmed is a combination of two further arguments.

The first is an argument of a familiar sort, formulated in modern terms by Richard Routley [6.28], drawn from environmental ethics and (a little inconveniently) named the "Last Man Argument." This is a little inconvenient, because it is gendered in a way which is entirely unnecessary but also plays upon a problematically gendered strand of environmentalist thinking. So, let us think of it as a "Last Human Argument." This last human has outlived all others and, with his dying movement, wantonly chooses to cause untold environmental devastation. Forests are blown up, canyons flooded, and so on. Surely, he or she does something wrong? Yet the wrongness is difficult to make sense of without allowing that something is actually harmed. We can also remove various candidates for such harm from the thought experiment, e.g., other creatures. This takes us as far as the original version went. The point was to drive the intuition that ecosystems and the environment matter in ways which go beyond their usefulness to us, or to any other particular creature.

The thought experiment can, however, be extended in ways which also remove all sentient life, and leave only the non-sentients. In such a case, the destruction still seems wrong and it is still difficult to say that wrong is done, but there is no harm. Harm again seems separable from an awareness of harm. A variant of this Last Human Argument has been run by Charles Cockell, in order to tease out the idea that microbes may have value in their own right [6.6]. We are invited to imagine someone who wantonly kicks through a microbial mat. Surely, they are doing something wrong? Surely, harm is done, and what harm presupposes is something to which the harm attaches. We can also strengthen the case in various ways: make the mat the only one of its kind, allow that all relevant scientific inquiry has already been conducted, and so on. It still seems that wrong is done, and that the wrong specifically involves harm. In some respects, this kind of argumentative move also overlaps with the choice between universes in the case above. In both, there is a presupposition in favor of life over non-life. Life is also taken to be the kind of thing that can be harmed, hence also the kind of thing that might be protected.

The second argument concerns ideas of directedness, flourishing or "telos." Our language for human well-being often draws from non-sentients, and in particular from the world of plants. There is something for a plant, or a tree, to *flourish*. Its development has direction. Trees are usually the exemplar which is appealed to here [6.32]. They have a "good of their own" [6.3]. This is a good which may be blocked, and the blocking is a form of harm to trees. This matches well with our ordinary ways of thinking about things. When someone sets fire to a tree, the tree is harmed (unless it is one of those forest trees which are part of a cycle in which fire helps propagation of offspring). The trees are harmed without being aware of the harm, and the harm takes the particular form of blocking their directedness or development. Something similar may be said about micro-animals and microorganisms. Indeed, while we may have no universal definition of life by contrast with non-life, we have multiple working definitions that we can use more locally, and they tend to include functional activity. Having something like directedness is, or is close to being, partly definitional for life. All life may be harmed in this sense.

These reflections may draw upon a kind of moral sensibility which is not shared by all readers, yet they do not require us to posit anything outlandish.

They do not require us to claim consciousness or sentience where we have no evidence for it, in order to underpin value talk in the relevant contexts. But, more importantly, they do not imply that we should adopt any impossibly demanding attitudes towards microorganisms. By this, I mean that an account of microbial value should not fall into the trap that Albert Schweitzer [6.29] fell into: the idea that all life should be regarded as equal, or as equal *through and through*, i.e., equal in more than some elusive ultimate metaphysical sense. (The universe is equally lacking in any concern about any of us. In that respect, we are all equal.) A restricted variant of this same mistaken move, which extends only to sentient life, is a familiar feature of classic animal rights theory [6.26], but not of more pragmatic forms of animal rights advocacy [6.4]: the idea that all value bearers are equal. Value is then thought of in binary terms. You can have it or not, and it always involves exactly the same moral standing.

A commitment to equality of this sort is not a psychologically available standpoint for beings such as ourselves. Nor is it at all required. We cannot treat microorganisms as having the same importance as humans, and should not ordinarily treat them as having the same importance as other sentient beings, entire rainforests, rivers, and so on. Indeed, the idea that all value bearers are equal, or that all life is equal, fits better with an understanding of value talk which again treats it as a way of speaking about hidden inner gold, rather than as a way of talking about our reasons for action (and response). When we adopt the latter approach towards value it should be readily apparent that we have very different reasons for action in relation to different sorts of creatures and different life forms. In ordinary life, my reasons for action in relation to a microbial mat and in relation to my wife Suzanne may in some small respects overlap, but they will also in some very large respects diverge. Even if I were to notionally claim to treat them equally, this would not accurately track the pattern of my actions. The latter would still show significant differences in my practices of valuing.

The crucially important political language of egalitarianism, which fits so well and does so much work in certain areas of our lives, and in some animal and environmental contexts, has little useful role to play in such deliberations about microbial value. We are not then called upon to value microbial life in impossible, or excessive, ways. Yet, to say this is not to revert to a "no value" view, or to say that terrestrial microbes and microbial life *in situ* elsewhere in space are much the same, from an ethical point of view. Again, we will have very different (legitimate) reasons for acting in relation to them. We will have reasons for protection in the case of Martian microbes that we do not have in relation to the microbes on the kitchen chopping board or in the bathroom sink. We will legitimately worry about biocontamination in relation to the former, but not the latter, even if decontamination protocols themselves involve the destruction of terrestrial microbes.

6.4   Contamination and Discovery

To recap, we *can* make sense of the widely shared intuition that some kind of special protection ought to be extended to even the most primitive forms of life, if we should ever find them elsewhere in the Solar System. Talk about their having value or (more precisely) intrinsic value can be understood in terms of our having reasons to act in ways which show a regard to avoid harm, but which are not driven by our own interests. Of course, the idea of "having reasons" of this sort is itself something that might be analyzed in various different ways. It is something else that we might argue about. However, for convenience, I will take it that a reason for action is something that plays a role within a practice of justification. Such practices have to end somewhere, at a point where we are satisfied that enough has been said. (We do not mimic children who simply say "But why?" no matter what they are told.) And the place where they end varies from context to context. A consideration may be justified in one context, but not in another context.

So, what then are the kinds of things that we might say, by way of justification, when considering the protection of microbial life from the dangers of biocontamination? Here are some examples: "It is the only life indigenous to this planet," "It comes from a second and entirely distinct genesis," "It has been around for billions of years as part of a process that we do not want to end," "We should not be the kind of beings who think only of our own interests," and so on. None of these are the kinds of bedrock considerations that our 19th century predecessors might have sought as a solid foundation to which every ethical justification could ultimately be traced. But very little of our patterns of valuing draws upon foundations of that sort. Such foundations do little in terms of lived ethical agency. Rather, we are almost invariably caught up in the different task of weighing one inconclusive consideration against another, while accepting that these considerations really do matter even though they are not conclusive.

Weighing up microbial protection is like this too. I have already committed, above, to the view that value bearers should not, and cannot, be treated equally except in some limited contexts. Our reasons for protecting microbial life *for its own sake*, will not be the same as our reasons for protecting one another from biohazards. And this applies here, or on any planet we happen to go to. And if we were ever to discover complex, sentient, life somewhere, in a place which also had microbial life, we would have reasons to prioritize protection of such complex and sentient life over the microbial life. Or, to regard the latter as part of a larger ecosystem which supports the more important kind of life. The fact that one kind of life is sentient, while the other is not, would itself be a reason that we could obviously offer in any justification of our actions. In some cases, it might be enough, in others a little more might need to be said. The general point here is, however, a simple one: all life *cannot* be treated equally all of the time, but what we offer by way of a justification for treating different life forms differently should at least be intelligible to others, and it should stem from something other than any simple form of anthropocentrism, fantasies about human destiny, or some similar ideologically-driven consideration. Moreover, it is life that matters greatly, not microbial life that matters greatly. Microbial life matters because it is life, and not because of the rights of individual microbes, or because of any impossible extension of admirable political ideas about equality.

Where this leaves us is with some easy choices, and some much harder choices which will shape the conclusion below. Among the easy choices there is the basic assigning of priority in comparisons between microorganisms and more complex beings. That will generally favour matters of life and death for sentient organisms and for beings like ourselves in relevant ways. We can, of course, imagine peculiar cases in which the life of a single human can only be saved at the expense of all known microbial life in some place. And that could be a genuine dilemma of a peculiar sort. I do not think we could ask humans to sacrifice their lives to save the second genesis, but we could understand why they might do so, and might even admire their doing so. Note also, I revert here to a language of historic depth (i.e., second genesis) to grasp a sense of the psychological state of a human caught up in such a predicament, and fully able to realize the enormity of the decision they face.

On the simpler side, and as an extension of this same life or death thought experiment, there is also the default commitment to preservation rather than destruction. If we find life elsewhere, even microbial life, we should not destroy it or expose it to risks of biocontamination, unless there is some absolutely compelling reason to do so. We should not be like the wanton agent in the Last Human Argument, and it is difficult to imagine someone who would consider it acceptable to act in such a way, unless they happened to be damaged by their own past. This default commitment to preservation rather than destruction extends to reasonable protocols concerning forward contamination. The assumption that the biochemistry of life elsewhere will be so different that it will always be safe from whatever we bring with us may be a convenient assumption, but it is not a safe assumption. We do not need to work with anything so optimistic, even though going would be considerably less expensive if it happened to be true. Protocols to avoid forward contamination, until we know what we are dealing with, should remain robust. And this is already covered in planetary protection protocols, although a further appeal to planetary environmental protection will add a further rationale for what we already try to do.

On the simpler side, there is also the question of the permissibility of experimentation. Generally speaking, experimentation only becomes an ethical problem in the case of experiencing subjects such as humans and other animals. Accepting microbial value is also not the same as attributing rights in the traditional way, i.e., to *individual* right bearers. There are few contexts in which we ever encounter microbes on an individual basis. Rather, except under laboratory conditions, we encounter them in aggregates. Hence, Charles Cockell's choice of examples, i.e., the wanton destruction of a microbial mat [6.6]. Valuing microbial life is largely about how we behave in relation to microbial life in aggregates and *as a whole*, or *overall*. And this is not something that would ordinarily be impacted by experimentation. Nor does it entail any consequentialist maximization, favoring a default of having more individual microbes rather than fewer on the grounds that every microbe is precious, so the more there are the more value will exist. (I do not think that value talk works this way with humans, much less with microbes.)

There is one possible qualification to this. The Viking lander experiment which delivered a false positive in 1976 involved testing for biosignatures in a regolith sample, then superheating the sample to destroy any microbial life, and retesting the sample. There may well be something wrong about arriving on a planet and *immediately*, as first priority, trying to kill indigenous life. Admittedly, it is difficult to explain why it would nonetheless be permissible later, but not *in that moment*. However, the intuition that it is not a good way to begin looks like it might have a justification. The justification could, of course, appeal not to microbial value, but to "who we want to be." Yet, a point of the latter sort, a virtue or character-focused point, would probably presuppose something like microbial value in order to make sense of the idea that harm has been done, and done in a way which should have been avoided.

What all this points towards or suggests is an ethical default of "protecting what we find," in the sense of "protecting whatever overall level of microbial presence we find elsewhere in the Solar System," and the requirement that we must have a good reason for disrupting this presence in significant ways. (And yes, this leaves questions over the distinction between "significant" and "non-significant" disruption.) In its favor, as a point about political realizability, this ethical approach will synchronize well with planetary protection for scientific inquiry. And neither involve anything akin to a microbial version of the prime directive in which noninterference with life is outlawed. As a qualifier, I use the language of "points towards" and "suggests," in order to emphasize the provisional nature of these conclusions, and the fact that they are not logically entailed by some set of first principles in combination with the available evidence. Arrival at this point has instead drawn from something less rigorous than that, but also from the kind of enquiry which may be well suited to the elusive nature of the difficulties of talking about ethics in relation to microbial life forms elsewhere, i.e., in a context which is relatively new.

A final simple side consideration, and one which aligns well with this ethical default of "protecting what we find," is that we should not add to the complexity of our ethical dilemmas unnecessarily, and without a good reason. There may, for example, be good reasons to take microbial life to Mars, or even tardigrades to the Moon *at some point in time*, but when we do so, we alter the playing field and the range of considerations that we must then take account of. This should not be done for trivial reasons, or by accident. Once it is *in situ,* microbial life may be on the way to having a different standing from the standing it has on Earth. A dropped Petri dish may be picked up or the disk with imprinted tardigrades may be picked up at this point in time. Later, may be different. On this view, what was wrong about crashing tardigrades onto the Moon was not just the "go it alone approach," which did not pay due regard to the need for an international consensus, it was also the risk of "changing of the playing field" for no good reason.

6.5 Conclusion

The final substantive section above considered some of the easier implications of accepting microbial value and doing so in a qualified and non-demanding way. As indicated, such an acceptance of microbial value will help to make sense of our shared interest in discovery. An interest which extends beyond scientific curiosity and is rooted in a sense that life matters to us. The best way to value life is, of course, to value living things when we can, and in the ways that we can. And this is about our relation to the rest of life, rather than simply a matter of recognizing inherently valuable structures or properties. Value talk works best when understood as a shorthand for our relation to others, to ourselves, and to what is non-human, i.e., as something relational. (Although it is not the relation itself that is valued, or at least not only this relation.)

On the harder side, we are left with the difficult question of specifying the extent of protection that any discovered microbial life would deserve, given that it would be an instance of life and that life is something we value. There may be no context-free answer to this. It might depend upon just how rare we consider life to be in the universe. If peculiar local conditions within the Solar System have given rise to two independent cases of life's evolution, but the peculiar conditions really are peculiar, I suspect that we would have to regard our reasons for microbial protection as stronger than they would otherwise be. Beyond that, any attitude towards microbial protection as part of a societal-level ethic would have to meet the requirements that any societal-level ethic of any sort must meet: it would have to be psychologically available to ordinary agents; politically viable (given some reasonable, if ordinarily flawed, set of political structures); and stable enough for some approximation to it to be accepted across multiple generations. The exploration of space is, after all, a multi-generational project.

These requirements suggest that a viable form of microbial protection from the risks of biocontamination, or from other harms, could not interfere with otherwise justifiable human activities such as infrastructure construction, scientific enquiry, and resource extraction. We can imagine cases in which it may make sense to say, "Don't mine there," because of the local presence and patterning of microbial life. Cases of this sort might be the "sharp end" of the ethical dilemmas we eventually face in the light of an acceptance of microbial value. (Rather than the fictionally extreme scenario considered above, where the human must die if the microbial life is to survive.) But, given that our attitude towards microbial protection is an aggregate attitude, and draws from a concern with life as such, it would be much harder to say, "Don't mine anywhere," on similar grounds. There might always be other reasons for us to restrict mining, but that is a different matter.

References

[6.1]   Aldiss, B., and Penrose, R., *White Mars*, Little, Brown, London, 1999.
[6.2]   Aquinas, T., *Questions of the Soul*, J.H. Robb trans., Marquette University Press, Milwaukee, 1984.
[6.3]   Attfield, R., The Good of Trees, *J. Value Inq.*, 15, 35–54, 1981.

[6.4] Clelland, C. and Wilson, E.M., Lessons from Earth: Towards and Ethics of Astrobiology, in: *Encountering Life in the Universe: Ethical Foundations and Social Implications of Astrobiology*, C. Impey, A.H. Spitz, W. Stoeger (Eds.), University of Arizona Press, Tucson, 2013.

[6.5] Cockell, C.S., Environmental Ethics and Size, *Ethics Environ.*, 13, 1, 29–37, 2008.

[6.6] Cockell, C.S., Jones, H.L., Advancing the Case for Microbial Consideration, *Oryx*, 43, 4, 530–26, 2009.

[6.7] Dunayer, J., *Speciesism*, Ryce Publishing, Derwood, MD, 2004.

[6.8] Eiseley, L.C., *The Unexpected Universe*, Harcourt, San Diego, 1972.

[6.9] Grau, C., Love and History, *South. J. Philos.*, 48, 3, 246–71, 2010.

[6.10] Haraway, D., *When Species Meet*, University of Minnesota Press, Mineapolis, 2008.

[6.11] Hepburn, R.W., Wonder, *Aristotelian Soc. Suppl. Vols.*, 54, 1, 1–24, 1980.

[6.12] https://www.hou.usra.edu/meetings/leag2017/presentations/tuesday/conley.pdf

[6.13] https://www.wired.com/story/a-crashed-israeli-lunar-lander-spilled-tardigrades-on-the-moon/

[6.14] Kant, I., Groundwork of the Metaphysic of Morals, in: *Practical Philosophy*, M. Gregor (Eds.), Cambridge University Press, Cambridge Mass, 2006.

[6.15] Leopold, A., *A Sand County Almanac and Sketches from Here and There*, Oxford University Press, Oxford, 1968.

[6.16] Lupisella, M., Ensuring the Integrity of Possible Martian Life, *49th IAA Congress*, Amsterdam, 1999.

[6.17] McDowell, J., Values and Secondary Qualities, in: *Mind, Value, Reality*, J. McDowell (Ed.), Harvard University Press, Cambridge Mass, 2002.

[6.18] McKay, C., Astrobiology and Society: The Long View, in: *Encountering Life in the Universe: Ethical Foundations and Social Implications of Astrobiology*, C. Impey, A.H. Spitz, W. Stoeger (Eds.), University of Arizona Press, Tucson, 2013.

[6.19] McKay, C., Does Mars have Rights, in: *Moral Expertise*, D. MacNiven (Ed.), Routledge, London, 1990.

[6.20] Milligan, T., *Nobody Owns the Moon: The Ethics of Space Exploitation*, McFarland, North Carolina, 2015.

[6.21] Milligan, T., The Duplication of Love's Reasons. *Philos. Explor.*, 16, 3, 315–23, 2013.

[6.22] Parfit, D., *Reasons and Persons*, Oxford University Press, Oxford, 1986.

[6.23] Passmore, J., *Man's Responsibility for Nature*, Duckworth, London, 1974.

[6.24] Persson, E., The Moral Status of Extraterrestrial Life, *Astrobiology*, 12, 10, 976–84, 2012.

[6.25] Pitcher, G., The Misfortunes of the Dead, *Am. Philos. Q.*, 21, 2, 217–225, 1984.

[6.26] Regan, T., *The Case for Animal Rights*, University of California Press, Berkeley, 2004.

[6.27] Rolston, H., The Preservation of Natural Value in the Solar System, in: *Environmental Ethics and the Solar System*, E.C. Hargrove (Ed.), Sierra Club Books, San Francisco, 1986.

[6.28] Routley, R., Is There a Need for a New, An Environmental, Ethic?, *Proceedings of the XVth World Congress of Philosophy*, vol. 1, pp. 205–10, 1973.

[6.29] Schweitzer, A., *The Philosophy of Civilization*, Prometheus Books, Amherst, NY, 1987.

[6.30] Singer, P., All Animals are Equal, *Philos. Exch.*, 5, 1, 103–116, 1974.

[6.31] Smith, K., The Curious Case of the Martian Microbes: Mariomania, Intrinsic Value and the Prime Directive, in: *The Ethics of Space Exploration. Springer*, J. Schwartz and T. Milligan, (Eds.), Springer, Heidelberg, London and New York, 2016.

[6.32] Stone, D., Should Trees Have Standing? Toward Legal Rights for Natural Objects, *South. Calif. Law Rev.*, 45, 450–501, 1972.

[6.33] Vidal, C., *The Beginning and the End: The Meaning of Life in a Cosmological Perspective*, PhD. Thesis, Vrije Universiteit Brussels, http://arxiv.org/abs/1301.1648, 2013.

7

Astrobiology Education: Inspiring Diverse Audiences with the Search for Life in the Universe

Chris Impey

Department of Astronomy, University of Arizona, Tucson, Arizona, USA

Abstract

The search for life beyond the Earth is one of the most compelling quests of modern science. Astrobiology appeals to a wide audience, including those who might not otherwise be engaged with science. This chapter gives an overview of the modes of astrobiology education and outreach that operate in the United States. Astrobiology is one subfield of astronomy, but it has connections to geology, chemistry, biology, and even sociology. This interdisciplinary context creates both challenges and opportunities for educators. Starting with professional development, there are just a modest number of Ph.D., M.Sc., and B.Sc. programs in astrobiology. Most professionals who identify as astrobiologists were trained in one science discipline and learned other subject matter and skills as needed through their careers. For non-science major undergraduates, "Life in the Universe" is a popular niche class that is offered as a follow-up to an introductory astronomy course. Astrobiology is offered to worldwide audiences of adult learners in the form of a MOOC, or massive open online class. Examples of teaching and outreach materials are given, and it is a good bet that the demand for astrobiology education and outreach will increase strongly if scientists achieve their goal of detecting life beyond the Earth.

Keywords: Astrobiology, education, pedagogy, degree programs, online courses, video lectures

Email: cimpey@arizona.edu

Chris Impey: http://chrisimpey-astronomy.com/, https://scholar.google.com/citations?hl=en&user=OrRLRQ4AAAAJ, https://en.wikipedia.org/wiki/Chris_Impey

Octavio Alfonso Chon Torres, Ted Peters, Joseph Seckbach and Richard Gordon (eds.) Astrobiology: Science, Ethics, and Public Policy, (135–156) © 2021 Scrivener Publishing LLC

7.1 The State of Astrobiology

The search for life in the universe has reached a pivotal stage. For a long time, astronomers have believed it to be likely that life exists elsewhere, based on a number of indirect lines of reasoning. These include: the fact that life formed on Earth within a half billion years, when the physical conditions were extreme; the ease with which life's basic ingredients combine naturally to form more complex molecular building blocks; the adaptation of extremophiles to a variety of ecological niches; and the abundance across time and space of the chemical ingredients needed for biology, along with the abundant habitable locations in the Solar System and beyond where biology could evolve. Refer to the Astrobiology Primer v2.0 for an overview [7.31]. None of these indications guarantee that biology exists beyond the Earth, but they have helped fuel the growth of astrobiology as a profession, and create excitement with public audiences. The long history of astronomy since the Copernican revolution has displaced us in centrality and importance in the universe. It would therefore be a surprise if biology were unique to our own habitable planet.

Figure 7.1 shows the research results from the past few decades that fuel optimism that life is widespread in the galaxy and the universe. Current strategies for discovering life beyond Earth are also indicated. We still don't know the full envelope of physical and chemical conditions

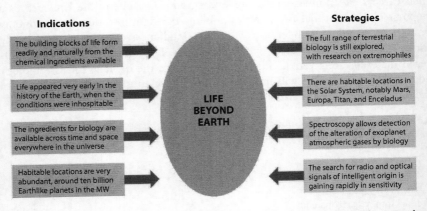

Figure 7.1 On the left are research insights that lead to an expectation that life on Earth is not unique. On the right are some of the main strategies being pursued to better understand the limits of biology on the Earth and search for it in the Solar System and beyond. (Copyright Chris Impey).

for terrestrial biology, so we have no basis for estimating the probability of biology on exoplanets with different conditions [7.62]. Even our definition of habitability is unreasonably tied to the situation of the Earth. Within the Solar System, Mars understandably gets the lion's share of attention, but there are a dozen locations with life's basic requirements: carbon-rich material, liquid water, and a local energy source. Most of these are moons of giant planets far from the traditional habitable zone [7.15] [7.42]. Spectacular success in detecting exoplanets has yielded hundreds of Earth-like worlds, and some are within a few dozen light years of Earth [7.76]. An upcoming generation of large ground-based and space-based telescopes will take spectra of the reflected light from the brightest of these planets, to look for signs that the atmosphere has been altered by biology [7.12]. These spectral lines are called biomarkers. Finally, astronomers are also placing a bet that some distant forms of life will have advanced to the level of intelligent and technology [7.71]. The tools of the search for extraterrestrial intelligence (SETI) are advancing rapidly in their sensitivity and reach [7.87].

Astrobiology is highly interdisciplinary. The practitioners are mostly astronomers, but the ranks of astrobiologists also included people trained in planetary science, geosciences, biology, and chemistry, along with a small cohort drawn from philosophy, sociology, and anthropology. We still only know of one place in the universe with life: this planet. Yet the excitement of the search has fueled growth, especially since the turn of the century. The emblem of astrobiology's success is the growth in the number of exoplanets, now standing at over four thousand [7.34]. The first was discovered using the Doppler method in 1995, leading to the award of the Nobel Prize in Physics in 2019 to co-discoverers Michel Mayor and Didier Queloz [7.59]. Less well-known is a precursor discovery of planets around a pulsar using radio timing methods in 1992 [7.86]. In 2000, the census was 50, all found with the Doppler method. In 2010, that had grown to 600. The surge in the last decade to 4000 was driven by NASA's Kepler satellite, detecting exoplanets as small as the Earth by partial eclipses that dimmed the parent star [7.13]. The number of exoplanets has increased exponentially, doubling every 27 months [7.43]. With new surveys and technical innovation, this trend might continue, projecting to 100 million exoplanets by 2050.

Figure 7.2 gives a snapshot of how astrobiology has grown in the 21st century. Four different metrics are considered, each normalized a year close to the turn of the century. The Astrobiology Science Conference (AbSciCon) began in 2000, and is the largest international meeting

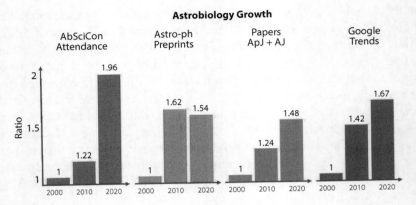

Figure 7.2 Graphs showing different aspects of the growth of astrobiology in the past two decades. From left to right is attendance at the biannual AbSciCon meetings, the number of arXiv preprints in astrophysics using the keyword astrobiology, the number of astrobiology papers in the two major AAS journals, and the number of Google searches astrobiology, as normalized to the word astronomy. (Copyright Chris Impey).

devoted to the subject [7.3]. Attendance nearly doubled from 500 in 2000 to just under 1000 at the most recent meeting in 2019. By comparison, attendance at the winter AAS meeting has not grown by more than 30%, using the four meetings in Seattle during this interval as a baseline. The second example shows the factor by which preprints with astrobiology as a keyword have outstripped all other preprints on the astrophysics section of the arXiv server [7.9]. The third example shows the growth of papers classified as astrobiology in *The Astrophysical Journal* and *The Astronomical Journal*, relative to the more modest overall growth of papers in those two flagship journals of the American Astronomical Society [7.7] [7.8]. The last example attempts to gauge the interest of the general public using the Google Trends tool [7.39]. Google searches for astrobiology have grown relative to searches for astronomy by two thirds since 2004 (the earliest date for which Google Trends data is available).

7.2 Astrobiology as a Profession

The context for astrobiology education is the intrinsic appeal of the subject. Humans have pondered their place in the universe for thousands of years. Modern cosmology places the Earth as an insignificant part of a vast and ancient universe. There is nothing special about our planet, our star, our galaxy, or our region of the universe. Stellar evolution has generated the chemical ingredients for life and dispersed them widely in space and time. Although we do not understand the path by which molecules in solution evolved into the first cells, parts of that path have been demonstrated in the lab, and it occurred by purely natural processes. But if life exists elsewhere, it need not use the same biochemistry or the same genetic code as life on the Earth. There is even less reason to think that complex organisms elsewhere, should they exist, will share the function and form of advanced terrestrial life forms. Uncertainty encourages speculation; the anticipation that we might soon detect life beyond Earth creates excitement. Education in astrobiology spans a range of modes from professional training of the next generation of astrobiologists to inspiring the general public with the question, "Are we alone in the universe?"

The pioneering generation of astrobiologists working in the second half of the 20th century, did not train to be astrobiologists. They began working on different aspects of life in the universe after training in a traditional science discipline. In the United States, the subject began to emerge in the 1950s. In 1953, the term astrobiology was coined by Russian astronomer Gavriil Tikhov [7.23], and Stanley Miller and Harold Urey carried out their famous laboratory experiment to try and simulate pre-biotic chemistry [7.64]. In the late 1950s, Joshua Lederberg first used the term exobiology to refer to concerns over potential contamination of the Earth by extraterrestrial life and the converse, the possibility that we might contaminate pristine environments beyond the Earth with terrestrial microbes. In 1960, NASA started awarding grants through an Exobiology Program, and Frank Drake conducted the first SETI experiment, Project Ozma [7.88]. NASA was crucial in fostering the emerging field. During the 1970s, the agency began to fund SETI and it launched life detection experiments to Mars aboard the twin Viking landers.

Astrobiology advanced further in the late 1990s with the discovery of the first exoplanets, its recognition as a future research area by the European Space Agency (ESA), and the formation of the NASA Astrobiology Institute, a virtual framework for collaboration and funding of research groups [7.65]. Since then, the profession of astrobiology has grown worldwide, with expansion beyond the United States, Canada, and Europe, into Asia, South America, Australia, and New Zealand. The International Astronomical Union (IAU) formed a Bioastronomy commission in 1982, and it was rebranded as Commission F3 Astrobiology in 2015 [7.45]. It currently has 206 members and operates a working group for Education and Training in Astrobiology. Detailed data are not available but in general

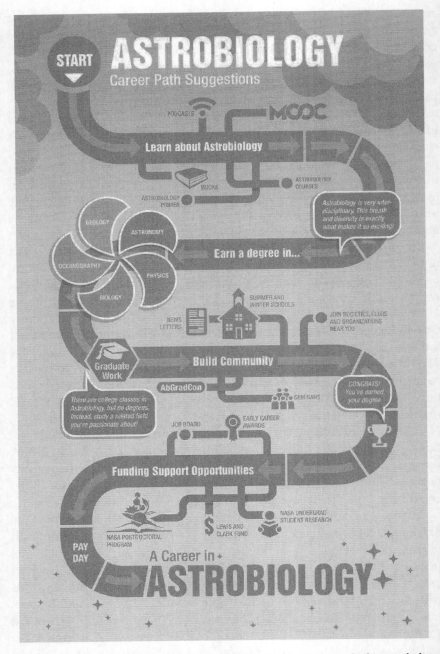

Figure 7.3 A simplified flowchart for professional development in astrobiology, including informal and formal education, building community, and finding financial support. (Courtesy NASA Ames Research Center).

astrobiologists have two types of employment: as university faculty doing teaching and research, and as researchers working for government agencies and research institutes. University culture has been traditionally organized in silos corresponding to departments and college. This creates obstacles to interdisciplinary degree programs and research activities [7.61] [7.72]. The situation is changing, but slowly, so research institutes have generally been able to better respond to the evolving landscape of astrobiology. Figure 7.3 shows a possible flowchart leading to a career in astrobiology.

7.3   Graduate Programs

The purest form of advanced credential in astrobiology is a graduate degree. Despite growth in the profession, astrobiology graduate programs are still quite rare. The difficulties arise when students admitted to an astronomy or astrophysics program, taking mostly classes in astronomy and physics, have to take graduate level courses in biology and chemistry. If they have little or no undergraduate preparation in the life sciences, this curriculum is very challenging. There may also be a perception among graduate students that an astrobiology degree, despite the breadth of science training involved, is more limiting in terms of subsequent career options. This will only be resolved when astrobiology has a highly visible place in more university departments. Many current astrobiologists followed a difficult path, taking life science courses as an overload, and sometimes working in near isolation with few mentors.

The longest-established program to offer a Ph.D. in astrobiology, dual-titled with a science department, is at the University of Washington. Their program is home to fifty graduate students, faculty, and researchers, spanning eight departments and four colleges. Penn State University also has a dual-title Ph.D. program. Other universities offer graduate minors, certificates, and concentrations, with fewer courses required. They include the University of Arizona, Arizona State University, the University of Colorado, and Georgia Institute of Technology. The first astrobiology graduate program in Canada was established in 2014 by the Origins Institute at McMaster University. A larger group of institutions and organizations offer graduate students research projects in astrobiology, often with specialized courses added to the existing curricula. An incomplete list is: the UK Centre for Astrobiology in Edinburgh, members of the European Astrobiology Campus, and the Australian Centre for Astrobiology. Opportunities for research and training are also beginning to appear at institutions in Asia and South America. NASA hosts links with job postings and information about employment in astrobiology [7.66] [7.67].

7.4   Undergraduate Programs

Given the breadth of knowledge required to master astrobiology, the earlier a student is able start, the better. Full degree programs are still rare. The first and so far only B.S. in astrobiology was created by the Florida Institute of Technology. Undergraduate minors are currently offered by the University of Arizona, the University of Kansas, Penn State University, and Rensselaer Polytechnic Institute, while Princeton University offers an undergraduate certificate, and Arizona State University offers a concentration in astrobiology. Astrobiology is difficult to configure as an undergraduate degree because physical sciences and life sciences each have layered curricula, where courses are taken in a specific sequence and build on each other. Once a student signs up for General Education or Foundation

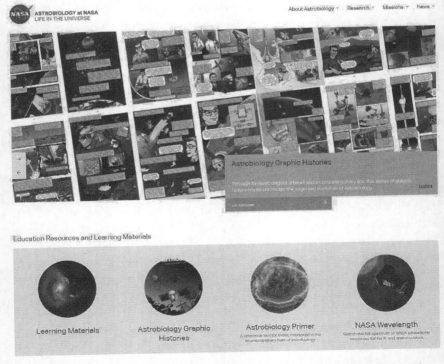

Figure 7.4 Examples of educational materials available at the NASA Astrobiology web site. This view highlights a set of eight graphic histories, but there are teaching materials of various kinds, a link to the Astrobiology Primer, information on careers and funding, and a link to the "Ask an Astrobiologist" show. (Courtesy NASA).

courses and required courses in their major, they have little latitude to add a sequence of courses from a different science major. Only 10% of students with double majors combine two science subjects [7.29]. As a good single resource, NASA has a web site with links to graduate and undergraduate degree programs, as well as individual courses and massive open online classes [7.68]. The NASA Astrobiology web site has extensive listings of scholarships, fellowships, and funding opportunities for undergraduates, graduate students, and postdoctoral fellows [7.70]. See the examples in the Figure 7.4 screen shot. The site also has a web form to sign up for their newsletter with regular updates on education programs. Jobs, including those in astrobiology, at every level from pre-doctoral to tenure-track faculty are kept updated monthly with the American Astronomical Society Job Register [7.1].

7.5 Conferences and Schools

For students intending to become astrobiologists, important educational opportunities are provided by schools, workshops, meetings and conferences. The pandemic has caused massive disruption to the flow of these events, with most in 2020 either cancelled on being held online. This situation is likely to continue into 2021 and possibly beyond. While online meetings have been successful, often with higher attendance than face-to-face meeting, some essential aspects of networking and informal interactions are lost when scientists can no longer gather. The major astrobiology conferences are the aforementioned biennial AbSciCon [7.3], the biennial European Astrobiology Conference [7.33], the biennial Astrobiology Australasia Meeting [7.6], the triennial International Conference on the Origin of Life [7.54], the annual Gordon Research Conference on the Origin of Life [7.40], and the annual Conference on Artificial Life [7.52]. One very valuable meeting for young professionals has been the annual Graduate Astrobiology Conference [7.2]. Organized and attended by graduate students, postdocs, and selected undergraduates, many attendees of this conference have gone on to successful careers in astrobiology [7.60].

Many young people who aspire to study astrobiology work in places without support or with no local community of astrobiologists. This professional isolation can be discouraging. Schools and workshops are an important means for learning and being part of a larger community. The pandemic has forced some schools to cancel their 2020 programs and others to shift to an online format. The most prominent annual schools are the International Summer School in Astrobiology held in Santander,

Spain, and open to students based in the U.S. and the E.U. [7.53], the Sagan Exoplanet Summer Workshop, supported and hosted by NASA [7.69], the Chinese Academy of Sciences Summer School in Planetary Sciences [7.18], the Keio Astrobiology Camp in Japan [7.55], and the Physics of Life Monsoon School in India [7.77]. The SETI Institute operates an annual Research Experiences for Undergraduates program, funded by the NSF [7.81]. Occasionally, the IAU International School for Young Astronomers is on the subject of astrobiology. At a more informal level, the AbGradCon web site [7.2] lists regional and international networks where astrobiologists can stay connected.

7.6   Courses for Non-Science Majors

Astronomy is a very popular subject for non-science majors looking to satisfy a requirement as part of their General education program. Surveys have found enrollments of about 250,000 per year in introductory astronomy courses nationwide. The breakdown is 20% in astronomy degree-granting universities, 60% in physics degree-granting universities, and 20% in two-year colleges and community colleges [7.4]. These numbers have been stable over a decade [7.37]. Astrobiology has become a popular option for a second General Education course for non-science majors, but there are no published statistics on enrollments in the United States and other countries. Using a Google search to do a non-comprehensive inventory, over a hundred courses are listed, at a wide variety of private universities (e.g., Cornell, Harvard, Princeton, USC), public universities (e.g., Arizona, Michigan, Ohio State, Pittsburgh, UCI, UCSD, Washington), and community colleges (e.g., Anne Arundel, LaGuardia, MiraCosta, Sierra, Snow). Astrobiology courses are also popular in Canada (e.g., Humber, McGill, Toronto, Western, York), the U.K. (e.g., Cambridge, Imperial, Edinburgh, Open University, Southampton), and Australia (e.g., Monash, Newcastle, New South Wales, Swinburne). Other countries do not generally have the science breadth requirements that create demand for an astrobiology class.

The astrobiology course for non-science majors has been a place where instructional experts have been creative in designing pedagogy that is innovative and takes full advantage of the range of disciplines involved in understanding life in the universe. Several examples are highlighted. The course "Habitable Worlds" was developed ten years ago at Arizona State University to try and incorporate active learning into a technologically advanced framework. Students learn by doing and they get adaptive feedback that customizes their path through the content. There are immersive, virtual field trips and the learning environment is data-rich so research can be done on how students learn. Figure 7.5 shows how this course maps onto the scientific method. The students make observations and build models to explain their observations. They then use their models to make predictions and then adjust the models accordingly. Finally, they integrate their models across disciplines to solve complex problems. The detailed feedback on where students succeed and fail was shown to create a cycle of improvement, both for the course and for instructors. Over 3000 non-science majors have taken this course [7.44].

The Center for Astronomy Education, hosted by the NASA Jet Propulsion Laboratory, has extensive astronomy materials, including many on astrobiology [7.17]. The research group led by Ed Prather has pioneered the use of lecture tutorials to enhance student learning in a traditional face-to-face classroom. Lecture tutorials are worksheets of carefully designed questions to make students think about challenging subjects. After a brief lecture, students work together in groups of two or three to complete the worksheet. Lecture tutorials scaffold learning and specifically address typical misconceptions [7.14] [7.77]. They often use a debate strategy where hypothetical students express opinions about a topic, and students in the class must say who they agree with, and why. They have been used extensively in geosciences and physics as well as in astronomy. Lecture tutorials for an introductory astrobiology class have been published in workbook format [7.78]. Figure 7.6 shows an example where the characteristics of a Solar System environment are matched to an extremophile organism, and Figure 7.7 shows part of the worksheet that makes an analogy of the factors in the Drake equation with a typical college life situation. Lecture tutorials

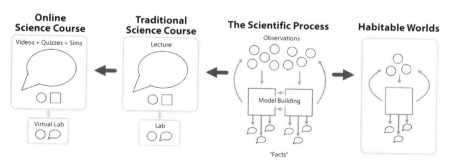

Figure 7.5 Diagram of the pedagogy of the "Habitable Worlds" online course from Arizona State University, showing how it maps to the scientific process and how the instructional model differs from traditional science courses and most online courses. (Courtesy Lev Horodyskyj).

TITAN'S ENVIRONMENT
Characteristics
Atmosphere containing nitrogen, methane, ethane, argon, and hydrogen
Very cold temperatures
Possible lakes/oceans of ethane/methane
Abundant light
May have continents

HYPOTHETICAL EXTREMOPHILE
Characteristics
Thrives in environments with low moisture/water
Prefers low pH
Uses oxygen
Uses organic carbon for energy
Prefers very cold environment

Figure 7.6 Examples of part of a lecture-tutorial on the interactions between hypothetical extremophile organisms and extreme environments. In this case, characteristics of a particular microbe are matched to Saturn's moon Titan. (Courtesy Ed Prather).

VARIABLE	ESTIMATED VALUE	NOTES
n – total number of students at your college/university		
f_f – fraction of females at your college/university		
f_p – fraction of those students that are psychology majors		A reasonable estimate might be that 1/5 of all students are psychology majors.
f_L – fraction of female students with long hair		
f_t – fraction of students at the library between 6:00 p.m. and 7:00 p.m.		
f_R – fraction of students in the reference section of library		
f_F – fraction of students in the reference section reading a journal article		For the students in the reference section, a reasonable estimate might be that 1/8 of them are reading a journal article
f_m – fraction of journal articles by a male author		
$F_{f,p,L,t,R,F,m}$ – number that represents the product of all the fractions above		
T – number of female psychology majors with long hair at the library between 6:00 p.m. and 7:00 p.m. in the reference section reading a journal articles by a male author ($F \times n$)		
CLASS AVERAGE, T_{avg}		

Figure 7.7 Part of a Drake equation lecture-tutorial. To teach estimation, an analogy is made between a specific situation, applying it to a subset of a large college population, and the fractional factors involved in estimating intelligent life in the galaxy. (Courtesy Ed Prather).

are a low-technology solution to learner-centered engagement in large classes. They can be used to convey complex concepts, such as exoplanet detection via microlensing, to an audience of non-science majors [7.85]. Although the use of portfolios to assess student performance in K-12 classrooms and to monitor the training of pre-service science teachers is increasingly common, their implementation in an undergraduate science classroom is limited [7.26]. For large introductory classes like "Life in the Universe" the assessment is usually multiple-choice tests and perhaps term papers. Portfolios offer a more authentic mode of assessment as well as a way to build multiple writing assignments into the class. Learning goals were to be able to assess scientific claims in the popular media, to recognize science as a dynamic and iterative process, and to communication science content to lay or peer audiences. The four inter-related units of the assessment are shown in Figure 7.8. Each one echoes in some way the process scientists follow when they do research and gain new knowledge. Where appropriate, students worked in pairs and small teams. The assignment is designed to

Portfolio menu.

Getting Your Feet Wet (60 points)	Weighing the Evidence (60 points)	Science in the Real World (60 points)	Outside the Box (40 points)
Part 1 (30 points total) Select two of the following activities to complete: 1. Make a scale model of the solar system or an atom scaled to a campus map. 2. Theories of Evolution 3. Intro to Spectoscopy 4. Intro to Extrasolar Planet Detection 5. Inverse Square Law **Part 2** (30 points total) Prelude to Weighing the Evidence: In Weighing the Evidence you will be asked to work extremely with one scientific article. In Preparation, select a related article in *Astrobiology Magazine* that covers some of the background science. After reading the article, write a synopsis, create a glossary of relevant scientific terms, and identify concepts that are still "muddy."	**Part 1** (30 points total) Select one article from those provided. Identify the types and numbers of pieces of evidence presented. Comment on the validity of the evidence presented. Note any evidence that is missing or that would contribute to a stronger argument. **Part 2** (30 points total) Select one of the following to complete: 1. Find a current media article about astrobiology. Identify the types and numbers of pieces of evidence presented. Comment on whether the conclusions of the article are justified based on the evidence presented. 2. Go to one of the listed talks. Outline the types and numbers of pieces of evidence presented. Comment on whether the conclusions are justified.	**Part 1** (30 points total) Select one of the scientists' interview trancripts to read. Identify the parts of the scientific method. Create a flowchart of the scientific process that the scientist went through. **Part 2** (30 points total) Select one of the following: 1. Given a set of data, create a list of questions you could ask and answer using the data set. Use the data to answer one question. 2. Given an instrument, discuss the types of questions you could answer. What limitations are there? 3. Given a set of observations, evidence, and snippets from conversations, what research questions or hypotheses can you arrive at?	Students must choose one of the following to complete for his/her portfolio. (40 points total) 1. Create an advertisement or marketing scheme for a new planet. 2. Create a brochure advertising a bogus investment opportunity. Use compelling "evidence" to ectice investors. 3. Write a children's story that introduces a key concept from astrobiology. 4. Create a painting for an astrobiology exhibit. 5. Design a museum exhibit on astrobiology that would fit in Steward Lobby. 6. Create a piece of curriculum on some aspect of astrobiology. 7. None of the above. Propose your own creative project.

Figure 7.8 The menu of options for a portfolio used in an introductory astrobiology class. Students get a lot of choice. They have to critique science articles that they find online, and they have to study how science actually works to address questions, and they get a chance to be creative and utilize their other skill sets. (Courtesy Chris Impey).

attach to the highest three levels of a Bloom taxonomy [7.57]. This type of course construction is quite labor intensive; two GTAs are needed for grading 1600 pieces of written work from 100 students. The first time this class was offered, in 2006, more than half the class agreed that portfolios provided a better assessment of learning than standard exams [7.74].

An even more "out of the box" way of teaching astrobiology involves the use of virtual worlds. The virtual world with the most traction for education is Second Life, launched in 2003 by Linden Lab. After rapid growth, it had one million users by 2013, but since then popularity has ebbed [7.10]. Second Life is similar to massive multiplayer online role-playing games, but it's not a game with objectives and manufactured conflicts. Rather, it is a world where a person's virtual representation, or avatar, can interact with places, objects, and other avatars. Astronomy and space applications in Second Life include an International Spaceflight Museum, a telescope training game, a recreation of the Apollo 11 landing site, and a planetarium

Figure 7.9 A virtual exhibit built in the virtual world Second Life for an astrobiology class. The layout was a spiral timeline of Earth's history (top), with animated exhibits at appropriate locations, such as the Cambrian explosion (bottom) in this example. (Courtesy Chris Impey).

that is operated by Tony Crider of Elon University [7.58]. In 2008, our education research group set up an island in Second Life and used it for students to build creative projects for an introductory astrobiology class [7.38]. Figure 7.9 gives a sense of what they created. Their projects were part of an interactive timeline of the Earth's history, where students placed their projects at appropriate locations along the spiral. They could use text, graphics, images, and animated 3D objects in their exhibits. They enjoyed the storytelling and communication possibilities of the virtual environment.

7.7   Massive Open Online Classes

Massive open online classes, or MOOCs, are a growing part of the education landscape [7.83]. They are typically free, and not for college credit, but completion certificates are available for a fee. The MOOC phenomenon has grown rapidly to 13,500 courses offered by 900 universities worldwide (excluding China, which has many MOOC learners, but no public statistics), reaching over 110 million adult learners [7.82]. We have investigated the motivations of MOOC learners in a large astronomy course and find that these learners are typically over 30, most already have an undergraduate degree and often a graduate degree, and they are motivated by a personal interest rather than by career advancement [7.35] [7.50]. Coursera has two astrobiology courses, one offered by the University of Edinburgh [7.24], and one offered by our group at the University of Arizona [7.51]. A Third course is on the emergence of life on Earth [7.36]. A MOOC has been spun off the successful astrobiology course Habitable Worlds, at Arizona State University [7.41]. The Santa Fe Institute offers an online course on the Origin of Life [7.27]. There is a unit on life on Earth and in the universe in the Khan Academy course on cosmology and astronomy [7.56]. The University of Central Lancashire has an online module on astrobiology leading to a Certificate in Astrobiology [7.84]. These online courses are suitable for anyone who wants to get the flavor of the search for life in the universe without committing to a degree program. A number of popular astrobiology books written for trade presses would be suitable to accompany such a course.

7.8   Teaching Materials and Books

In addition to full courses and instructional modules for astrobiology, a diverse array of teaching materials can be found online. Many of these are hosted by, or linked from, the main NASA Astrobiology web site. They include videos, games, posters, lesson plans, interactive materials, and podcasts. NASA also hosts a set of materials for teachers to implement learning progressions in astrobiology. Learning progressions are very effective in building knowledge within a theoretical framework of how people learn [7.32]. Another collection of astrobiology resources is hosted by Carleton College as part of the National Science Digital Library, funded by the National Science Foundation [7.63]. SAGANet is another source of educational materials for STEM teachers, parents homeschooling their children, camp counsellors, and leaders of after-school programs [7.80]. The group is an informal network of young scientists who conduct a live monthly broadcast of "Ask an Astrobiologist" and who routinely answer questions on Twitter and in their web site chat space.

The field of astrobiology is well-served by books at various levels. Unfortunately, the higher-level books suitable for graduate students and advanced undergraduates are no longer up to date. Jonathan Lunine's *Astrobiology: A Multi-Disciplinary Approach* was released in 2004 by Benjamin Cummings with no second edition yet, and *Planets and Life: The Emerging Science of Astrobiology*, edited by Woody Sullivan and John Baross, has also not been updated since a 2003 first release by Cambridge University Press. The best recent book is *An Introduction to Astrobiology*, edited by David Rothery, Iain Gilmour, and Mark Sephton [7.79]. Articles

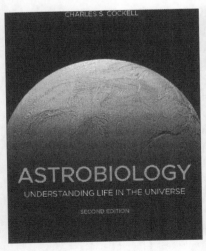

Figure 7.10 Two of the most prominent textbooks for astrobiology written at an introductory level for non-science major undergraduates. Roughly 30,000 students per year take a Life in the Universe course at universities around the United States, often as a science requirement for their degrees in other subjects. (Left: Courtesy Wiley. Right: Courtesy Pearson).

from leading research scientists are gathered in *Frontiers of Astrobiology*, edited by Chris Impey, Jonathan Lunine, and José Funes [7.47]. A very good primer can be found in David Catling's *Astrobiology: A Very Short Introduction* [7.16]. The two major textbooks at the introductory level are *Astrobiology: Understanding Life in the Universe* by Charles Cockell [7.25] and *Life in the Universe* by Jeff Bennett and Seth Shostak [7.11], illustrated in Figure 7.10. An excellent book for high school students, that comes with a companion teacher's guide, is *Astrobiology: An Integrated Science Approach*, by Jodi Asbell-Clarke *et al.* [7.5].

Astrobiology is intrinsically interdisciplinary, and there are books that echo the strategy of this volume by considering its broader implications. *Talking About Life* collects three dozen interviews with leading astrobiologists, but also includes the perspectives of poets, writers and philosophers [7.46] (Figure 7.11). This scope is broadened to encompass theology in *Encountering Life in the Universe: Ethical Foundations and Social Implications of Astrobiology*, edited by Chris Impey, Bill Stoeger, and Anna Spitz [7.48] (Figure 7.11). Steven Dick presents a historical overview in *Astrobiology*,

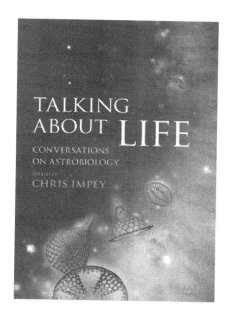

Figure 7.11 Two of the less conventional approaches to conveying elements of astrobiology to a general audience. *Talking About Life* is a set of three dozen conversations with leading research astrobiologists, and *Encountering Life in the Universe* considers ethical and social implications of the search for life in the universe. (Left: Courtesy Cambridge University Press. Right: Courtesy University of Arizona Press).

Discovery, and Societal Impact [7.30]. Two philosophers give their perspectives in *The Quest for a Universal Theory of Life*, by Carol Cleland [7.22] and *The Astrobiological Landscape*, by Milan Ćirković, [7.21]. There are other books that address topics in astrobiology, with many on exoplanets and others on niches such as the Drake equation, the Fermi paradox, extremophiles, and the origin of life on Earth.

This abundance of writing in an indication of the vibrant state of astrobiology as the search for life in the universe continues. As demonstrated elsewhere in this volume, the search for life beyond Earth resonates with profound issues of how humans view their place in the universe, and the ethical choices we make as we live our lives on Earth and contemplate a time when we will live off-Earth [7.19] [7.49]. Surveys of students in countries as diverse as Brazil [7.28], Peru [7.20], Sweden [7.75], and the United States [7.73] show that young people are intrigued and often inspired by the subject, even if their grasp on its scientific principals is weak. For educators at all levels, it is a challenge and a pleasure to teach astrobiology.

References

[7.1]   AAS, *American Astronomical Society Job Register*, 2020, https://jobregister.aas.org/.

[7.2]   AbGradCon, *Graduate Astrobiology Conference*, 2020, https://www.abgradcon.org/home.

[7.3]   AbSciCon, *Astrobiology Science Conference*, 2021, https://www.agu.org/abscicon.

[7.4]   American Institute of Physics, *Physics Enrollments in Two-Year Colleges*, AIP Statistical Research Center, American Institute of Physics, Washington, DC, 2013, https://www.aip.org/sites/default/files/statistics/undergrad/tyc-enrollme nts-p-11.pdf.

[7.5]   Asbell-Clarke, J.E., Barstow, D.W., Edwards, T.E., Larsen, J.L., *Astrobiology: An Integrated Science Approach*, TERC, Cambridge, Mass, 2005.

[7.6]   Astrobiology Australasia Meeting, *Astrobiology Australasia Meeting*, 2020, https://www.aam2020.org/.

[7.7]   Astronomical Journal, *Home page*, 2020, https://iopscience.iop.org/journal/1538-3881.

[7.8]   Astrophysical Journal, *Home page*, 2020, https://iopscience.iop.org/journal/0004-637X.

[7.9]   Astrophysics, *The arXiv server*, 2020, https://arxiv.org/archive/astro-ph.

[7.10] Axon, S., *Returning to Second Life*, Ars Technica, 2017, https://arstechnica.com/gaming/2017/10/returning-to-second-life/.

[7.11] Bennett, J.O., Shostak, S., *Life in the Universe*, Fourth Edition, Pearson, London, England, 2016.

[7.12] Birkby, J.L., Spectroscopic Direct Detection of Exoplanets, in: *Handbook of Exoplanets*, H.J. Deeg and J.A. Belmonte (Eds.), pp. 1485–1508, Springer, New York, 2018.

[7.13] Borucki, W.J. *et al.*, Characteristics of planetary Candidates Observed by Kepler. II. Analysis of the First Four Months of Data. *Astrophys. J.*, 736, 1, 19–41, 2011.

[7.14] Brogt, E., A Theoretical Background on a Successful Implementation of Lecture-Tutorials. *Astron. Educ. Rev.*, 1, 6, 50–58, 2007.

[7.15] Castillo, J., Vance, S., The Deep Cold Biosphere? Interior Processes of Icy Satellites and Dwarf Planets. *Astrobiology*, 8, 2, 344–346, 2008.

[7.16] Catling, D.C., *Astrobiology: A Very Short Introduction*, Oxford University Press, Oxford, England, 2014.

[7.17] Center for Astronomy Education, *Astronomy 101*, 2020, https://astronomy101.jpl.nasa.gov/.

[7.18] Chinese Academy of Sciences, *Summer School in Planetary Sciences*, 2020, http://planet.ustc.edu.cn/en/index/.

[7.19] Chon-Torres, O.A., Astrobiology and its Influence on the Renewal of the Way We See the World from the Teloempathic, Educational, and Astrotheological Perspective. *Int. J. Astrobiology*, 2020, https://doi.org/10.1017/S1473550420000087.

[7.20] Chon-Torres, O.A. *et al.*, Attitudes and Perceptions Towards the Scientific Search for Extraterrestrial Life Among Students of Public and Private Universities in Peru. *Int. J. Astrobiology*, 2020, https://doi.org/10.1017/S1473550420000130.

[7.21] Ćirković, M.M., *The Astrobiological Landscape: Philosophical Foundations of the Study of Cosmic Life*, Cambridge University Press, Cambridge, England, 2012.

[7.22] Cleland, C.E., *The Quest for a Universal Theory of Life: Searching for Life as We Don't Know It*, Cambridge University Press, Cambridge, England, 2019.

[7.23] Cockell, C.S., Astrobiology and the Ethics of New Science. *Interdiscip. Sci. Rev.*, 26, 2, 90–96, 2001.

[7.24] Cockell, C.s., *Astrobiology and the Search for Extraterrestrial Life*, 2020, https://www.coursera.org/learn/astrobiology.

[7.25] Cockell, C.S., *Astrobiology: Understanding Life in the Universe*, Second Edition, Wiley-Blackwel, Hoboken, NJ, 2020.

[7.26] Collins, A., Portfolios for Science Education: Issues in Purpose, Structure, and Authenticity. *Sci. Educ.*, 76, 451–463, 1992.

[7.27] Complexity Explorer, *Complexity Explorer Santa Fe Institute, Origin of Life*, 2020, https://www.complexityexplorer.org/courses/95-origins-of-life.

[7.28] de Souza, R.F., de Carvalho, M., Matsuo, T., Zaia, D.A.M., Study on the Opinion of University Students of the Origin of Universe and Evolution of Life. *Int. J. Astrobiology*, 9, 2, 109–117, 2010.

[7.29] Del Rossi, A.F. and Hirsch, J., Double Your Major, Double Your Return? *Econ. Educ. Rev.*, 27, 375–386, 2008.

[7.30] Dick, S.J., *Astrobiology, Discovery, and Societal Impact*, Cambridge University Press, Cambridge, England, 2018.

[7.31] Domagal-Goldman, S.D. and Wright, K.E., The Astrobiology Primer v2.0. *Astrobiology*, 16, 8, 561–653, 2016.

[7.32] Duschl, R.A., Learning Progressions: Framing and Designing Coherent Sequences for STEM Education. *Discip. Interdscip. Sci. Educ. Res.*, 1, 4, 2019. https://doi.org/10.1186/s43031-019-0005-x.

[7.33] EANA, *European Astrobiology Conference*, 2020, http://www.eana-net.eu/.

[7.34] Exoplanet Catalog, *NASA Exoplanet Catalog*, 2020, https://exoplanets.nasa.gov/exoplanet-catalog/.

[7.35] Formanek, M., Buxner, S., Impey, C.D., Wenger, M., Relationship between Learners' Motivation and Course Engagement in an Astronomy Massive Open Online Course. *Phys. Rev. Phys. Educ. Res.*, 15, 020140, 2019.

[7.36] Fouke, B.W., *Emergence of Life*, 2020, https://www.coursera.org/learn/emergence-of-life.

[7.37] Fraknoi, A., Enrollments in Astronomy 101 Courses. *Astron. Educ. Rev.*, 1, 1, 121–123, 2002.

[7.38] Gauthier, A., Impey, C.D., ASTR202, Exploring Life in the Universe. *EDUCAUSE Rev.*, 43, 5, 1–4, 2008.

[7.39] Google, *Google Trends*, 2020, https://trends.google.com/trends/?geo=US.

[7.40] Gordon Research Conference, *Gordon Research Conference on the Origin of Life*, 2020, https://www.grc.org/origins-of-life-conference/2020/.

[7.41] HabWorlds Beyond, *HabWorlds Beyond*, 2020, https://www.habworlds.org/.

[7.42] Heller, R. *et al.*, Formation, Habitability and Detection of Extrasolar Moons. *Astrobiology*, 14, 9, 798–835, 2014.

[7.43] Heller, R. and Kiss, L.L., Exoplanet Vision 2050, 2019, https://arxiv.org/pdf/1911.12114.pdf

[7.44] Horodyskyj, L.B., Mead, C., Belinson, Z., Buxner, S., Sernken, S., Anbar, A.D., Habitable Worlds: Delivering on the Promise of Online Education. *Astrobiology*, 18, 1, 86–99, 2018.

[7.45] IAU, *International Astronomical Union Commission F3 Astrobiology*, 2020, https://www.iau.org/science/scientific_bodies/commissions/F3/.

[7.46] Impey, C.D. (Eds.), *Talking About Life: Conversations on Astrobiology*, Cambridge University Press, Cambridge, England, 2010.

[7.47] Impey, C.D., Lunine, J., Funes, J., *Frontiers of Astrobiology*, Cambridge University Press, Cambridge, England, 2012.

[7.48] Impey, C.D., Impey, C., Stoeger, B., Spitz, A. (Eds.), *Encountering Life in the Universe: Ethical Foundations and Social Implications of Astrobiology*, Cambridge University Press, Cambridge, England, 2013.

[7.49] Impey, C.D., *Beyond: Our Future in Space*, Norton, New York, 2016.

[7.50] Impey, C.D., Wenger, M., Formanek, M., Buxner, S., Bringing the Universe to the World: Lessons Learned from a Massive Open Online Class on Astronomy. *Communicating Astronomy Public*, 21, 20–30, 2016.

[7.51] Impey, C.D., *Astrobiology: Exploring Other Worlds*, 2020, https://www. coursera.org/learn/astrobiology-exploring-other-worlds.

[7.52] International Society for Artificial Life, *Conference on Artificial Life*, 2020, https://alife.org/conferences/.

[7.53] International Summer School on Astrobiology, *International Astrobiology Summer School*, 2020, https://astrobiology.nasa.gov/career-funding/ astrobiology-summer-school/.

[7.54] ISSOL, *International Conference on the Origin of Life*, 2020, https://issol.org/.

[7.55] Keio Astrobiology Camp, *Keio Astrobiology Camp*, 2020, https://sites. google.com/view/kac2020/home.

[7.56] Khan Academy, *Unit: Life on Earth and in the Universe*, 2020, https://www. khanacademy.org/science/cosmology-and-astronomy/life-earth-universe.

[7.57] Krathwohl, D.R., A Revision of Bloom's Taxonomy: An Overview. *Theory Pract.*, 41, 212–218, 2002.

[7.58] Krider, A., *Space Science and Astronomy Education in Second Life*, 2020, https://facstaff.elon.edu/acrider/acrider/Second_Life.html

[7.59] Mayor, M. and Queloz, D., A Jupiter-mass Companion to a Solar-type Star. *Nature*, 378, 6555, 355–359, 1995.

[7.60] McGonigle, J.M., Motamedi, S., Rapf, R.J., Astrobiology Graduate Conference: A 15-Year Retrospective. *ACS Earth Space Chem.*, 3, 12, 2675–2677, 2019.

[7.61] Merino, N. *et al.*, Living at the Extremes: Extremophiles and the limits of Life in a Planetary Context. *Front. Microbiol.*, 2019. https://doi. org/10.3389/fmicb.2019.00780.

[7.62] Microbial Life Educational Resources, *Astrobiology in the Classroom*, 2920, https://serc.carleton.edu/microbelife/extreme/astrobiology/educators.html.

[7.63] Miller, S.L. and Urey, H.C., Organic Compound Synthesis on the Primitive Earth. *Science*, 130, 3370, 245–251, 1959.

[7.64] NAI, *NASA Astrobiology Institute*, 2020, https://astrobiology.nasa.gov/nai/.

[7.65] NASA Astrobiology, *Learning Materials*, 2020, https://astrobiology.nasa. gov/classroom-materials/.

[7.66] NASA Careers, *NASA Astrobiology Careers and Employment*, 2020, https://astrobiology.nasa.gov/careers-employment/.

[7.67] NASA Courses, *NASA Astrobiology Courses and Programs*, 2020, https:// astrobiology.nasa.gov/careers-employment-courses/.

[7.68] NASA Exoplanet Science Institute, *Sagan Exoplanet Summer Virtual Workshop*, 2020, https://nexsci.caltech.edu/workshop/2020/.

[7.69] NASA Funding, *NASA Astrobiology Scholarships and Fellowships*, 2020, https://astrobiology.nasa.gov/career-funding/.

[7.70] Nature, SETI at 50. *Nature*, 461, 7262, 316–317, 2009.

[7.71] Obradovic, S., Publication Pressure Creates Knowledge Silos. *Nat. Hum. Behav.*, 3, 1028–1029, 2019.

[7.72] Offerdahl, E., Prather, E.E., Slater, T.F., Students' Pre-Instructional Beliefs and Reasoning Strategies About Astrobiology Concepts. *Astron. Educ. Rev.*, 2, 1, 5–27, 2003.

[7.73] Offerdahl, E. and Impey, C.D., Assessing General Education Science Courses: A Portfolio Approach. *J. Coll. Sci. Teach.*, 41, 5, 19–25, 2012.

[7.74] Persson, E., Capova, K.A., Li, Y., Attitudes Toward the Scientific Search for Extraterrestrial Life Among Swedish High School and University Students. *Int. J. Astrobiology*, 18, 280–288, 2018.

[7.75] Petigura, E.A., Howard, A.W., Marcy, G.W., Prevalence of Earth-size Planets Orbiting Sun-like Stars. *Proc. Natl. Acad. Sci.*, 110, 48, 19273–19278, 2013.

[7.76] Physics of Life Monsoon School, *Simons-NCBS Physics of Life, Annual Monsoon School*, 2020, https://theory.ncbs.res.in/physlife.

[7.77] Prather, E.E., Slater, T.F., Adams, J.P., Bailey, J.M., Jones, L.V., Dostal, J.A., Research on a Lecture-Tutorial Approach to Teaching Introductory Astronomy for Non-Science Majors. *Astron. Educ. Rev.*, 3, 2, 122–136, 2004.

[7.78] Prather, E.E., Offerdahl, E., Slater, T.F., *Life in the Universe: Activities Manual*, Pearson Prentice Hall, Upper Saddle River, NJ, 2006.

[7.79] Rothery, D.A., Gilmour, I., Sephton, M.A., *An Introduction to Astrobiology*, Cambridge University Press, Cambridge, England, 2018.

[7.80] SAGANet, *Classroom Astrobiology*, 2020, http://saganet.ning.com/page/saganet-ed.

[7.81] SETI Institute, *Research Experiences for Undergraduates*, 2020, https://www.seti.org/research-experience-undergraduates.

[7.82] Shah, D., *MOOCs: By the Numbers*, Class Central Report, 2019, https://www.classcentral.com/report/mooc-stats-2019/.

[7.83] Spector, J.M., A Critical Look at MOOCs, in: *Open Education: From OERs to MOOCs*, M.J. Kinshuk and M.K. Khirbi (Eds.), pp. 135–147, Springer, New York, 2016.

[7.84] University of Central Lancashire, *Certificate in Astrobiology*, 2020, https://studyastronomy.com/courses/university-certificate-in-astrobiology/.

[7.85] Wallace, C.S., Chambers, T.G., Prather, E.E., Brissenden, G., Using Graphical and Pictorial Representations to Teach Introductory Astronomy Students About the Detection of Extrasolar Planets via Gravitational Microlensing. *Am. J. Phys.*, 84, 5, 335–343, 2016.

[7.86] Wolszczan, A. and Frail, D., A Planetary System Around the Millisecond Pulsar PSR1257+12. *Nature*, 355, 6356, 145–147, 1992.

[7.87] Zeeya, M., Search for Extraterrestrial Intelligence gets a $100 Million Boost. *Nature*, 523, 7561, 392–393, 2015.

[7.88] Zuckerman, B. and Tarter, J., Microwaves Searches in the U.S.A. and Canada, in: *Strategies for the Search for Life in the Universe (Proceedings)*, vol. 83, pp. 81–92, Astrophysics and Space Science Library, Springer Nature, London, 1980.

8

Genetics, Ethics, and Mars Colonization: A Special Case of Gene Editing and Population Forces in Space Settlement

Konrad Szocik[1]*, Margaret Boone Rappaport[2] and Christopher Corbally[3]

[1]*Department of Social Sciences, University of Information Technology and Management in Rzeszow, Rzeszów, Poland*
[2]*The Human Sentience Project, LLC, Tucson, AZ, USA*
[3]*Vatican Observatory, Department of Astronomy, University of Arizona, Tucson, AZ, USA*

Abstract

Future human space missions are challenging for astroethics for many reasons—political, social, economic or environmental. The ethical challenge increases when human presence in space is discussed in relation to population forces such as natural selection and genetic drift. In our chapter we discuss possible trajectories of population forces in different variants of future human space settlements. One of the discussed issues is the idea to control—at least to some extent—possible negative outcomes of population forces in space, or to accelerate some of them. We argue that the idea of human modification by genetic editing becomes clearer and more reasonable when it is discussed in the specific terms of population biology.

Keywords: Human space missions, natural selection, genetic drift, speciation, gene editing, population, demography, genetic engineering

Corresponding author: kszocik@wsiz.edu.pl
Konrad Szocik for work on this text received the Bekker Fellowship (3rd edition) funded by the National Agency for Academic Exchange (Decision No. PPN/BEK/2020/1/00012/DEC/1) for a research stay at Yale University (USA) in the academic year 2021/2022.

Octavio Alfonso Chon Torres, Ted Peters, Joseph Seckbach and Richard Gordon (eds.) Astrobiology: Science, Ethics, and Public Policy, (157–176) © 2021 Scrivener Publishing LLC

8.1   Introduction

8.1.1   The Complex Relationship Between Population Forces and Ethics

Fertility, mortality, and migration are the major concepts in demography and population studies. Their discussion can seem oddly impersonal at times. Take fertility: The sum total effect of so many deeply personal decisions about sex, reproduction, and family. Mortality can also seem dry and unfeeling. It captures the loss of people, and those people had families, are mourned, and their bodies are handled in special, culturally appropriate ways. This chapter addresses primarily the third major concept: migration. We focus on a type of migration that has never occurred before, and our generalizations are based on theory and observations based on terrestrial populations. In the future we will have a new type of entity: An off-world population. Here, we address some (not all) of the known population forces that determine human population characteristics and influence their relationships to astroethics.

All these human actions that comprise fertility, mortality, and migration are guided by human ethics—the offshoots of human theological and philosophical systems that guide behavior, including intimate behavior. Ethics are a special type of rule whose sum total guides decisions about "doing the right thing" [8.10]. Humans often follow ethics without even knowing they are doing so. They often constitute a large part of what anthropologists call "covert culture." There are efforts to state ethics in codified form—in literature, religious documents, art, film, and even politics. Still, so many of the intricately detailed ethical decisions that humans make are private. They are difficult for field interviewers to elicit. Respondents often resist their exposure because they are so close to the core of how they think, feel, and why they act as they do.

The reader should realize at the outset that we will address population forces, such as natural selection and genetic drift, which biologists and demographers have demonstrated in exquisite detail. The forces are a reality. They do exist. They are very difficult to "see" sometimes because they occur at the level of a population. Tabular analyses are useful. Charts are useful. Literature is useful. They all help us verify that the human population changes in regular, scientifically verifiable ways, and will continue to do so. Ethics function to guide many of these changes. They justify actions, organize our thinking about others, and test us daily to judge our own behavior—when we stop and bother to do so. Usually ethics operate at a group level that is obscure, except to the demographer, the priest, the historian, the philosopher, and others like those bothersome anthropologists! Three of those types are represented here as co-authors.

8.1.2   Humans Evolving on Earth and Mars

There is growing evidence that modern humans are continuing to evolve [8.2], and therefore, a new species of the genus *Homo* may already be form-ing. The colonization of our solar system will produce a variety of relatively small, off-world populations and some strong population forces that could speed genomic change. Population genetics [8.4] and human genomics [8.3] are now much better able to precisely define differences between related groups, so we should be able to test the divergence of off-world populations from the earthbound. We know that the terrestrial population will continue to become more genetically similar in the future because of the strong forces of natural selection in large populations. However, the population on Mars may differentiate from the large Terran population. Here, we ask: How? Why? Why is it important?

Mutations have been captured by the human genome over evolutionary time and the forces behind those changes have not ceased, but they have changed since our species emerged between 300,000 and 400,000 years ago, in relatively small groups of ancient humans all over the African con-tinent [8.5]. Eventually, cognitive "feedback loops" emerged that favored specific genetically based cognitive capacities, like reading, mathematics, and religious thinking; and thus humans became involved in their own evolution naturally [8.10]. Now, genetic engineering involves humans even more deeply, and with greater precision.

Factors that may speed up the evolution of *Homo sapiens* include: (1) population forces such as natural selection and genetic drift, and (2) the nature and stability of existing genes in the human genome and how they interact with extraterrestrial environments. The human species will change when it ventures to Earth's moon, Mars, and farther out to the Jovian moons. The task is to anticipate how much and the consequences in terms of ethics and society. Those projections will enable the engineering of the human genome so that it fares just as well or better on a new plan-etary body as on Earth. Our genome adapted over millions of years to the atmosphere, gravity, and abundance of water and sunlight on Earth. All those factors change, for example, on Mars or the Jovian moon Europa.

While we can soften the impact of the Martian environment, for exam-ple, we cannot erase it. Human settlement of Mars constitutes a population bottleneck in which humans break off from a larger group and do not con-tinue to breed freely with it. If the human species can reproduce on Mars, then classic inbreeding in this smaller population will ensue. If technologies emerge to gestate humans outside a maternal body—and they are already in development—then inbreeding may be controlled. However, as more and more humans come to Mars, the population will continue to change, but the direction of change may, itself, change. The human Martian population will remain small in comparison to the Terran population, for the foreseeable future. Mars cannot serve as a "safety valve" for overpopulation because it is literally impossible to even "make a dent" in the huge terrestrial population with current interplanetary transportation systems. Martians will become their own unique population, related to Mother Earth, but different.

8.1.3   Bioenhancements: Science, Technology, and Ethics

In order to anticipate the ethical issues that derive from the inevitable change of the human genome on Mars, it is useful to examine the forces at work. The human genome is stable but also wonderfully changeable. It is made of stuff that reacts with its environment to create usually small, but sometimes momentous, changes that are expressed in the human phenotype. Most natural mutations of our genome are disadvantageous, but a few are beneficial. As we anticipate the settlement of Mars, we need to change that balance so that more are advantageous. That may require both genetic engineering and selection of crew who can tolerate Mars best. Few have breathed this basic truth, but selection of crew and settlers will steer the population profile of future Martians.

There is good reason to assume that humans today may have strong arguments against the genetic engineering of future Martian crew and settlers. There is no less reason to assume that genetic intervention may be needed and so, justified. It may seem like space exploration will provide the first time in a human history when humanity may be obligated to intervene into natural genetic order. In fact, humans have been selectively forming subpopulations throughout their long history. Genetic engineering will be more obvious, rationalized, purposeful, and targeted, but its purpose and endpoint are not new.

Still, some features of genetic engineering are unique for space colonization and they should not be confused with other human colonization projects that we know from our history. What is specific to future space colonization is the fact that humans possess for the first time in their history effective tools for scientific human modification. We assume here, for the sake of argument, that effective technologies of gene editing will be widely available in the near future, when human interplanetary missions become a reality. We can find a rationale for gene editing, and other kinds of biomedical human enhancement, in the history of settlers colonizing Earth. European colonists wanted to be resistant to some local diseases that they had never encountered in Europe. Or, they wanted to be better adapted to new local climatic conditions. They were not able to achieve their goals because there were no biomedical tools that could help. Eventually, many populations of settlers did acclimate naturally to new and different environments because of culture, and because of the flexibility, adaptability, and plasticity of the human body and brain, and, the genome's ability to change over time. Genetic engineering takes a "short cut" to off-world adaptation that was not available to them.

As we plan for off-world exploration, we should realize that progress in science and technology does not automatically open doors for bioenhancement like genetic engineering, and that some of the resistance is due to human ethics and moral systems that were adapted to a different time and place. We assume that any colonization that would be realized today on Earth, for example, of the few remaining places on Earth that have not been colonized, would not be supported by human enhancement technologies. However, we assume that when humanity starts at some point in the future to explore and settle new locales beyond Earth, humans not only can, but under some conditions, should make use of bioenhancement technologies.

From an ethical point of view there is nothing inherently morally wrong with the human attempt to control genetic variation and reduce the number of disadvantageous mutations. This is one of the most salient features of social, religious, and moral systems: To define who should reproduce with whom. Rules guiding human reproduction, which are embodied in kinship systems and other social structural rules, usually have a goal—to retain certain genetics that are deemed desirable and avoid other genetics that are deemed undesirable. This is part of the history of humans on Earth. Therefore, one could rightly ask why human bioenhancements should be viewed as taboo and excluded, whereas efforts at human control over the natural world, including non-human animals and natural events like weather, are not taboo. It is difficult to see how this is different from the "speciesism" that is widely criticized by ethicists. Animal ethicists and environmental ethicists are at the forefront of the critics.

8.1.4 A Set of Astrobioethical Guidelines for Off-World Exploration

We propose a model of space astrobioethics that states that humans have a responsibility to reduce the risk and mitigate the hazards caused by radiation in space and microgravity, and wherever it is so high as to cause damage to humans, as it would on Earth's moon and Mars without substantial protection. That can be accomplished by modification of the targeted environment. Terraforming Mars is one possible option. Artificial adaptation of humans to the space environment is another option. These examples provide only a general outline, and needed modifications are highly context-dependent. Factors that should be considered are the rationale for the human space mission that may require gene editing, and the reasonable choice of an available technology. Bioethical considerations of technologies used in space should not be developed without any reference to the purpose of a space mission and the policies for space exploration held by government agencies and private companies. Societal and even international codes of ethics also form a context for consideration of bioenhancements.

Our conclusion is that human enhancements for space are not inherently morally wrong or right. Their use by humans is. They are neutral in moral terms and may be evaluated as morally wrong or right only in terms of extrinsic, instrumental moral assessments. If this is true, a bioethics for space that addresses the desirability of gene editing should not give undue weight to arguments that forbid gene editing. This conclusion is relevant for groups of scholars and clerics who recommend extending human space colonization to some point in the future, when hazardous factors in spacefaring could be reduced by non-enhancement technologies. There is a moral repugnance among some observers to application of gene editing techniques. Their reaction is not consistent with science but seated in a faulty interpretation of the human genome as "perfect" and immutable. There is a kind of faulty logic behind this type of reasoning that is not ethical, itself, especially when one considers the large number of both highly and slightly deleterious genes in the human genome, some of which wreak havoc on human life and health [8.3] [8.8].

Pre-judgments about use of human enhancements in space exploration are inconsistent with the value we place on human life. It is comparable to some parents' denying medical treatment to their gravely ill children because all modern medicine is deemed "ungodly." To avoid reducing risk to human life by avoiding gene editing is illogical if one accepts the premise that gene editing is not inherently morally wrong, or always dangerous. We suggest that there are not sufficiently strong reasons to defend an assumption that humanity should avoid gene editing for space and find other ways to achieve desirable space goals. There may be no such ways to avoid risk to humans. Space exploration—like all human exploration in history and back into prehistory—is dangerous, and therefore the context of space ethics must include assessment of dangers for humans in new and very challenging environments. Humans have *never* avoided all risk. It is not possible to do so.

Future human long-term space missions, including colonization programs, create a specific set of ethical questions that derive from the nature of space exploration, itself, and from the goals of specific missions (some of which are more dangerous than others). First, the hazardous aspects of space, such as high radiation exposure and reduced gravity, cannot be compared to terrestrial conditions—at least within an ethical context—because humans seek out and use terrestrial and space environments for entirely different purposes. Second, strong rationales are developing to avoid bone breakage due to the substantial declines in bone density in space. Scientific studies have demonstrated the dangers of even short-term missions in space [8.12]. Once those rationales are in place, space policies and programs will be able to change more easily. There is already a well understood rationale for human interventions to control disadvantageous mutations in space. Radiation is dangerous. Even now, some humans freeze eggs or sperm because they will be undergoing radiation treatment.

Gene editing on Earth for terrestrial populations will create a different set of ethical questions. For space exploration, our present concerns about gene editing and other human enhancements center on their cost and risk of side effects. If cost or risk or both become too great, gene editing may cease to be a reasonable option. In the future, these very general guidelines will be replaced with more finely tuned and strategically targeted assessments of the appropriateness of individual bioenhancements.

8.2   Population Forces and the Ethical Issues They Raise

8.2.1   Natural Selection and Genetic Drift on Mars

Martian environmental characteristics will exert the pressures of natural selection on the human genome, but it depends on the size of the Martian population whether these forces are "strong" or "weak." Gravity will be lower, like sunlight, and the atmosphere will be poisonous and unbreathable. Radiation will hit Mars almost unabated by a thin atmosphere. As far as we know, there is little water on Mars, and it is dry and dusty because of the absence of organic material and water. Martian dust is powder fine and it will get into everything and make dust abatement a major engineering and health issue. Natural selection involves the "pressure" of all these environmental factors, and they and others may cause the human species to change naturally.

The speed of anticipated change has been a subject of debate. However, while genomes usually change slowly, there are many examples of rapid change. Genetic engineering may also create rapid changes, which could soften the impact of environmental factors and change the human genome to accommodate the new planet. The reader should remember that natural selection is strongest when populations are large, and the human population on Mars will be, for the foreseeable future, small.

Another population force may exert a stronger effect than natural selection on this small Martian population. The force of genetic drift has a much stronger effect in small populations, in comparison to natural selection. It has been suggested that genetic drift was especially forceful in shaping the genes for neurocognitive human traits in small populations of very early humans [8.9] [8.10].

8.2.2   Contrasting and Convergent Population Forces on Earth and Mars

A great deal of the human genome is "neutral" in comparison to our nearest biological relative, the chimpanzee [8.6]. Humans have relatively few "selective sweeps" in their genome, which can be identified by inspection. This suggests genomic material was retained *not* because of its adaptive quality and the operation of natural selection, but because of other mechanisms, including random genetic drift.

As Mars is colonized in the near future, natural selection will have a stronger force on Earth, in the very large human population there. On Mars, genetic drift will have a stronger effect, responding to "population bottlenecks" and resulting inbreeding. The possibility that natural human reproduction may not be as easy on Mars as on Earth introduces very careful, selective, and perhaps risky human reproduction on Mars, which may further accentuate the effect of genetic drift, reducing the size of the reproductive population even further, thus accentuating the effects of drift. Additional migration from Earth, followed by crossbreeding with the Martian population, will reduce the effects of genetic drift, but probably not much. It is in the early decades of Martian settlement that we anticipate the largest changes in the human genome, through a combination of genetic drift and genetic engineering.

On Earth, there will be increasing homogenization of the large population there, with worldwide migration, resettlement, and continuing mixture of major geographical groups. Humans are all very much alike genetically, and they will become more so if the present trends prevail. If war, environmental catastrophes, or pandemics divide humans into reproductively isolated groups, this homogenization will be curtailed, at least temporarily. At the present, it is going full throttle, while the entire world population slows its growth and continues to age. Eugene Harris has suggested that humans may be due for an "evolutionary tune-up" [8.3], because natural selection will tend over time to eliminate deleterious alleles in our very large human population.

For example, human genes on Earth that place people at high risk for diabetes, obesity, and alcoholism will become less widespread on Earth, while individuals who carry these diseases succumb to their ravages, live shorter lives, and reproduce less. Eventually, populations carrying these deleterious alleles will shrink, as they complete their transition to genetic resistance to them (which began with the advent of human agriculture and intensive farming of grains). However, where diseases are caused by multiple alleles, like schizophrenia, the process of eliminating these genes through natural selection may take centuries—without genetic engineering. If genetic engineering is implemented, then care must be taken not to affect nearby genes that contribute to brain size and human intelligence. At the present, the set of genes impacting schizophrenia is not fully identified, including regulatory genes, which may exert important effects. However, good work has begun, and there is hope that genetic engineering can help to reduce the impact of schizophrenia.

There are strong cultural forces at work to provide robotic, cyborg, and genetic-engineering enhancements for humans with all types of disabilities in earthbound populations. Some disabilities will be more easily engineered than others. There will be an important flow of medical expertise from Earth to Mars in terms of genetically engineered enhancements, and eventually, an equally important flow of expertise from Mars to Earth, as research capacities mature and target their work on Martian problems. Proportional change of the Martian population will potentially be much greater because it is so much smaller. And, while medical expertise will back-flow from Mars to Earth, the Terran population will be so large, in comparison, that introduced changes will probably not be noticed. Whether a Martian population will eventually be recognizable by casual human inspection will depend on which genes are changed.

8.2.3 Population Forces When Humans Colonize Mars, the Asteroids, and Outer Planets

Early in our species' history, genetic drift was extremely important in small groups of evolving African hominins. Within the next few hundred years, countervailing forces to natural selection will likely begin to operate once more when humans colonize Earth's Moon, Mars, some of the larger asteroids, and some of the moons of the gas giants, such as Europa.

The settlement of these bodies raises the possibility of the reproductive isolation of small human populations once more, and therefore the renewal of the strong forces of genetic drift. When early colonies are cut off from a regular infusion of genes from the large human population on Earth, new advantageous *and deleterious* genes may arise and become widespread in the small, isolated populations, especially if settlers are exposed to higher levels of radiation on the trip out, or after arrival, or both, and if they can reproduce in these higher-radiation environments, at all [8.13].

Again, we note the eventual backflow of new (or existing) alleles from these off-world settlements to Earth or possibly to Mars or Earth's Moon. If infusions occur on Earth, mutations will likely not spread widely. The differences in population size will be very great, and any mutation introduced from off-world settlements would have to be extraordinarily advantageous to gain a foothold in the large Earth population. If very deleterious, i.e., deadly, then quarantine measures will be instituted and the populations will be further prevented from reproduction and gene flow.

If new alleles are introduced into another small population, for example, from the population on the moon of a gas giant, to Mars, then the possibility exists that new mutations could take hold in that secondary environment. The forces of genetic drift would likely predominate on Mars. However, the forces of natural selection could then predominate if conditions are particularly rigorous. On Mars, difficult living conditions will prevail in the early decades of settlement, and natural selection could exert renewed force on all pioneers from Earth. As Matute notes, genetic drift and natural selection can operate at the same and in different directions at the same time, or sequentially [8.7]. The operation of one does not obviate the operation of the other.

A more likely scenario is one in which humans in off-world settlements remain isolated, and, if new and useful mutations arise, they will remain in that settlement. In this way, it is possible that new subspecies of human will arise, if isolation continues for a sufficiently long time. Evolution in small populations can occur very rapidly, and if a community is cut off without new settlers, then mutations could spread fast. However, we feel that it may be more likely that space travel (especially propulsion) technologies will improve, and new mutations on Mars will be swamped by the later arrival of larger numbers of humans from Earth. If this happens, then the isolation required for a new subspecies to develop will disappear in a relatively short time.

8.3   Ethical Issues Implied by Population Forces and Genome Modification

8.3.1   Selection of Interplanetary Migrants Based on Invasive Genetic Procedures

These facts about population forces, especially migration between planetary bodies, open up many ethical questions, which will be considered, solved, discarded, and argued throughout the following centuries of the colonization of our solar system. Gene editing of colonists or settlers, which is accomplished with the specific goal of accentuating certain genetic effects and avoiding other mutations that are deleterious, raises ethical issues for migration to all off-world destinations.

Dystopian images come to mind of humans who are accepted as settlers only if they undergo invasive procedures. And yet, we wonder how different that is from the straightforward selection of new settlers for inclusion in extraterrestrial populations who have the best chances of survival in a harsh environment. There is clearly the possibility that new settlers will be selected for some part of their genome, some phenotypic traits, or it may be required for the settlers to modify some part of the genome before they are sent to a space colony. In that case, they would have a choice, and if they chose not to undergo the procedures, then migration would not be possible. Tailor-made populations are not beyond the imagination. Will this be acceptable to some migrants, at some point in the future? Indeed, there are many occupations that a person cannot now seek if certain genetic defects are present, like asthma, severe bipolar disorder, and sickle cell anemia.

8.3.2   Required Pre-Settlement Genetic Remediation

The opposite but related issue is that space agencies or companies may require and guarantee desirable changes in settlers or workers (for example, miners), and that selection might be very practical indeed. The population of a space colony could be changed in proportion to the size of the new gene pool introduced by new settlers, and this entire process could be purposefully implemented. This fact raises political as well as ethical issues for agencies, companies, or countries purposefully changing the composition of a targeted off-world population. It might be difficult, but not impossible. This possibility is both conceivable and real. Indeed, one can imagine a benign example in which all colonists destined for Mars would be required to undergo genetic modifications to counter bone loss. From a certain perspective, this is extremely practical. The gravity of Mars is lower, and the journey to Mars, itself, could reduce bone density. If settlers arrive who cannot work without fracturing bones, then no one gains. A broader issue is that if nefarious powerful forces arise on Earth, they could indeed change the genetic profile of a space colony in different directions, perhaps with the goal of power or profit. This type of development is surely conceivable, and it could have consequences, both political and ethical, for generations because once a genetic selection occurs, future generations will reflect that selection.

8.3.3   Moral Context for Genetic Engineering for Space

We conclude that gene editing for space program purposes is not intrinsically morally wrong. A purposeful act that is not intrinsically wrong or right is usually rooted in another act or phenomenon and its value. If we accept that gene editing for space may be morally wrong or right mainly in an extrinsic sense, we must identify the external factors that could shape its moral value in both right and wrong directions. Thus, for each and every space mission, the moral deficits and moral rectitude of genetic engineering must be assessed. The knowledge base on the safety of specific engineering techniques is still very modest, but it will be augmented in the future. Therefore, whether a specific procedure is acceptable for a specific type of mission or a specific type of off-world colony would involve a fairly detailed analysis. The question of whether to use genetic engineering is very complex, and we should acknowledge that complexity.

8.4   Case Types for Off-World Population Change and Their Ethical Implications

Challenging astroethical issues confront the human race and the human space enterprise from a long-term perspective that is connected to the population forces discussed earlier. Here, we first consider ethical issues implied by an isolated population—an isolated space colony.

8.4.1   The Case of the Isolated Space Colony

Because evolution in small populations can sometimes occur rapidly due to genetic mutations followed by isolation and genetic drift, confined human space settlements could result. This type of isolated community is challenging not only from a biological but also an ethical point of view, and decisions about the nature, functions, and destiny of the space colony will be based on both biology and ethics. As in cases of reproductively isolated populations on Earth, decisions by community leaders (or earthside sponsors of a space colony) must take into account the health of the population from the perspective of inbreeding. If genetically based diseases become common, the colony may want to resort to either natural or artificial means to correct this genetic distribution. In the past, there are examples of very isolated Amazonian tribes that, for example, had a high proportion of club feet. There was a great deal of inbreeding because of the small size of the group and its isolation from other groups.

The result was that this unfortunate distribution could not be naturally avoided. Now, this type of unfortunate distribution can be altered with genetic engineering or further with planned in-migration. The use of genetic engineering in such a case would be a positive, fruitful, and ethical decision to take. It could be seen as morally right, and one that would prevent human suffering.

Given a longer time frame, the possibility exists that the genetic profile of a space colony would change so much that a new subspecies of human would be taking hold. This could result in lower success of reproduction if new settlers from Earth were to join the colony. (Recall the common definition of "species" is that reproduction produces "fertile offspring.") That biological knowledge should be the foundation of political decisions that have serious, long-term ethical implications. In the case of a small isolated space settlement—possibly more than in many cases in the history of Earth—political decisions have much more serious biological and ethical consequences. On Earth, infusion of new settlers would often be available as people migrated, but in a small isolated space colony, it would be less likely that new settlers would arrive unless planned as a long-term settlement project. Decisions would focus on community goals, sustainability, desirable population size, and the consequences of directly addressing the subspecies issue. Genetically, there is no firm definition of what a subspecies would be. However, in a small colony, inbreeding could well bring out deleterious genes whose distribution would need to be considered, and for which solutions would need to be found.

8.4.2   The Case of an Inclusivist or Exclusivist Space Colony: Science, Research, Intelligence

Consider the following case of the settlement of a space colony that is planned to be exclusivist or inclusivist. For example, a mission to establish a colony primarily for scientific research purposes. It would be a small, exclusive, science outpost with a rotated crew. None of the population or evolutionary effects discussed above would be operating. The colony would be for research and not to build a long-term human community—which includes the idea of replacement through reproduction or in-migration. A mission aimed at science would not pose ethical issues related to population forces.

Nevertheless, there are other challenges that make the mission problematical and that involve astroethical issues. The rationale (and therefore funding) for such a mission is always at stake. At some point, there will be genetic modifications that are known, safe, and even required for off-world scientific personnel. Would scientists be required to have these modifications in order to do their work?

Finally, it is conceivable that outposts established for science evolve into communities that are more or less permanent. If human reproduction is successful, the community can then begin to plan its future and engage in a new cultural mix. However, given the founding generation, this would be a highly exclusivist type of colony—smart and talented children and grandchildren of scientists. The only eventual problem would then again be inbreeding, if the colony persisted. It is possible that the colony would want to remain exclusivist, and that would entail ethical questions if they prevented in-migration according to certain exclusivist principles. In parallel examples on Earth, "out-marrying" might be forbidden, i.e., reproduction with someone who did not fit specific requirements. Ethical issues related to freedom of choice, freedom to marry, and freedom to reproduce might ensue.

8.4.3   The Case of the Space Refuge as an Ethically Expensive Option

Consider the following thought experiment, by assuming that humans will attempt to create a space settlement designed as a "space refuge." If selective political decisions determine who can leave Earth, they will also determine who can survive an existential catastrophe on Earth—environmental or agricultural collapse, pandemic, or endemic warfare with or without Artificial Intelligences, as just a few examples. The space settlement would be founded as a means to ensure human survival as a species, but under some conditions These conditions might include what appears as a rational consideration of other types of refuges, such as an aquatic or subterranean refuge. Planners could assess the desirability of these types of refuges using cost/benefit and risk/benefit analysis, and they could assess the types of refugees who would be "most appropriate" for the refuge.

If the planned number of human survivors is relatively small, the dominant population force in a future space refuge will be genetic drift because genetic drift is stronger in small groups. The obverse effect would be that in a larger population of such a colony, the stronger impact would be from natural selection.

Planning for transportation off Earth in the case of a catastrophic event would be a major undertaking and plans would run into severe limitations. "Evacuation of Earth" is not a realistic possibility, and selection of who goes and who stays is fraught with ethical implications. Furthermore, ethical issues will likely arise between the space refuge and terrestrial populations. These are populations that will have coexisted on two planets—Earth and the refuge. Planning may require interplanetary trips, with preferences for who travels and who does not. These requirements would likely be enforced as long as the space refuge does not obtain status as an autonomous settlement.

Full autonomy would be required for any effective space refuge, and this means no dependence on deliveries of supplies from Earth. Those supplies would cease in any case, when the catastrophic event occurs. The fragility of interplanetary transportation is an important factor in determining whether Earth can be a lifeline, and whether the space refuge can survive independently. The human populations on Earth who possess financial and technological means to organize transport of people and early supplies for survivors will determine the profile of the population that eventually stabilizes in the space refuge. Whether resourcing a space refuge would be encouraged or even allowed is yet another matter. Prohibition of interplanetary trips might be justified in order to keep the refuge population "pure" or conforming to certain criteria. If this happens, no arrivals from Earth would be accepted. If survival of humans as a species is considered as the main goal of a space refuge, limitations on who survives in the space refuge become ethically difficult choices, and for many, unacceptable.

Justifications based patently on biology are particularly onerous and recall "survival of the fittest" rationales. Still, in a post-apocalyptic scenario, it could be deemed justifiable to select strong, healthy migrants, just as today it is reasonable to select strong, healthy astronauts to work on the International Space Station. Finally, there is the issue of what happens once the refuge is "full," or supply trips are unneeded because the space refuge has become autonomous. Should trips continue with additional survivors from Earth? Should migrants be preferred or prohibited?

8.4.4 The Case of the Formation of a New Species of Human

Let us return to a case suggested earlier: the possibility of formation of new species of the genus *Homo* in a space settlement of the future. It appears to us that only two possible ethical choices have been considered, and we see the situation as much more complex. The initial question is: Should we avoid a new speciation if it is within the realm of possibility as a result of off-world exploration? Or, should we encourage it, or at least view it as a result of normal evolutionary processes? The three co-authors are in disagreement as to whether there are good reasons to have only one human species. One argument is this: "It makes things simpler." However, has it? The human species has divided itself into many types of groups based on genetic, social, and economic factors. The groups disagree, but they work together. It is difficult to see how being one species has made things simple or convenient. The fact is that the human species as it exists today is one in a series of species in the genus *Homo*. We are all very similar genetically because of repeated bottlenecks during our evolution. And still, is a new species of *Homo* necessarily unfortunate? Among the co-authors there are varying opinions. Some say, it is a natural occurrence. Others say, a new species would be unfortunate and we should stay as we are. A strong counter-argument to this is the proven fact of a large array of deleterious genes in the human genome, which cause untold damage and suffering. Changes of the human genome should not be categorically rejected.

There is a scenario in which speciation may well occur, given sufficient time, separation, and lack of crossbreeding between Earth and a space settlement. Indeed, this may happen if humans set up a space refuge. Designated survivors who live in the space refuge will, at some point in the future, become increasingly differentiated genetically from the large Earth population from which they descended. Population forces operating, such as genetic drift and natural selection, will have led to evolution of new traits in the refuge population. While this is not necessarily a scenario which we want and seek, under some conditions in the future, it may be impossible to avoid. Humans will be modified genetically, but this will be the result of natural population processes. The major question will be whether this is a genetic, social, economic, and/or ethical problem. Again, the co-authors disagree.

We turn to religious ethics and population forces next. That will provide a somewhat different perspective on the seemingly "impersonal" population forces of genetic drift and natural selection.

8.5   Religious Ethics and Population Forces

Migration is an early theme in the Bible, and many of the reasons for migration were the same as we have mentioned in this chapter, although oral traditions endow them with religious motivations and rationales. Indeed, when space colonization becomes easier and more frequent, it is very likely that some space colonies will be established for religious reasons of avoiding persecution or establishing a special and isolated community.

In one story from the Bible, Abraham is told by God to take his family, servants, and flocks and move to Canaan, which was located in the area now occupied by modern Lebanon (Genesis Ch. 12 & 13). In another tale, Jacob's favorite son Joseph found himself in Egypt, thanks to the machinations of his elder brothers, and he settles there after several changes in fortune (Genesis Ch. 37ff). A third example is found in the story of the Israelites in Egypt, who became a thriving community, although eventually an enslaved one until Moses, again under God's command, moves them into the desert and finally to the "promised land" where modern Israel is located (Exodus Ch.1ff). While the writers of Genesis and Exodus saw the hand of God in all these migrations, the social roots were overcrowding, famine, and political persecution. These are still important factors in migration three thousand years later, and they are likely to continue to be so, while humans establish space colonies and even refuges.

We read in the Bible that the physical survival of the Israelites was behind these migrations, but the survival of this and other groups' strong religious beliefs has been a compelling motivation time and time again: the Puritans who settled in New England; the French Huguenots in Louisiana; and the Mormons in Utah. They brought with them their religious beliefs and ethics. Social and ethical rules for migrating groups in the Bible tended to be strongest in what mattered most to the survival of the group's identity, and so they concerned who could marry whom and produce offspring, and how family life was organized. The backbone of ethics is usually enshrined in the group's religious beliefs, which were very effective in maintaining the kind of practices that ensured that their identity persisted.

However, the Bible is full of examples of ways in which Israelites found it hard to maintain their religious practices after the exodus from Egypt and after they had settled in the "promised land." In Israel, they came under the less stringent Canaanite influence and their beliefs in multiple gods. The only solution to this type of "religious infection" was rigid isolation of the group. We see some of the same isolation practices in the case of the pandemic today, and indeed we can see some of the backlash as dissatisfaction with the inability of some evangelical groups to proselytize. Few new converts can be attracted if everyone stays home.

Our expectation is that any new Martian settlers who will make the long journey for strong religious reasons will similarly try to become as isolated as possible. Kim Stanley Robinson's Mars Trilogy [8.11]—*Red Mars, Green Mars*, and *Blue Mars*—has several good examples of both traditional and new religious groups attempting to maintain isolation from the remainder of the Martian population.

How might this affect the population forces and so the survivability of the migrating group? Again, looking back at the Bible, we see that Isaac told Jacob to marry one of his first cousins and to expect God's blessing, so Jacob ended up marrying two of them (Genesis 28:2). On Mars, as in some cultures on Earth, it may be preferable to marry first cousins [8.1], despite the danger of recessive genes. Does this mean that the isolated Martian settlers will revert to so-called "primitive" customs? We can predict that marriage customs will likely follow those of earthbound, small isolated communities, but the genetic knowledge guiding those customs will be far from primitive. Depending on how difficult human reproduction is on Mars, we may see the drafting and use of an entirely different set of religious principles and ethics. If the group cannot reproduce, then it would be difficult to maintain isolation of it, because members will die off without replacements.

Thus, we have to ask: Will the religion of the Martian settlers be forced to change, along with their ethics? Yet again, we look back to the Bible for examples. Just as the change from a nomadic to an agricultural society changes the approach to the deity, so will the religion of the new Martian settlers incorporate their new experiences, practices, and health. However, we believe that fundamental changes in religions will not likely occur quickly, except in isolated cases of crew or settlers. The relative stability in religion will be helped by initially keeping contact with their fellow believers on Earth, despite the many minutes lag in communication between the two planets. We will expect families on both planets to keep contact in some measure. The same happens on Earth between families and religious groups on different continents.

Over the years, the "new Martians" will become distinct from the Earth dwellers, in culture and religion, and in genetics, through forces we have discussed in this chapter, especially natural selection and genetic drift. Bioenhancements that may become necessary to sustain life and reproduction in an un-Earthlike environment will change the meaning and cultural content of what is considered "natural" and "normal." With those changes will come a change in ethical principles—a change in the rules. Whatever the underlying philosophy, ethics must follow local conditions in organizing its values for people who face new challenges. Those values may also find a grounding in a new, developing religion. It is a development that will happen through the expanded vision that surviving and thriving on Mars will need and provide.

8.6   Conclusions

Main evolutionary processes and the human genome as such are sources of different ethical issues and dilemmas in a natural environment on Earth. The context of future space missions introduces new challenges for the ethics and bioethics of human biology. The ethics of population biology in space is an important subcategory of space ethics. As we showed, the structure and size of a future human space outpost, but also its relation to

Earth and possible new migrations, will affect the way the population forces work. If a future space settlement is designed as a confined colony, and no new immigrants from Earth are welcome, we expect that genetic drift will become a dominant population force which may increase a number of hazardous mutations. This is a risk that can be avoided not necessarily through genetic engineering but—at least to some extent—through proper space policy. Genetic drift is not a trivial issue because one of the possible long-term consequences may be a further speciation of a human population living in space. An important issue which arises here is a possibility of artificial and intended human control over the dynamics of population forces, which is ethically very controversial and might be called "genocide."

Genetic changes will happen naturally in a space settlement but some of them may be artificially applied and shaped by genetic engineering to accelerate and direct the process. While this is also ethically controversial as long as any genetic engineering—both on Earth and for purposes of space missions—is one of the main issues in bioethics, we argue that there are some reasons which may justify genetic engineering for space. There are some situations where humans should reduce randomness and limit the risk. Because space missions as such are very risky human projects, the precautionary principle would imply that we should reduce all possible fields of risk, including survival of humans in space both at the individual and population levels. Some natural genetic mutations may be disadvantageous for a human population living in a space settlement. Even if genetic engineering as such still remains a controversial issue, challenges of population biology enforced by a hazardous space environment may be a good reason to treat future human space missions as a kind of specific astroethical situation.

Acknowledgement

Konrad Szocik for work on this text received the Bekker Fellowship (3rd edition) funded by the National Agency for Academic Exchange (Decision No. PPN/BEK/2020/1/00012/DEC/1) for a research stay at Yale University (USA) in the academic year 2021/2022.

References

[8.1]   Bhopal, R.S., Petherick, E.S., Wright, J., Small, N., Potential social, economic and general health benefits of consanguineous marriage: results from the Born in Bradford cohort study. *Eur. J. Public Health*, 24, 862–869, 2014.

[8.2]   Cochran, G., and Harpending, H., *The 10,000 Year Explosion: How Civilization Accelerated Human Evolution*, Basic Books, New York, NY, 2009.

[8.3]   Harris, E.E., *Ancestors in Our Genome; The New Science of Human Evolution*, Oxford University Press, Oxford, 2015.

[8.4]   Daniel, L. and Clark, A.G., *Principles of Population Genetics*, Fourth Edition, Sinauer Associates of Oxford University Press, Sunderland, MA, 2006.

[8.5]   Hublin, Jean-Jacques, Ben-Ncer, A., Bailey, S.E., Freidline, S.E., Neubauer, S., Skinner, M.M., Bergmann, I., Le Cabec, A., Benazzi, S., Harvati, K., Gunz, P., New Fossils from Jebel Irhoud, Morocco and the Pan-African Origin of *Homo sapiens*, *Nature*, 546, 289–292, 2017.

[8.6]   Lachance, J., and Tishkoff, S.A., Population Genomics of Human Adaptation, *Annu. Rev. Ecology, Evolution, Systematics*, 44, 123–143, 2013.

[8.7]   Matute, D.R., The Role of Founder Effects on the Evolution of Reproductive Isolation, *J. Evolutionary Biol.*, 26, 2299–311, 2013.

[8.8]   O'Bleness, M., Searles, V.B., Varki, A., Gagneux, P., Sikela, J.M., Evolution of Genetic and Genomic Features Unique to the Human Lineage, *Nat. Rev. Genet.*, 13, 853–66, 2012.

[8.9]   Rappaport, M.B. and Corbally, C., Cultural Neural Reuse, Re-Deployed Brain Networks, and Homologous Cultural Patterns of Compassion, in: *Biological Systems from a Network Perspective*, C. Timoteo, R. Cazalis, R. Cottam, (Eds.), pp. 111–134, Namur University Press, Namur, Belgium, 2019.

[8.10] Rappaport, M.B. and Corbally, C., *The Emergence of Religion in Human Evolution*, Routledge, Oxford, 2020.

[8.11] Robinson, K.S., *Red Mars, Green Mars, Blue Mars*, Spectra reprint editions, New York, 2003.

[8.12] Sibonga, J.D., Spaceflight-induced bone loss: is there an osteoporosis risk? *Curr. Osteoporos. Rep.*, 11, 2, 92e98, 2013.

[8.13] Konrad, S., Marques, R.E., Abood, S., Lysenko-Ryba, K., Minich, D., Kędzior, A., Biological and social challenges of human reproduction in a long-term Mars base. *Futures*, 100, 56–62, 2018.

9

Constructing a Space Ethics Upon Natural Law Ethics

Brian Patrick Green

*Markkula Center for Applied Ethics, Santa Clara University,
Santa Clara, California, USA*

Abstract

For centuries, natural law ethics served as a starting point for the consideration of the universal aspects of ethics across human cultures. While natural law ethics is currently mostly talked about among Roman Catholic ethicists, natural law discourse is the historical origin of contemporary human rights discourse, and ethical naturalism itself is experiencing something of a revival in secular philosophy, and shares many commonalities with Catholic discussions of natural law. An updated version of natural law can provide a foundation upon which space ethics can be built. This space ethics seeks to choose justly which purposes are better or worse when conflict is unavoidable. This ethic respects all living things, but especially demands that we protect humanity from extinction, since humans are the only known moral creatures in existence (if we discover other moral creatures, we should protect them as well). Lastly, the disadvantages (lack of immediate popularity, potential particularist opposition), ambiguities (difficulties in achieving moral neutrality and in knowing and interpreting nature correctly), and advantages (alignment of humanity with the universe, intuitive psychological appeal, and practicality) of this ethic are explored.

Keywords: Space, ethics, space ethics, natural law, astrobiology, astroethics, astrobioethics, ethical naturalism, teleology

Email: bpgreen@scu.edu

Brian Patrick Green: https://www.scu.edu/ethics/about-the-center/people/brian-green/, https://scholar.google.com/citations?user=NjrARt4AAAAJ&hl=en, https://scu.academia.edu/BrianGreen

Octavio Alfonso Chon Torres, Ted Peters, Joseph Seckbach and Richard Gordon (eds.) Astrobiology: Science, Ethics, and Public Policy, (177–192) © 2021 Scrivener Publishing LLC

9.1   Introduction

"Natural law" in the context of outer space may evoke notions of gravity and electromagnetic radiation. But in the context of ethics, it evokes quite different notions related to human nature, human needs, and human activities based on our shared humanity.

While some might now consider natural law ethics to be parochially Roman Catholic or anachronistic, at its core natural law ethics seeks to be a "mere" secular ethic, a naturalistic approach to ethics based on human nature, no theology required (and I have argued previously that this ethic can be built upon the basic metaphysical assumptions underlying the scientific method) [9.23]. Theology can form natural law's context, but the idea of natural law originates in Greek and Roman thought [9.5] [9.27], predating Christianity, and was adopted into Catholicism for the sake of facilitating moral conversations upon a common moral ground with all people, whether Roman Catholic or not.

Continuing this objective, natural law did, over time, gradually convert into the natural rights discourse and from there into the human rights discourse, thus somewhat fulfilling this ancient role in the contemporary world [9.50]. But, with its focus on humans to the exclusion of the environment (and stretches of the imagination attempting to amend this, such as the idea of "nature's rights"), the human rights discourse is a bit of a thin starting point for space ethics, particularly since nearly all of space is completely without humans.

Reaching beyond human rights, into the deeper philosophical context of the roots of human rights in natural law, adds adaptive flexibility and strength to the discourse, even to the point of being able to fruitfully engage such further "beyonds" as space. Additionally, recent work in virtue ethics and multicultural ethics of technology [9.53], when contextualized into a natural law/naturalistic framework, show the advantages of this approach for engaging across cultures in our pluralistic world—a distinct advantage when it comes to the future of humanity in space.

In this chapter I will consider the relationship of space and natural law ethics, and in particular will argue for how natural law ethics might help with the construction of an ethics for space exploration, sometimes called "astroethics." First, I will examine what space ethics and what natural law ethics are. Second, I will delve more deeply into how to construct a space ethics upon natural law. Third, I will discuss the disadvantages, ambiguities, and advantages of thinking about space ethics from the perspective of natural law.

9.2 Space Ethics and Natural Law Ethics

There have been many explorations into the ethics of space, though the field is certainly more nascent than related fields such as bioethics and engineering ethics. Such landmark volumes as those by Arnould [9.6], Bertka [9.11], Hargrove [9.24], Schwartz and Milligan [9.42], Smith and Mariscal [9.47], and Vakoch [9.52]; many articles in such journals as *Astrobiology* [9.31] [9.38] [9.41], *Futures* [9.7] [9.46] [9.48], *The International Journal of Astrobiology* [9.32] [9.36] [9.39], and *Space Policy* [9.15] [9.45] [9.54]; as well as other books, journals, and untold numbers of conferences and conference papers, show the field has developed and formed a coherence over the last few decades. Yet we would be remiss to forget the countless science fiction stories that have explored ethical questions in space as well, many of which have laid out ethical problems in great detail, such as those by Asimov [9.9] [9.10], Liu [9.28], and Robinson [9.40].

However, among all of these works, the intersection of space ethics and natural law ethics is still quite unusual. To understand their relationship, it is first important to know what each one is.

First, space ethics is the study of ethical issues related to space exploration, use, and settlement. It includes everything from the basic risks of spaceflight to human beings and the justice of allocating resources in this way, to the ethics of terraformation and interstellar travel [9.17]. Space ethics gives the opportunity to take traditional Earth-bound systems of ethics and stretch them to their largest scales and potentials—both to the benefit and detriment of those systems. For example, Mill's utilitarianism operates in a certain way on Earth—but at the interstellar scale it operates a little bit differently, as do Kantian deontology and Aristotelian virtue ethics [9.22]. Space stretches and even breaks ethical systems, showing us aspects of them that we might previously have missed, as well as successes, failures, and opportunities for improvement. Space ethics is merely Earth ethics extended into space, surely no small task, but most likely a comprehensible one to readers of this volume—so I will spend more time discussing natural law.

Second, at its most basic level, natural law ethics states that what something is says something about what it should[1] do, in a prescriptive, normative sense. For example, multicellular plants typically grow upwards, away

---

[1]The use of the word "should" here could be interpreted descriptively or prescriptively, which gets into the question of relating "ought" and "is," a question too big for this chapter, but one which has been addressed numerous times (e.g., three examples: [9.13] [9.25] [9.43]). Suffice it to say that Hume's is-ought problem is not as robust an obstacle to naturalistic ethics as it once was construed to be [9.23].

from accelerating forces (e.g., gravitational fields), and toward light (e.g., stars). Animals generally have sensory, metabolic, and motor capacities that allow them to move more than rooted plants do (e.g., fins, legs, wings, etc.) as well as behave in novel ways (e.g., swimming, running, flying). Plants which do not grow towards light, or animals which refuse to move (excluding sessile ones such as barnacles), may be experiencing difficulties of some sort that eventually damage their flourishing. The idea that being and action are related can be summed up in the Medieval Latin phrase *agere sequitur esse,* which means "action follows being." This phrase is found in the writings of Thomas Aquinas [9.1]; however, the general sense of the phrase is traced originally to Aristotle's notions of hylomorphism: that beings are made of matter and form, existence and essence, combined.

This is one thing when it comes to plants and animals, but what about natural law ethics as it applies to people? We do expect the average human to perform certain types of activities, there is no doubt. We expect the average human to drink liquids. We expect them to eat food. We expect them to breathe, move, speak, and so on. In fact, we have human rights based upon these expectations, and if we deny people drink, food, air, movement, and speech, we are ethically culpable for the harm that may come to them. The entire medical profession (in a very broad sense, including speech, occupational, and physical therapists, etc.) can in some way be seen as an homage to a normative understanding of nature, trying to make sure that human bodies operate in particular ways and are capable of expressing certain capacities such as language, self-care, and movement.

And what about human agency and freedom? Humans can choose to lie to each other or steal from or rape or murder each other. In each of these, one natural and good human power is distorted into something evil. Speech is good and makes possible human cooperation and truth-seeking, but lies manipulate and distort cooperation for the sake of evil and obstruction of the truth. Material objects, such as tools and food, are necessary for human life, and we have hands to create such objects and carry them, but stealing unjustly acquires these objects for one person, while depriving another. Sexual intercourse is good and promotes loving relationships, procreation, and family, but distorted by power and violence it treats others as dehumanized objects and commits grave bodily and psychological violations. And the application of force can be a good thing—fighting off dangerous animals or other humans in order to protect oneself or the innocent, but unjustly directed against the innocent, this ability to exert force destroys the foundation of all other human goods: life itself. The good capacities of human life need to be directed towards good ends, and distortions of those capacities need to be limited for the sake of protecting others. And nearly all human cultures recognize this (at least in some fashion) in their customs, norms, and taboos.

The efforts we humans take to help and protect ourselves and each other, in order to live in a typically human manner, demonstrate that human nature has action-guiding teleologies built into it from which moral imperatives can be derived. For example, we care for babies and raise them into people who are capable of, in turn, caring for themselves and others [9.30]. And these built-in teleologies, "entelechies," are found in all life forms: every living thing is on a trajectory through time, seeking fundamental and basic goals such as energy, self-repair, and reproduction. Exactly what those operations require for any particular creature can vary considerably: photosynthesizing light or eating plants, oozing sap or clotting blood, budding clones or reproducing sexually. But the overarching ethical imperative is towards the flourishing of the organism by permitting it to seek its natural ends—with one exception: if those ends might be destructive to other organisms, then ethical discernment may be necessary in order to determine which side of the conflict has moral priority.

For example, if a human is preyed upon by a mosquito carrying a deadly virus, there is a serious conflict of ends. The human endeavors to survive and flourish, as does the mosquito, as does the pathogenic virus, though each in their own way. Not all ends can be respected in this case: least of all the pathogen's, which is not only lacking in self-consciousness, acting on chemical-level dynamics, but is also intrinsically destructive in its ends. The pathogen is capable of very little in the way of flourishing, merely existing to copy itself mindlessly, unlike humans, who (while quite capable of mindless copying) are also capable of science, art, philosophy, love, and so much more. In this case, the human takes moral priority over the pathogen because humans have vastly more possible virtuous ends than a pathogen.

If the pathogen can be stopped, this helps the human being, and possibly the mosquito as well. But what if the pathogen cannot be stopped, is it all right instead to stop the mosquito in order to protect the human? Once again, the mosquito either lacks self-consciousness or is very minimally conscious, acting on instinctive ends, which, while they are not intrinsically destructive to humans, are irritating and risk the side effect of infection by a deadly pathogen. The mosquito is effectively a material, though not a formal, aggressor—one which causes harm not because it intends to (for a mosquito cannot "intend" in any human sense), but merely by its action. The mosquito is also, once again, not capable of much in the way of virtue. A virtuous mosquito is one that can suck blood and reproduce without getting swatted. Mosquitoes have no science, no art, no philosophy, and very little in the way of love. In short, the mosquito can justifiably be stopped in its activities relative to other creatures. In fact, humans have some ethical priority over other creatures in many cases specifically because of one thing: we possess the capacity to think ethically and thereby *be* ethical. This is not a blanket approval for all human desires, but insofar as those ends inescapably conflict, it gives humanity some priority.

Humans are the only creatures in the universe that we know of who are capable of ethics. As such, we have a special place, as the "foothold" of morality in the universe [9.25]. If we encountered other ethically intelligent beings in space we would of course have to respect them as well. And just because we are more capable than other creatures, it does not mean that those creatures have no value at all; indeed, every living thing should be permitted to seek its own ends to the greatest extent possible, without interference, providing that those ends do no harm other creatures. (Note that non-intervention in other organisms' affairs is generally presumed. Humans are not to go around being interstellar do-gooders, stopping all predators, parasites, and pathogens, but rather to aid in the spread of life to new places, and to step in when morally significant creatures are at serious risk, such as with extinction [9.21].)

Here are three principles that can be derived from this respect-for-*telei* natural law ethic:

1. Respect *telei*, both human and non-human, promote their fulfillment and minimize their thwarting, and if possible, prevent cross-purposes. Discriminate between cross-purposes via capacity for virtue.
2. Respect life itself, as the ground of all *telei*.
3. Humans (and other intelligent and moral creatures, if they exist), as creatures capable of respecting *telei*, ought to exist.

While the progression of thought runs from first to third, the order of moral priority is third to first [9.18] [9.23].

These rules, as far as I can tell, should be just as valid on Earth as they are anywhere in space. For humanity, right now, it means particularly that we ought to make sure that our existence continues into the far future by exploring and settling beyond the Earth [9.19].

9.3   A Natural Law Ethics Including Space

When it comes to space ethics, natural law seeks to discover and facilitate (within reasonable limits) the purpose and meaning of living creatures.

These entelechies are demonstrations that living creatures are more than just dead matter, they are matter with goals, and these goals have significance, not only to the creature itself (which is likely trying to survive and reproduce), but possibly to other life in the universe as well. And we should not deny the additional possibility that these goals might also in some way reflect goals for the universe itself. This is of course one of the great mysteries of existence: What's it all about? Why is there anything at all? What is my role in the universe?

While religious and other worldviews typically have answers to these questions, coming at them from a more secular materialistic perspective leaves the answers as something of a mystery, and even the question itself is sometimes dismissed as nonsense (typically as a conclusion extending from the assumption of materialism). Secular materialism holds sway in most conversations on space, and we could easily continue in that vein, except for the fact that with all of these odd life forms going about their entelechies, and the presumption that they just pop into existence from abiotic origins whenever given a chance, it does look somewhat suspiciously like the universe itself is *doing something*. Life seems to be a unique and intrinsic part of the development of the universe, and not an aberration. Furthermore, *life has goals*, and oddly, by extension, therefore *the universe has goals* in these particular instantiations of the universe called life forms. The question is whether all these tiny individual goals of individual life forms build up into something significant at the aggregate scale, or even grander, at the scale of the universe, or multiverse. The goals of life forms are expressing something inherent in the universe (since the possibility for purpose is intrinsic to our material universe), but the deeper question is "What is the universe trying to do?" And then for humanity the question of space ethics: "*Should we help* the universe in its efforts or not?"

All of the above is still a vast simplification. There are many dead ends and poor choices possible in such a natural law space ethic. But for the sake of simplicity, here are a few necessary assumptions for a natural law space ethic.

First, teleology is real. For several centuries this has been a somewhat controversial assertion, but there is more than enough evidence to make the case for the real existence of teleology in the universe [9.20], even if limited to the mere existence in human minds or human social conventions, such as the meanings of words [9.44]. If words have meaning, then teleology exists in the universe, and so much more so if biology has intrinsic purposes [9.13], or the universe itself [9.23] [9.25]. By reading this chapter, you demonstrate the existence of teleology, as the purpose of my words pass through writing and into your mind (hopefully to beneficial effect).

Second, some teleologies are more morally valuable than others. This gets into potentially horrifying terrain if done wrong, prioritizing some goals over others can be terribly abused, but in reality it is very basic. Spending one's entire life playing mindless video games is likely to be less worthwhile than trying to improve the world by feeding the poor, educating children, or healing the sick. If one has a chance to do one or the other, one ought to choose to do something beneficial to humankind and our fellow living creatures in the universe.

Third, this beneficence can be expressed at larger scales in space than on Earth. If it is correct that the universe is "doing something," then we can choose to either go along with this action or oppose it. We could choose to exterminate humanity and every living thing on Earth as a horrible mistake on the part of the universe, dead matter gone terribly wrong. Still, after Earth's biosphere and humanity are dead (no mean feat), life could evolve again elsewhere and our will to oppose life would have been thwarted. Ultimately, opposing life seems to entail opposing the "metaphysical grain" of the universe, and is therefore futile. It also goes against some of the most basic human instincts to survive, love, raise children, live in society, and learn about the world [9.3]. Choosing the way of death is the wrong choice: it is not only anti-human, it seems to be anti-universe, and ultimately futile. What if we chose the way of life instead?

Rather than trying to exterminate humanity and all life, what if we tried to help all humans to reach their greatest potential? What if we also attempted to do this for the biosphere, and the plants and animals in it (within reasonable limits)? What if we took this task beyond the Earth, expanding Earth life to lifeless places, or what if we found life on other planets and "helped" that life develop whatever potential it might have?[2] Perhaps the goal of the universe is life itself, and whatever further things life might develop into. This is a more noble task than the futile one of negation and destruction. This is an accomplishment for which humankind could be pleased, for which our descendants could feel joy in their hearts to have had such ancestors. Non-conscious life forms, both Earthly and perhaps from other origins of life, would likewise flourish, but never know the beneficence from which they grew.

From an outsider's perspective on the Roman Catholic Church, this perspective could seem odd. Where is God? Where is Jesus? Where is the

---

[2]As considered in David Brin's Uplift series, starting with *Sundiver* [9.12].

Second Coming? They are all still there. Humankind is still loving God and neighbor, including all of our neighboring humans, Earthly life forms, and potential extraterrestrial life. Jesus is still there, living in a humanity that gives forth its life for each other and for the rest of creation. The Second Coming is God's prerogative and will happen no matter what humanity might do, whether we are on Earth twiddling our thumbs or across the universe tending the garden of creation. There is nothing in the history of the Church that directly opposes such thinking, and in fact much that supports it.

Of note is the very basic choice between the way of death and the way of life, which are explicitly considered in the Bible in the book of Deuteronomy 30:19: "I have set before you life and death, blessings and curses. Choose life, so that you and your descendants may live" [9.33]. This is emphasized again centuries later in the *Didache* of the first century after Christ [9.26].

With space exploration and settlement, humankind can choose life, and choose it on the grandest scale. Our descendants can choose life as well, and cultivate the flourishing of other living things as well, thus promoting "the richness and diversity of life" in the universe [9.39]. And one need not be religious in any sense to see the rationale of this project—it is merely aligning the actions of humanity with the natural grain of the universe as it appears in a relatively neutral interpretation of secular reality.

In my chapter in Smith and Mariscal's *Social and Conceptual Issues in Astrobiology* [9.47], I argued that several major proposed ideas for space ethics—those of Lupisella and Logsdon [9.29], Smith [9.45], and Randolph and McKay [9.39]—are in fact, at their foundations, natural law ethics [9.18]. I then combined these with insights from René Girard and Peter Thiel concerning the avoidance of competition (aligned with the above idea of avoiding conflicting *telei*) [9.16] [9.49]. Lastly, I extracted some commonalities across these approaches and condensed them to the three principles listed above. In my view, a space ethics constructed upon the basic assumptions of natural law ethics should have these traits. However, with this perspective there are disadvantages and ambiguities, as well as advantages.

9.4 The Disadvantages, Ambiguities, and Advantages of a Natural Law Space Ethics

This particular description of a natural law ethics for space has some advantages, disadvantages, and ambiguities. Let me begin with two disadvantages:

possible poor reception due to traditionalist associations and possible conflict with exclusivist/particularist ethical systems.

First, particularist ethics, such as those presented by some historical religions, charismatic leaders, and metaphysical and/or political movements (e.g., Marxist communism), have demonstrated that they are highly motivating and enlivening of the human spirit, while natural law ethics—as a more universalist approach, and one with much historical baggage—does not seem to motivate with such fervor, nor get much beyond academic discussions. While in the guise of "human rights," natural law ethics has become quite widespread; in and of itself, the name "natural law ethics" is stodgy and often makes people think of poorly argued traditionalist critiques of progressive norms. This could perhaps be remedied by some efforts to re-describe and update the tradition.

Second, as a universalistic perspective, natural law ethics may be anathema to more particularistic and exclusive approaches to ethics, such as those of certain religions, political movements, or other worldviews. While I think that the best ideas of various worldviews could reach consonance with this ethic, thus permitting forward movement together (as with human rights discourse), no doubt some will disagree. In a pluralistic world, this could be a significant impediment to the use of this ethic in space; and yet, a pluralism of ethics in space could itself be a major impediment to human movement into space.

There are also some ambiguous aspects to this natural law space ethic, and I will present two related ones: the impossibility of absolutely certain moral neutrality, and the difficulty of establishing a perfectly clear interpretation of nature.

First, an absolutely certain and neutral moral perspective is impossible. We can attempt to achieve it, but it will always be in need of revision and updating, as we learn more things about the universe, and recognize more perspectives that should be included in this ethic. Even then, there could be significant errors embedded into the ethics that we would not recognize until later in its development. This is a disadvantage yet also an advantage, in that flexibility is possible within this ethic. It has the benefit of aligning with reality as well, since we really cannot understand ethics deeply without knowing the impacts of our choices at the very largest scales of space and time. Numerous thinkers have noted this limitation in ethics, ranging from Aristotle considering the impossibility of wisdom among the young (for in relation to the universe we are all young) [9.4], to Aquinas considering the limitations of human capacities when considering the good [9.2], to J.R.R. Tolkien's assertion, via the character Gandalf, that "even the very wise cannot see all ends" [9.51], to various utilitarians stating that moral perfection is impossible [9.8] and therefore some uncertainty is a require-ment for ethics (including ethics for AI [9.14]).

Second, even within the notion of natural law there could be many approaches that could be taken, some which could be viewed as (and be) quite immoral to other potentially plausible versions of natu-ral law. That is to say that natural law is, in the words of Jean Porter, "underdetermined"—i.e., nature and morality are not only loosely con-nected epistemologically, but perhaps also socio-culturally, and even biologically [9.37]. There could be dramatically different approaches to natural law, more than one of which might work as a space ethic. For example, in Cixin Liu's *Three-Body Problem* trilogy, space civilizations must make fundamental choices between kindness towards other species and brutality towards other species, manifesting as either deterrence or genocide [9.28]. While the ethic I have laid out above is clearly on the side of kindness, if other species exist who do not accept the path of kind-ness and instead choose the brutal path, then deterrence becomes nec-essary to hold those species at bay. And if weaker species are incapable of adequate deterrence, then they could be exterminated (unjustly from the "kind" perspective, justly from the "brutal" perspective). A natural law ethic of brutal exterminationism would be horrifying to encounter [9.21]. This brutal ethic appears in other science fiction novels too (not to mention in the genre of existential horror), thus showing that the idea is not so unusual [9.34] [9.35]. Indeed, in the history of humankind, it seems that human groups have not infrequently driven other groups of humans extinct, just because we could. And by "we" I mean "us" because we are the descendants of the survivors and the perpetrators. I find this brutal interpretation of natural law to be incorrect, because it is the path of death; but that is a much longer argument that can be made here, hence the ambiguity can only be pointed out.

There are also several advantages for basing space ethics on natural law, of which I will list three: moral alignment with the universe, intuitive appeal, and practicality.

First, this is an ethic which is, I think, fundamentally aligned with life and the universe. The universe spontaneously evolves life from abiotic chemicals, and all indications are that this should not happen infrequently. This life then develops over time, eventually, in some cases, developing technology and perhaps becoming capable of space travel, and thereby can spread life and evolution to the places in the universe which are as yet without life. This is a pro-life perspective in the deepest sense, in that is seeks to increase the overall amount of life in the universe, and it is human-ity which is uniquely suited to this goal [9.19]. Those who oppose life or who are anti-natalist will obviously not like this perspective, but they have already chosen their end, and it is self-limiting. This insight leads us into astrobioethics, that branch of space ethics which focuses specifically on issues surrounding life on Earth and off-Earth.

Second, such an ethic based on natural law has an intuitive aspect to it. It seems to appear over and over again as the foundation for other ethical systems, and therefore does seem to originate repeatedly and in various forms in human psychology and culture. And human psychology and culture, as manifestations of the directionality of the universe, should have a say in this. Humankind, as a species and as individuals, recognizes the reality of good and evil, and that we should choose good and flourishing not only for ourselves but for others as well. Humanity does not need to choose an ethic which is opposed to our flourishing, because such a choice is actually less plausibly accurate; even intrinsically self-defeating.

Third, this ethic is imminently practical. It gives guidance, unlike some other ethical systems which are too theoretical or removed from reality. For example, deontological ethics is too formal for most humans to follow in any real sense; rather insofar as people follow deontology, it tends to act as a rationalistic reinforcement of their own intuitions. Likewise, consequentialism requires capacities of ratiocination that exceed human ability, especially when considering the long term and large scale. Natural law ethics, on the other hand, due to its intuitive connection to human psychology and alignment of that psychology with a meaning and purpose in the universes, up to and including the possibility of God, ought to be practical in a way that these other philosophies are not.

The above are not the only possible advantages, disadvantages, and ambiguities, but they are some of the more significant ones.

9.5   Conclusion

A space ethics constructed upon the foundation of natural law has several basic aspects, such as respecting natural *telei*, respecting life in general, and respecting human life (and the lives of all moral agents) in particular. This ethic has relationships to questions of space especially when considering the existence of teleology, the relative values of conflicting teleologies, and the eventual possible scale of ethics in the universe. Its drawbacks are related to its possible poor reception and possible conflict with exclusivist ethical systems. Its ambiguities include the impossibility of the absolute certainty of moral neutrality and the difficulty of establishing a perfectly clear interpretation of nature. Its advantages are its seeming moral alignment with the universe, its intuitive appeal, and its practicality. These are only the first steps towards a natural law space ethics; if this path proves to be fruitful, hopefully there will be many more.

References

[9.1]   Aquinas, T., *Summa Contra Gentiles*, O.P. Joseph Kenny, (Ed.), Hanover House, 1955-57, 69.20, 97.4.

[9.2]   Aquinas, T., *Quaestiones Disputate de Veritate*, Html edition, R.W. Mulligan, S.J., J.V. McGlynn, S.J., R.W., Schmidt, S.J. (transl.), O.P. Joseph Kenny, (Eds.), Henry Regnery Company, Chicago, IL, USA, 1952-54, 25.1 (Responso, para. 4) https://isidore.co/aquinas/english/QDdeVer25.htm.

[9.3]   Aquinas, T., *Summa Theologica, Complete English Edition in Five Volumes*, vol. II, Fathers of the English Dominican Province (transl.), Ave Maria Press, Inc., Notre Dame, IN, USA, 1948, 1009 (I-II, 94.2).

[9.4]   Aristotle, Nicomachean Ethics, in: *The Basic Works of Aristotle*, W.D. Ross (transl.). R. McKeon (Ed.), pp. 1027–1030, Random House, New York, NY, USA, 1941, (1141a17-1142a31).

[9.5]   Aristotle, Rhetoric, in: *The Basic Works of Aristotle*, W. R. Roberts (transl.), R. McKeon (Ed.), pp. 1317–1451, Random House, New York, NY, USA, 1941, (1373b2-8).

[9.6]   Arnould, J., *Icarus' Second Chance: The Basis and Perspectives of Space Ethics*, Springer, Wein, NewYork, New York, NY, USA, 2011.

[9.7]   Arnould, J., The emergence of the ethics of space: the case of the French space agency. *Futures*, 37, 245–254, 2005.

[9.8]   Arrhenius, G., An Impossibility Theorem for Welfarist Axiologies. *Econ. Philos.*, 16, 247–266, 2000.

[9.9]   Asimov, I., *Foundation*, Bantam, New York, NY, USA, 1991[1951].

[9.10] Asimov, I., *I, Robot*, Bantam, New York, NY, USA, 1991[1950].

[9.11] Bertka, C.M. (Ed.), *Exploring the Origin, Extent, and Future of Life: Philosophical, Ethical, and Theological Perspectives*, Cambridge University Press, Cambridge, UK, 2009.

[9.12] Brin, D., *Sundiver*, Bantam Books, New York, NY, USA, 1980.

[9.13] Deacon, T., *Incomplete Nature: How Mind Emerged from Matter*, W.W. Norton & Company, New York, NY, USA, 2011.

[9.14] Eckersley, P., *Impossibility and Uncertainty Theorems in AI Value Alignment or why your AGI should not have a utility function*, 2019, arXiv. org. https://arxiv.org/abs/1901.00064.

[9.15] Fogg, M.J., The ethical dimensions of space settlement. *Space Policy*, 16, 205–11, 2000.

[9.16] Girard, R., Mimesis and Violence: Perspectives in Cultural Criticism. *Berks. Rev.*, 14, 9–19, 1979.

[9.17]   Green, B.P., *Space Ethics*, Rowman & Littlefield, London, UK, forthcoming.

[9.18]   Green, B.P., Convergences in the Ethics of Space Exploration, in: *Social and Conceptual Issues in Astrobiology*, K.C. Smith and C. Mariscal, (Eds.), pp. 179–196, Oxford University Press, Oxford, UK, 2020.

[9.19]   Green, B.P., Self-Preservation Should Be Humankind's First Ethical Priority and Therefore Rapid Space Settlement Is Necessary. *Futures*, 110, 35–37, 2019. https://www.sciencedirect.com/science/article/pii/S0016328718303173?dgcid=author.

[9.20]   Green, B.P., Habitus in the Roman Catholic Tradition: Context and Challenges, in: *Habits in Mind: Integrating Theology, Philosophy, and the Cognitive Science of Virtue, Emotion, and Character Formation*, G.R. Peterson, J.A. Van Slyke, M.L. Spezio, K.S. Reimer (Eds.), pp. 41–57, Brill, Leiden, Netherlands, 2017.

[9.21]   Green, B.P., Astrobiology, Theology, and Ethics, in: *Anticipating God's New Creation: Essays in Honor of Ted Peters*, C.R. Jacobsen and A.W. Pryor (Eds.), pp. 339–350, Lutheran University Press, Minneapolis, MN, USA, 2015.

[9.22]   Green, B.P., Ethical Approaches to Astrobiology and Space Exploration: Comparing Kant, Mill, and Aristotle, in: *Special Issue: Space Exploration and ET: Who Goes There?*, J. Arnould (Ed.), 2014, Ethics: Contemporary Issues 2(1): 29–44. http://bit.ly/1Umz6Qr.

[9.23]   Green, B.P., *The Is-Ought Problem and Catholic Natural Law*. Dissertation, Graduate Theological Union (ProQuest), Berkeley, CA, USA, 2013.

[9.24]   Hargrove, E.C. (Ed.), *Beyond Spaceship Earth: Environmental Ethics and the Solar System*, Sierra Club Books, San Francisco, CA, USA, 1986.

[9.25]   Jonas, H., *The Imperative of Responsibility*, University of Chicago Press, Chicago, IL, USA, 1984.

[9.26]   Knight, K. (Ed.), The Didache, The Lord's Teaching Through the Twelve Apostles to the Nations. M.B. Riddle (transl.), in: *The Ante-Nicene Fathers*, vol. 7, A. Roberts, J. Donaldson, A. Cleveland Coxe (Eds.), Christian Literature Publishing Co., Buffalo, NY, USA, 2020[1886], http://www.newadvent.org/fathers/0714.htm.

[9.27]   Levy, E., Natural Law in the Roman Period. *Natural Law Institute Proceedings*, vol. 43, pp. 43–72, 1949.

[9.28]   Liu, C., *The Three Body Problem*. K. Liu (transl.), Tor, New York, NY, USA, 2014[2006].

[9.29]   Lupisella, M., Logsdon, J., Do We Need a Cosmocentric Ethic? Paper IAA-97-IAA.9.2.09. *Presented at the International Astronautical Federation Congress*, American Institute of Aeronautics and Astronautics, Turin, Italy, pp. 1–9, 1997.

[9.30]   MacIntyre, A., *Dependent Rational Animals: Why Human Beings Need the Virtues*, Open Court, Chicago and La Salle, IL, USA, 1999.

[9.31]   McKay, C.P., Marinova, M.M., The Physics, Biology, and Environmental Ethics of Making Mars Habitable. *Astrobiology*, 1, 1, 89–109, 2001.

[9.32] Milligan, T., Common origins and the ethics of planetary seeding. *Int. J. Astrobiology*, 15, 4, 301–306, 2016.

[9.33] New Revised Standard Version (Eds.), *The Holy Bible, New Revised Standard Version*, Oxford University Press, New York, NY, USA, 1989.

[9.34] Pellegrino, C., *Flying to Valhalla*, William Morrow and Company, New York, NY, USA, 1993.

[9.35] Pellegrino, C. and Zebrowski, G., *The Killing Star*, William Morrow and Company, New York, NY, USA, 1995.

[9.36] Peters, T., Does extraterrestrial life have intrinsic value? An exploration in responsibility ethics. *Int. J. Astrobiology*, 18, 4, 304–310, 2019.

[9.37] Porter, J., *Nature as Reason: A Thomistic Theory of the Natural Law*, pp. 19, 49, 117, 126–7, 131, etc., William B. Eerdmans Publishing Company, Grand Rapids, MI, USA, 2005,

[9.38] Race, M.S., Denning, K., Bertka, C., Dick, S., Harrison, A., Impey, C., Mancinelli, R., Workshop Participants, Astrobiology and Society: Building an Interdisciplinary Research Community. *Astrobiology*, 12, 10, 958–965, 2012.

[9.39] Randolph, R.O., McKay, C.P., Protecting and expanding the richness and diversity of life, an ethics for astrobiology research and space exploration. *Int. J. Astrobiology*, 13, 28–34, 2014.

[9.40] Robinson, K.S., *Red Mars*, Bantam, New York, NY, USA, 1993.

[9.41] Rummel, J.D., Race, M.S., Horneck, G., and the Princeton Workshop Participants. Ethical Considerations for Planetary Protection in Space Exploration: A Workshop. *Astrobiology*, 12, 11, 1017–1023, 2012.

[9.42] Schwartz, J.S.J. and Milligan, T. (Eds.), *The Ethics of Space Exploration*, Springer International, Switzerland, 2016.

[9.43] Searle, J., *Making the Social World: The Structure of Human Civilization*, Oxford University Press, Oxford, UK, 2010.

[9.44] Searle, J., What is an Institution? *J. I. Econ.*, 1, 1–22, 2005.

[9.45] Smith, K.C., Manifest complexity: A foundational ethic for astrobiology? *Space Policy*, 30, 209–214, 2014.

[9.46] Smith, K.C. and Abney, K.A. (Eds.), Human Colonization of Other Worlds, Special Issue. *Futures*, 110, 1–66, 2019.

[9.47] Smith, K.C. and Mariscal, C. (Eds.), *Social and Conceptual Issues in Astrobiology*, Oxford, UK, Oxford University Press, 2020.

[9.48] Szocik, K., Wójtowicz, T., Boone Rappaport, M., Corbally, C., Ethical issues of human enhancements for space missions to Mars and beyond. *Futures*, 115, 102489, 2020.

[9.49] Thiel, P. and Masters, B., *Zero to One: Notes on Startups, or How to Build the Future*, pp. 1–43, Crown Business, New York, NY, USA, 2014.

[9.50] Tierney, B., *The Idea of Natural Rights: Studies on Natural Rights, Natural Law and Church Law, 1150-1625*, Scholars Press for Emory University, Atlanta, Georgia, USA, 1997.

[9.51] Tolkien, J.R.R., *The Fellowship of the Ring*, p. 74, Houghton Mifflin, New York, NY, USA, 2004[1954].

[9.52] Vakoch, D.A. (Ed.), *Extraterrestrial Altruism: Evolution and Ethics in the Cosmos (The Frontiers Collection)*, Springer, Heidelberg, Germany, 2013.

[9.53] Vallor, S., *Technology and the Virtues: A Philosophical Guide to a Future Worth Wanting*, Oxford University Press, Oxford, UK, 2016.

[9.54] Williamson, M., Space ethics and protection of the space environment. *Space Policy*, 19, 47–52, 2003.

10
Two Elephants in the Room of Astrobiology

Jensine Andresen

Independent Scholar, Virginia, USA

Abstract

Video and radar data, together with recent actions taken by the Office of the Director of National Intelligence (ODNI), the Department of Defense (DoD) and the Senate in the United States (U.S.), indicate that the U.S. government knows that unidentified ariel phenomena (UAP) and unidentified submerged objects (USOs) are real and most likely are of extraterrestrial origin. This is referred to as the extraterrestrial hypothesis (ETH), which, broadly, argues that an extraterrestrial intelligence (ETI) representing an extraterrestrial civilization (ETC) is operating these objects. Astrobiologists should pay attention to this data and should bring their considerable scientific expertise to bear in analyzing it. Relatedly, astroethicists must actively oppose any attempts to use the reality of UAP, USOs, and the Intelligent Beings—most likely ETI—responsible for their operation on and around Earth—collectively, the phenomenon—as pretext to rationalize the militarization and weaponization of space. To ensure that space be maintained as a peaceful arena free of international conflict and to facilitate international, scientific cooperation in astrobiology, astroethicists actively must engage in the formulation of a space policy that is cooperative and constructive. Astrobiologists should create a robust, ethics-based community to oppose the militarization and weaponization of space and to engage in an open-minded manner with the reality of the phenomenon and any ETI and ETC responsible for it.

Keywords: Astroethics, extraterrestrial life, extraterrestrial hypothesis (ETH), extraterrestrial intelligence (ETI), extraterrestrial civilization (ETC), unidentified

Email: Jensine@alumni.harvard.edu

Jensine Andresen: https://jensineandresen.academia.edu/, https://www.researchgate.net/profile/Jensine_Andresen, https://scholar.google.com/citations?user=Q7mL-0YAAAAJ&hl=en

Octavio Alfonso Chon Torres, Ted Peters, Joseph Seckbach and Richard Gordon (eds.) *Astrobiology: Science, Ethics, and Public Policy*, (193–232) © 2021 Scrivener Publishing LLC

aerial phenomena (UAP), militarization of space, weaponization of space, space policy

Abbreviations

ASAT	anti-satellite
BAASS	Bigelow Aerospace Advanced Space Studies
CIA	Central Intelligence Agency
COMETA	Comité d'Études Approfondies
DEW	directed energy weapon
DIFAA	Department of Investigation of Anomalous Aerial Phenomena
DoD	Department of Defense
EMP	electromagnetic pulse
ETC	extraterrestrial civilization
ETH	extraterrestrial hypothesis
ETI	extraterrestrial intelligence
FOIA	Freedom of Information Act
FTL	faster-than-light
FY	fiscal year
HCM	hypersonic cruise missile
HGV	hypersonic glide vehicle
IC	intelligence community
MDA	Missile Defense Agency
NASA	National Aeronautics and Space Administration
NDAA	National Defense Authorization Act
NORAD	North American Aerospace Defense
NTSB	National Transportation Safety Board
ODNI	Office of the Director of National Intelligence
ONI	Office of Naval Intelligence
SBIRS	Space-Based Infrared System
SDA	Space Development Agency
SDI	Strategic Defense Initiative
SAP	Special Access Program
SOCOM	Special Operations Command
SOF	Special Operations Force
STRATCOM	Strategic Command
UAP	unidentified aerial phenomena
UAPTF	Unidentified Aerial Phenomena Task Force
UFO	unidentified flying object

UN   United Nations
USO   unidentified submerged object

10.1   Identifying the Two Elephants

There are two elephants standing silently, unfairly yoked together high up on a tightrope in the room of astrobiology. Even though many people walk past these two elephants every day, aware of their precarious perch, they often avert our eyes and pretend they are not there. What are they?

One silent elephant is the phenomenon[1], a term I use to describe the combined reality of Unidentified Aerial Phenomena (UAP), or what many people call Unidentified Flying Objects (UFOs), Unidentified Submerged Objects (USOs), and Intelligent Beings. When I capitalize "Intelligent Beings," I am referring to beings that are not human and are not members of intelligent animal species as we commonly know them on Earth. Following the extraterrestrial hypothesis (ETH), one may argue that these Intelligent Beings are an extraterrestrial intelligence (ETI)[2] that represents at least one extraterrestrial civilization (ETC) responsible for the operation of UAP on and around Earth.

If we define astrobiology as Lucas Mix does — "Astrobiology is the scientific study of life in space. It happens when you put together what astronomy, physics, planetary science, geology, chemistry, biology, and a host of other disciplines have to say about life and try to make a single narrative" [10.110] — then why don't astrobiologists study the phenomenon scientifically? That is the first elephant.

The second silent elephant—which is cruelly yoked to the first by the rough-and-ready circus of the U.S. military-industrial complex—is the weaponization of space. It requires the astrobiology ethicist, also referred to as the astroethicist, to weigh in on the relevance of the 1967 United Nations (UN) Outer Space Treaty.[3] This Treaty—formally called the Treaty on Principles Governing the Activities of States in the Exploration and Use of Outer Space, including the Moon and Other Celestial Bodies [10.164]— stresses that celestial locations can be used "exclusively for peaceful purposes" and explicitly prohibits the "placing in orbit around the Earth any

---

[1]A documentary film on the topic is entitled *The Phenomenon* [10.123] [10.107].
[2]For more on the implications of ETI for the academy and society, see [10.11].
[3]Although it was written in 1966, this treaty entered into force on October 10, 1967, which is why it often is referred to as the 1967 Outer Space Treaty.

objects carrying nuclear weapons or any other kinds of weapons of mass destruction" [10.163].

Ted Peters, among others, registers regret that no planetary community of moral deliberation yet has amassed sufficient clout to enforce the 1967 UN Outer Space Treaty [10.122]. Astrobiologists and astroethicists can facilitate the development of such a strong moral community. In the meantime, however, this second elephant remains silently standing. This means that astrobiology—a scientific discipline, not a military one—continues to unfold within the context of the militarizing and weaponizing of space.

Without delay, we must take these two elephants down from the tightrope and onto the bridge of astroethics before they either fall or the tightrope snaps. Further, as the subdiscipline concerned with ethics that complements the science of astrobiology, astroethics has a moral dimension and can help bring the two elephants down to safety. Because it is a human endeavor, science is never completely separate from ethics and morality, and astrobiology is no different than any other scientific discipline in this regard. Furthermore, astroethics also naturally connects scientific discussions to broader, societal ones regarding what human beings understand about space, the cosmos, and the phenomenon. When people engage in such conversations, they draw information from a diversity of sources—e.g., from the science of astrobiology, coverage of rocket launches in the media [10.48] [10.85], the informed perspectives of astronauts,[4] credible witness reports of UAP and USO sightings [10.113], including many in Alaska [10.8], reports of UAP in formations of as many as fourteen up to a stated thirty at one time [10.138], reports of direct,

---

[4]One such astronaut is Helen Sharma [1991, Soyuz TM-11], the first Briton to enter space [10.124]. Like Sharma, Edgar Mitchell also held the opinion that intelligent extraterrestrial beings were operating in and around Earth. Mitchell vocally opposed the militarization and weaponization of both Earth and space, and he suggested that UAP sightings were common near military bases because this ETI was warning human beings about the dangers of militarization. "Let's hope that is exactly what the ETs, extraterrestrials, are trying to show us," he said, "We don't need to be this warlike civilization" [10.120]. In a leaked email to John Podesta, White House Chief of Staff to former President Bill Clinton, Counselor to former President Barack Obama, and campaign chairman for Hillary Clinton, Mitchell wrote, "Remember, our nonviolent ETI [Extraterrestrial Intelligence] from the contiguous universe are helping us bring zero point energy to Earth .... They will not tolerate any forms of military violence on Earth or in space" [10.20]. In fact, many U.S. astronauts are reported to have witnessed UAP, including John Glenn [1962, Mercury-Atlas 6], Gordon Cooper [1963, Mercury-Atlas 9], James McDivitt [1965, Gemini 4], Buzz Aldrin [1966, Gemini 12], and Ron Evans [1972, Apollo 17]. Like Mitchell [1971, Apollo 14], some also have spoken about their sightings publicly [10.120]. Both Mitchell [10.109] and Cooper [10.36] explicitly, on camera have stated that extraterrestrials have visited Earth.

firsthand encounters with Intelligent Beings [10.86] [10.87], television programs [10.151], and movies. For example, one sighting of an unidentified aerial phenomenon is documented on television by Chad Horning, a Co-Executive Producer of a program on sightings and other experiences. Interestingly, in a meta way, Mr. Horning saw an unidentified aerial object when he was on location in Alaska working on a documentary television program on the same topic, an event which he describes during one episode of the series [10.65]. As they reflect upon this abundance of information, astroethicists can ensure that opinions formed occur in the context of sound, moral deliberation.

10.2   The Phenomenon Elephant

The phenomenon elephant is this: although many astrobiologists ask *if* extrater-restrial life exists, significant evidence suggests to witnesses, investigators, and aviation experts—and, increasingly, to many others, including many respected scientists [10.93] [10.143] [10.144]—that UAP and USOs are real. One such group comprises The Galileo Project, at Harvard University, which seeks "to bring the search for extraterrestrial technological signatures of Extraterrestrial Technological Civilizations (ETCs) … to the mainstream of transparent, validated and systematic scientific research" [10.52]. Given the length of time over which such UAP and USOs have been observed, which is decades at the least and, on my view, probably at least thousands of years, it makes logical sense that they represent technology of advanced ETI operating on and around Earth. In other words, many astrobiologists are working to determine if something exists while many other individuals in society have collected significant data showing not only that the phenomenon is real, but, also, that it is already here. As journalist Tom Rogan succinctly states, "…the credible witness statements are numerous, matched to data recordings, and associated with a far larger set of similar incidents currently not public" [10.133].

In 1950, Enrico Fermi is said to have posed the question, "Where is everybody?" [10.70]. Given the size of the cosmos, among other possible reasons, Fermi expected that our skies would be filled with extraterrestrial emissaries. But Fermi could not see them—even though many others could. Does this mean ETI, or extraterrestrial intelligence, is not here, because certain scientists have not had personal sightings themselves? When one phrases the question this way, the logical answer is, "No." Indeed, not long after Fermi made his comment, many people witnessed multiple UAP over Washington, D.C., in what "was known as the Washington UFO flap" of 1952 [10.25] or "the Washington, D.C. flap" [10.3]. A full decade before that, an estimated fifteen UAP appeared over Los Angeles, only to be shot at by U.S. military gunners [10.56]. Given these sightings together with almost complete certainty that no country possessed advanced propulsion in the 1940s and 1950s of the type witnessed, it makes logical sense to conclude that at least some UAP are extraterrestrial in origin.

However, many scientists have followed Fermi's lead in understanding the issue of UAP in the way that Fermi did [10.66], despite centuries of sightings of UAP by reliable witnesses [10.166] [10.167]. Probably because of Fermi's significant status in the scientific community, his quandary—which really was nothing more than a casual comment—attained the status of paradox, becoming known as "the Fermi Paradox." As Eric Korpala astutely notes, however, the Fermi Paradox "is neither a paradox nor a concept that originated with Fermi." It is only speculation. Korpala is impatient with speculation. The definition of success in science is not "Did I find what I was looking for?" but merely "Do I understand more than I did yesterday?" [10.77].

If astrobiologists would take the time to examine UAP data, they would understand more than they did yesterday. In fact, they likely would understand more than they ever have before.

The U.S. government is paying serious attention to UAP. The U.S. Department of Defense (DoD), often referred to as the Pentagon [10.19] [10.21] [10.34] [10.37] [10.59], issued a press release on August 14, 2020 that states: "On August 4, 2020, Deputy Secretary of Defense David L. Norquist approved the establishment of an Unidentified Aerial Phenomena (UAP) Task Force (UAPTF). The Department of the Navy, under the cognizance of the Office of the Under Secretary of Defense for Intelligence and Security, will lead the UAPTF" [10.156]. A former Pentagon official recently implied that the U.S. government possesses "stark visual documentation" of UAP [10.82].

While they served as U.S. Secretary of Defense and Director of National Intelligence, respectively to Mark Esper and John Ratcliffe, received briefing on UAP. Such briefings also have been given to the Senate Select Committee on Intelligence, the Senate Armed Services Committee, and several members of the Joint Chiefs of Staff. Government officials in Japan also state that they discussed the topic during a meeting with Mark Esper in Guam. A former Pentagon official stated, "We would not have moved forward without briefing close allies. This was bigger than the U.S. government" [10.82]. This constitutes a relatively rapid reversal of decades of denial regarding the reality of UAP (then called UFOs and USOs) among

U.S. government officials. In this direction, Joseph Gradisher, spokesperson for the Office of the Deputy Chief of Naval Operations for Information Warfare at DoD, recently stated that UAP have entered airspace designated for the military as often as multiple times per month. The *Washington Post* states that this "recent uptick of sightings of unidentified flying objects," or UAP, is what "prompted the Navy to draft formal procedures for pilots to document encounters" [10.121].

Prior to the Pentagon's action in creating a task force to study the phenomenon, on June 17, 2020, U.S. Senator and former Republican presidential candidate Marco Rubio, from the United States Senate Select Committee on Intelligence, submitted the Intelligence Authorization Act for FY 2021. In this, the annual intelligence authorization bill, the Committee included the provision that a public report be prepared regarding a "detailed analysis of unidentified aerial phenomena data and intelligence reporting collected...by the Office of Naval Intelligence (ONI), geospatial intelligence, signals intelligence, human intelligence, measurement and signals intelligence, and the FBI" [10.2] [10.43]. The Act is part of the recently signed $2.3 trillion spending bill, which provided DoD and the intelligence agencies in the United States with 180 days to produce a public report on what they know about UAP [10.38].

After the Senate Intelligence Committee submitted the Intelligence Authorization Act to the full Senate, on June 17, 2020, [10.77] Rubio made this comment during an interview with a CBS affiliate in Miami: "Frankly, if it's something from outside this planet, that might actually be better than the fact that we've seen some technological leap on behalf of the Chinese or the Russians or other adversary that allows them to conduct this sort of activity" [10.134]. Rubio continues to discuss the phenomenon [10.45] [10.82].

The Office of the Director of National Intelligence (ODNI) in the U.S. delivered its 'Preliminary Assessment: Unidentified Aerial Phenomena' to Congress on June 25, 2021. The unclassified portion of this report was posted online. With reference to UAP reports from 2004 to 2021, the report observes, "Most of the UAP reported probably do represent physical objects given that a majority of UAP were registered across multiple sensors, to include radar, infrared, electro-optical, weapon seekers, and visual observation." It also reports that some UAP "reportedly appeared to exhibit unusual flight characteristics" (p. 3). Further, certain UAP demonstrated what appears to be advanced technology, such that in 18 incidents described over 21 reports, "observers reported unusual UAP movement patterns or flight characteristics," such as remaining stationary in winds aloft, moving against the wind, maneuvering abruptly or at considerable speeds, all "without discernable means of propulsion." Also, some UAP demonstrate "accelerator of a degree of signature management" (p. 5) [10.119].

In addition to Rubio, other former and current U.S. senators also have shown interest in the phenomenon. Harry Reid is very involved in discussion of the phenomenon [10.17] [10.37] [10.123] [10.107] [10.128] [10.129] and has indicated that he thinks some UAP may not be of human origin [10.128] [10.129]. This position appears to be shared by John O. Brennan, former director of the Central Intelligence Agency (CIA), who states, "I think some of the phenomenon we are seeing is unexplained and we don't yet understand. It may be some type of activity that constitutes a different form of life" [10.22] [10.82]. Reid requested the support of Daniel Inouye and Ted Stevens to secure funding to study the phenomenon [10.37] [10.82] [10.94] [10.150]. On July 30, 2009, a prior organization, Bigelow Aerospace Advanced Space Studies (BAASS), issued a 494-page *Ten Month Report*, and monthly reports to the Pentagon and annual program updates to the DIA regarding the phenomenon [10.82] [10.94].

John McCain and Jack Reed [10.74], Mike Gravel [10.58], Mark Warner [10.16] [10.82] [10.94] [10.172], and other lawmakers [10.16] also have shown interest in the phenomenon. North American Aerospace Defense Command (NORAD) has tracked UAP [10.98] [10.135], including one instance in 2008 during which a squadron of fighter jets and subsequent alert fighters were scrambled after an unidentified aerial object that they were unable to locate [10.31]. In addition, an unidentified aerial object was detected in Washington, D.C. on November 26, 2019, causing the White House and Capitol buildings to be put on lockdown and fighter jets to be scrambled [10.88]. CNN's flighty cover story that a flock of birds caused this mighty hullabaloo [10.32] systematically and thoroughly was deconstructed by reporter Tom Rogan [10.132]. Even *The New York Times* has published articles discussing the role of the ONI in studying UAP [10.18] [10.19] and has commented on "the possible existence of retrieved materials" from UAP [10.18].

Although general reporting on U.S. government interest in the phenomenon often discusses the idea that UAP may be examples of advances in human technology, one former, high-ranking U.S. government employee and prominent member of the U.S. national security apparatus, Christopher K. Mellon, has made a number of clear statements indicating that he does not think UAP are human in origin [10.97] [10.105] [10.107], an opinion that is not contradicted by Mellon when he interviews Reid [10.102]. Mellon is a former U.S. Deputy Assistant Secretary of Defense for Intelligence and later for Security and Information Operations at the Pentagon, former Staff Director for the U.S. Senate Select Committee on Intelligence, and also the great-grandson of Gulf Oil co-founder William Larimer Mellon [10.173].

Mellon states: "There is a profound mystery here. Are we alone in the universe?" [10.101]. Mellon also makes this direct remark: "Every source that I've gone to and spoken with—uh, in the aerospace industry, in the Pentagon—has denied that we have a TR3B or anything matching the performance, uh, even remotely, of what's being observed" [10.104]. With respect to UAP seen by military witnesses, Mellon also states that with respect to the U.S. military: "We do not operate test aircraft in the vicinity of carrier battle groups without some prior coordination. It would be completely uncharacteristic of the military to operate in such a strange, haphazard manner ... The evidence we have suggests these are not U.S. vehicles" [10.101]. Mellon also states that the U.S. has designated test sites and does not need to use the airspace over civilian farms and ranches to test secret technology [10.103], and (in reference to the USS *Nimitz* incidents) involving UAP, that the U.S. military would not "blindside the commander of the carrier battle group" [10.105]. In other fora, Mellon reiterates his view that the truly unexplained UAP are not advanced U.S. craft [10.97] [10.103] [10.105]. Furthermore, Mellon states that the U.S. has a global intelligence collection apparatus from 22,500 miles [10.96] [10.102], "to virtually to the deepest levels of the ocean" [10.102]. This means that the U.S. military already has considerable data on UAP [10.96] [10.102]. After the unclassified portion of ODNI's 'Preliminary Assessment' was published on June 25, 2021, Mellon stated that the government still may be withholding some data: "It is far from clear how much information [the Pentagon task force] could elicit from organizations led by 4-star flag officers and agency directors." He added, "Undoubtedly some important information was not shared, potentially for a variety of reasons. Congress should inquire about that" [10.15].

More recently, Mellon has published two blog posts explicitly stating his position that the ETH is a valid possible explanation for certain UAP [10.99] [10.100]—in the most recent of these two posts, he writes, "non-human origin is presently the theory that best fits all the facts" [10.100]. As stated above, my view is that the ETH is the most plausible explanation for what has been and is being observed with respect to various UAP. I differ with Mr. Mellon in as much as I do not view this either as a national security threat to the nation-states of the world nor as an existential threat to humanity. I have a much more optimistic view regarding the intentions and actions of the ETI responsible for UAP operating on and around Earth.

In addition to in the U.S., developments regarding UAP also are occurring in other countries. Very soon after DoD's announcement on August 14, 2020, that it had set up a new task force to study UAP on August 4, 2020 [10.156], an English-language media outlet ran the story that Russian cosmonaut Ivan Vagner had captured a one-minute video of a UAP when he was aboard the International Space Station (ISS). This was confirmed by Russia's space agency, Roscosmos. Furthermore, "Vagner has reported that it [i.e., the video] has been submitted to Roscosmos experts for review," and Russian media outlet TASS confirmed that "[t]he video is being examined at the Space Research Institute of the Russian Academy of Sciences" [10.29].

Many countries actively investigate UAP. These include Brazil [10.14] [10.168], Canada [10.83] [10.84], France [10.27] [10.54], the United Kingdom (U.K.) [10.115] [10.118], Perú [10.127],[5] and Uruguay [10.51] [10.69], among many others. Also, "[t]he Canadian government hosts a publicly searchable archive of government records about UFOs dating back to the 1950s" [10.83]. Even the U.S. National Security Agency (NSA) has released historical files on UFOs/USOs/UAP following Freedom of Information Act (FOIA) requests and made them publicly available online in the "UFO Documents Index" [10.161]. Comité de Estudios de Fenómenos Aéreos Anómalos (CEFAA, Committee for Studies of Anomalous Aerial Phenomena), under the auspices of the Civil Aeronautics General Administration in the Republic of Chile, also actively studies the phenomenon [10.33], as does Centro Ufologico Nazionale in Italy [10.40]. Many countries possess "either declassified and published U.F.O. files," including France, "the U.K., Denmark, Brazil, Russian, and Sweden," while some countries "have formed their own official organizations dedicated to the issue," such as Chile, mentioned above, and, also, Perú [10.82].

If so many governments, organizations, and military officials take UAP seriously, why do the majority of astrobiologists forgo the topic? It is difficult to say, although there is good reason for astrobiologists to examine the video data of UAP recently released by the Pentagon [10.34]. Many

---

[5]The Peruvian government's Department of Investigation of Anomalous Aerial Phenomena (DIFAA), a branch of the Peruvian Air Force (FAP), reopened in 2013 [10.127]. The FAP recently confirmed a sighting near the Jorge Chávez International Airport in Lima [10.111].

aviation experts have been called in to do so, and they are unable to explain it away as human technology.[6]

Astrobiologists can follow the good, scientific lead of aviation experts and can analyze UAP data as they do any data, by examining it rigorously rather than by eschewing it out of hand. The phenomenon invites astrobiologists and others to inquire into whether the ETH to explain UAP and USOs can be confirmed or not. In fact, the ETH has traction in many circles. Indeed, Comité d'Études Approfondies (COMETA), a former organization comprised of retired French generals, scientists, and space experts, among others, concluded that the ETH "was the most logical explanation" for what was being observed. U.S. Air Force Project Sign investigators also "leaned in favor of the extraterrestrial hypothesis" in their top-secret "Estimate of the Situation" memorandum [10.82].

Given the size of the cosmos and the rapid discovery of exoplanets [10.46], including hundreds of millions of potentially habitable exoplanets in the Milky Way galaxy alone, together with the fact that faster-than-light (FTL) travel [sometimes referred to as superluminal travel] is

---

[6]One such expert is Steve Justice, former Director of Advanced Systems Development at Lockheed Martin Skunk Works [10.72]. Justice specifically compares the UAP seen in the Pentagon video to the performance of the X-43A scramjet, an unmanned hypersonic aircraft. Justice explains how the UAP exhibits more sophisticated technology than the X-43A [10.73]. Another aviation expert who analyzes the UAP videos is Dr. Travis Taylor [10.145] [10.146]. Taylor is highly credentialed scientifically, holding six university degrees: B.E.E. Electrical Engineering; M.S. Physics; Ph.D. Optical Science and Engineering; M.S.E. Mechanical and Aerospace Engineering; M.A. Astronomy; and Ph.D. Aerospace Systems Engineering [10.174]. For the past two decades, Dr. Taylor has worked on various programs for the U.S. DoD and the National Aeronautics and Space Administration (NASA) [10.126]. Taylor compares the Pentagon videos to footage of a UAP taken at Skinwalker Ranch in Utah, observing that in both cases, one sees evidence of a temperature differential such that the air around the craft is colder than the surrounding air [10.146]. Retired Lt. Col. Christopher Cooke, who spent his career as a pilot flying sophisticated aircraft, and Captain Ross ("Rusty") Aimer, who flew commercial planes for over four decades and who works as a consultant for the National Transportation Safety Board (NTSB) [10.4], also analyze the videos during the first episode of the television program *Unidentified: Inside America's UFO Investigation*. Cooke states, "I have no idea what that thing is, and it's not acting like a—" Aimer cuts in here and says, "aircraft." "It's not acting like anything I've ever seen," Cooke continues [10.35]. "Yeah," Aimer agrees. [10.6]. In another television program *Black Files Declassified*, Aimer states that the video footage is authentic, has been checked multiple times, and shows an object that defies conventional human understanding of physics with respect to thermodynamics, aerodynamics, and gravity. Given that the object has no flight surfaces and shows no heat signature indicating a source of propulsion, Aimer's conclusion is that it is not an example of human technology [10.5].

theoretically possible [10.140] [10.82]. Given recent evidence regarding liquid water on Mars [10.92] [10.10], thereby making it plausible that Mars could support life, it should come as no surprise that an ETI already is operating on and around Earth.

Even though it is clear that the phenomenon is not human, what still remains unclear whether it is "extraterrestrial" in the strict sense of the term. Nevertheless, while Intelligent Beings could have evolved on Earth, even alongside *Homo sapiens* for some period of time, the sophisticated propulsion technology demonstrated by UAP and USOs suggests to me that the Intelligent Beings responsible for the phenomenon are, in fact, extraterrestrial. This ETI most likely represents one or more ETCs.

The reality of the phenomenon presents astrobiology with an interesting fork in its disciplinary journey. Either the field simply can punt on exploring data relating to the phenomenon and can change its mandate to be the detection of—*per se*—extrasolar civilizations; or astrobiologists can say "Hello" to this enormous elephant in their midst and bring their considerable scientific expertise to bear in helping to understand it.

Having astrobiologists engaged in the study of UAP and USOs is they right path forward. Astrobiologists can continue to work on the detection of extraterrestrial life 'out there' at the same time that they can collect and consider quantitative and qualitative data and witness and experiencer reports any ETI already present on and around Earth. After all, the ETI responsible for the operation of UAP and USOs on and around Earth may represent only one ETC among myriad more from this vast cosmos we call home that the human species can look forward to meeting.

10.3   The Weaponization Elephant

While we safely may have taken one elephant down from the tightrope by paying long overdue attention to the phenomenon, there is a weightier elephant still left up there, dangling precipitously on the tightrope above. This second elephant is the weaponization of space.

On January 10, 1963, scientists in the U.S. asked the government to make the search for extraterrestrial life "the principal scientific objective of the nation's space program" [10.116]. This second elephant is the weaponization of space. In fact, there is evidence that this vision linking the U.S. space program to extraterrestrial life may in fact have been followed—just not publicly, and not precisely in those terms.

The U.S. government has possessed significant evidence regarding the reality of the phenomenon since at least the 1950s. Former President Harry S. Truman stated so directly on camera, in response to a question from a reporter, "Did the Joint Chiefs of Staff talk to you or concern you about the unknown or the unidentified flying objects? I'm speaking of the flying saucers." Truman replied, "Oh yes, we discussed it at every conference that we had with the military and they never had [*sic.*] been, never were able to make me a concrete report" [10.25].[7] Many government and military persons have gone on the record about the phenomenon [10.147], Truman also stated, "I can assure you that flying saucers, given that they exist, are not constructed by any power on Earth" [10.154] [10.147].

The historical implication of Truman's statement is that knowledge of the phenomenon has been caught up with—if not a reason for—the U.S. government's misguided actions to militarize and weaponize space. Said more directly, part of the impetus for weaponizing space appears to be projective and paranoid fear of the phenomenon—even though there is no evidence that the ETI responsible for UAP and USOs are a threat. To the contrary, the "threat" narrative is projective, since human beings constitute a threat to themselves, to others, and to the environment. For example, there is evidence that the ETI operating on and around Earth now may be quite concerned about human nuclear weapons and especially any use of nuclear weapons in space—a usage that they may be taking steps to forestall [10.152] [10.64] [10.139]]10.30] [10.12] [10.153] [10.135] [10.47].

The BAASS report mentioned above discusses "Project Northern Tier," which evidently involves "instances where dozens of UFOs flew over restricted airspaces of facilities housing nuclear weapons," and it mentions a "football field-sized disc" illuminating a "missile site," under "(Clear Intent, 27)" [10.94]. Mellon also discusses the connection between UAP and the Northern Tier [10.106], as does journalist Tim McMillan [10.94] [10.95].

---

[7]This source originally was available at https://www.amazon.com/UFOs-White-House-Files-Season/dp/B082FQ476D and appears to remain available to watch for some or all of those who purchased it on or prior to June 23, 2020. As of May 1, 2021, however, it appears to be unavailable to watch for at least some U.S. customers, with Amazon.com providing the following message, "This video is currently unavailable to watch in your location."

It is well known that over the course of decades, the U.S. government has pursued an aggressive agenda to militarize and weaponize space—part and parcel of its overall policy of American exceptionalism and concomitant projection of power [10.28] in pursuit of what is referred to frequently as 'full spectrum dominance.' Furthermore, there is evidence that because the new generation now deciding whether or not there should be military conflict in space experienced neither World War II nor the Cold War, certain individuals who are members of this generation do not view a "sanctuary approach" to space, such as that embodied by the 1967 UN Treaty, as particularly persuasive [10.61].

Unfortunately, inconsistencies in space law permit the militarization and weaponization of space to proceed forward despite the existence of the 1967 UN Treaty. Loopholes in space law are so extensive that governments and private companies are able to pursue their own agendas with respect to space regardless of how self-interested and/or ruthlessly aggressive they may be, with almost no checks whatsoever on these activities. This topic is encyclopedic, so I refer readers to an excellent article by Y. Zhao in *Oxford Research Encyclopedias* [10.180] and also to an excellent article published by H.T. Kim in Korea [10.75]. Aggressive actions by vested interests that push militarism and weapons development and deployment result in a world in which the scientific disciplines of astrobiology is compelled to contend with concrete actions taken on a daily basis to weaponize space, even if it may prefer to leave this heavy and difficult elephant on a tightrope in midair.

10.4   U.S. Government Spending on Weapons for Space

The U.S. military and intelligence community (IC) spend billions of dollars per year on the militarization and weaponization of space. Other countries such as China, Russia, France, etc., also invest heavily in space-based weapons systems, as do private companies such as defense contractors, and, also, ultrarich individuals who increasingly possess multinational footprints of their own. Unfortunately, international treaties and conventions provide inadequate guidance to hold back the current weaponization of space [10.89]. Such spending egregiously wastes resources that should be allocated towards moral, logical, and humanitarian ends.

The U.S. Air Force Space Command operated for years before organizational changes within DoD on December 20, 2019 elevated it into an independent branch of the military called the U.S. Space Force [10.177].[8]

---

[8]For more on the U.S. Space Force, see [10.142].

Legislatively, the U.S. Space Force was created by the National Defense Authorization Act (NDAA) for fiscal year (FY) 2020 [10.24] [10.57]. FY 2020 covers the time period from October 1, 2019 to September 30, 2020. The portion of the FY 2020 NDAA relating to the creation of the U.S. Space Force is "Subtitle C—Space Matters," 919-937, which falls within Title IX—Department of Defense Organization and Management. According to "§ 9091. Establishment of the Space Corp," "(a) ESTABLISHMENT.— There is established a United States Space Corps as an armed force within the Department of the Air Force" [10.1].

The U.S. Space Force is the first new armed service created in the U.S. since 1947 [10.53]. The plan put forth for the U.S. Space Force before it was created bears similarities to DoD's reorganization of its Special Operations Force (SOF), [10.50] which is part of the U.S. Special Operations Command (SOCOM) [10.44] [10.60].

Because the U.S. Space Force now is an independent branch of the U.S. Armed Forces, and, as such, can receive funds as line item allocations in the NDAA, more money can be spent on the militarization and weaponization of space. Trying to hone in on official DoD spending for the U.S. Space Force is not straightforward, however. The FY 2020 NDAA Executive Summary states, "The NDAA supports a total of $750 billion in fiscal year 2020 funding for national defense" (p. 5) [10.162]. Furthermore, news outlets carried the story that "only" $40 million (a drop in the bucket in the ocean of DoD expenditures) was being allocated to the U.S. Space Force in its initial year [10.24], as per the 2020 NDAA [10.142].

Publicly, the fanfare that attended the announcement of the formation of the U.S. Space Force—even though it is mostly an administrative change, as noted above—would have people think that the militarization and weaponization of space is only getting started. In reality, however, a massive proposed increase in expenditures for the U.S. Space Force from 2020 to 2021—from $40 million to a little over $15 billion [10.159]—indicates that the U.S. government's plans to weaponize space have been in the planning and implementation stages for decades that currently are accelerating.

Furthermore, whatever public documents state regarding the funding the U.S. Space Force do not tell the entire story. The so-called real money to weaponize space—by which I mean the big money—is hidden in what is referred to as "the black budget." In certain instances, such covert programs are referred to as Special Access Programs (SAPs) [10.91] [10.154], which often are outside the oversight of elected representatives.

Black budget, or "black files," funding to militarize and weaponize space has been part of the surreptitious U.S. military and IC landscape for decades. Recently declassified documents indicate that the U.S. military's program to weaponize space began in the 1960s [10.13]. Prior to the announcement on December 20, 2019 of the formation of the U.S. Space Force [10.177], declassified documents show that black budget spending for a space-based military program and "top-secret military research dedicated to the weaponization of space" already was an estimated $10 billion per year, with one top-secret mission already in orbit [10.13]. An article from *The New York Times* also reveals that there was a "coast-to-coast space complex" secretly operating in 1989 or earlier [10.23]. Said bluntly, what the NDAA reports is being spent on the militarization and weaponization of space is only the tip of the iceberg, with the real amount undoubtedly being much higher.

Space weapons programs have a long a history in the U.S., and the systems have become increasingly technologically sophisticated over time. When one factors in spending from the $90 billion per year that comprises clandestine government programs in the U.S. referred to collectively as "the black budget," U.S. Space Force black budget spending already may be around $10 billion per year [10.13].

On what is all of this money being spent? Although the scope of this topic is too broad for treatment in this chapter, I will mention a few examples here.

The Arms Control Association provides information on official U.S. government spending for what commonly is referred to as the U.S. missile "defense" program. The word "defense" could not be more misleading. Keep in mind that while the Arms Control Association cites official dollar amounts, the public is not being appraised regarding how much more is being allocated to this and to similar weapons systems in the black budget. Nevertheless, the Association's website states that for fiscal years 1985 to 2019, the Missile Defense Agency (MDA) estimates that $200 billion was appropriated for the Agency's programs, which "does not include spending by the military services on programs such as the Patriot system or the many additional tens of billions of dollars spent since work on anti-missile systems first began in the 1950s." Further, "the fiscal year 2020 request seeks $380 million over the next five years to develop and test by 2023 a prototype space-based laser weapon to destroy ICBMs during their boost and midcourse phases of flight." With respect to Space-Based Infrared System (SBIRS) High, "The most recent sensor, GEO-4, was launched aboard an Atlas V rocket on January 19, 2018." Now, "Lockheed Martin is under contract to produce GEO-5 and GEO-6, which will be launched in 2021 and 2022, respectively" [10.130].

In December 2020, the U.S. Space Force received the fifth satellite for its SBIS constellation, the SBIRS GEO-5, which was delivered to it from Lockheed Martin. This "is the first satellite in the constellation to be made with the company's LM 2100 bus, which is the new upgraded common platform for the system" [10.79]. Like other weapons systems, it is an example of a horrific application of human ingenuity to potentially lethal rather than life-enhancing ends.

In fact, the U.S. is and has been developing many different weapons for space. A partial list includes: anti-satellite (ASAT) weapons capable of destroying other satellites; whatever military and intelligence technologies were lofted into space by the space shuttle; technologies developed in conjunction with the Strategic Defense Initiative (SDI), which morphed into the MDA, including airborne lasers and other high-energy physics-based technologies; and the X-37B.[9]

What the X-37B is doing in space has not been made fully public [10.55]. Speculatively, it may be deploying satellites, setting up a Star Wars-like defense system, putting the elements of a U.S. Space Force into place, working on weapons and propulsion drives, carrying out missions for the U.S. Space Force, and, given that it is a reusable platform, transporting some form of technology into space. Indeed, the very existence of the X-37B in space indicates that the U.S. Space Force already was flying some time ago [10.13].[10]

As we consider current developments in U.S. space policy, Noam Chomsky's warning from almost twenty years ago regarding the U.S. agenda to use space offensively for military purposes to serve U.S. interests in its hegemonic pursuit of global dominance is instructive [10.28]. As observed by Patrick Tucker, Technology Editor, *Defense One*, the X-37B permits the U.S. military to create a persistent presence in space. The X-37B can be armed with lasers, missiles, and a small crew consisting of, or including, an advance strike team. Tucker states that since the X-37B is in constant orbit, it could "strike anywhere in the world at a moment's notice." Tucker observes further, "...but if you have a persistent presence in space potentially such a plane could carry a small crew, including, perhaps, an advanced [*sic.*] strike team, that could be put down, on Earth, at any particular spot, at a moment's notice....A plane such as that would have that potential use" [10.149].

---

[9]For more on the X-37B, see [10.171].
[10]For a discussion on the role of the X-37B in on the weaponization of space, see [10.55].

Tucker also discusses directed energy weapons (DEWs) such as high-tech lasers and neutral particle beams, the latter of which work with the electromagnetic spectrum, and he mentions that the U.S. military may have plans to put a laser on a F-35 stealth fighter [10.148] [10.149]. The U.S. recently tested a DEW that can destroy aircraft mid-flight [10.81]. It also should be mentioned that France is spending $4 billion to weaponize space, which includes arming ASATs with lasers and machine guns. The U.S. military may have similar programs [10.13].

Electromagnetics also can be weaponized [10.176], and electromagnetic pulses (EMPs) can be used in space. Starfish Prime is an example of a nuclear test conducted in outer space that "revealed the destructive impact of the Electromagnetic Pulse (EMP) produced by a nuclear explosion" [10.39].

Hypersonics are another major area of U.S. government spending on weapons for use in space.[11] There currently is a hypersonics race between the U.S., China, and Russia for unmanned hypersonic missiles, though there also are plans to try to create manned hypersonic craft. The official DoD annual budget for hypersonics was $100 million when the program was starting out, but it has jumped to $2.6 billion in recent years [10.9].

There are two types of missiles in the U.S. hypersonic missile development program, hypersonic glide vehicles (HGVs), which are "maneuverable glide vehicle[s] capable of speeds greater than Mach 5"; and hypersonic cruise missiles (HCMs), "[c]apable of sustained, powered and maneuvering hypersonic flight." According to George Nacouzi, Senior Engineer at RAND Corporation, HCMs travel at extremely high speeds—Mach 5, 6, or even 10 or so—they are intensely destructive, and they are almost impossible to defend against. They are so accurate and move so quickly that they do not need to be armed with explosives, since the vehicles themselves have so much kinetic energy that they cause significant damage when they hit their targets [10.114].

---

[11]Hypersonics are discussed in depth in episodes 1 and 2 of *Black Files Declassified* [10.9] [10.136].

10.5   The Military-Industrial Complex Operates Under Euphemisms Citing "Government-Industry" Linkages

The weaponization of space is literally unfolding under the noses of scientists and members of the broader public. This occurs in the context of what today are referred to euphemistically as "government-industry alliances" and "government-industry initiatives." In his Farewell Address of January 17, 1961, then President Dwight D. Eisenhower referred to these relationships explicitly as "the military industrial complex" [10.90].[12] Although Eisenhower warned against the dangers of this nexus of power, not only does it persist, but it has grown significantly since the 1960s. It is the main driving force pushing forward the U.S. government's aggressive space policy.[13]

The euphemisms of government-industry "alliances" and "initiatives" describes those relationships often but not exclusively forged within the Beltway, i.e., at gatherings in Washington, D.C. itself and within the I-495 ring around D.C. encompassing parts of Maryland and also the opulent yet traditionally understated area known as NOVA (Northern Virginia). What may surprise some readers is that such government-civilian partnerships do not refer only to partnerships between the U.S. military and defense contractors such as Lockheed Martin and Boeing, but they also refer to alliances between DoD and companies such as SpaceX, Virgin Orbit, and Blue Origin.

For those who are unfamiliar with how industry and government come together to further mutual interests such as the weaponization of space, I describe here one example of how this occurs. Although I did not attend this event, I attended one very much like it. Ironically, the settings for such events are similar to academic conferences, except that there is a tremendous amount of money involved and the people are discussing how to create lethal weapons systems rather than how to further the state of knowledge on behalf of humankind.

On December 3, 2019, some few weeks prior to the legislative creation of the U.S. Space Force later that month, a government-industry meeting occurred at the U.S. Chamber of Commerce in Washington, D.C. It was called "LAUNCH: The Space Economy." C-SPAN3 broadcast this conference on January 6, 2020, under the banner, "Space Policy & Commerce."

---

[12]For an excellent historical contextualization of Eisenhower's speech, see [10.165].
[13]For an overview of space policy in the U.S., see [10.141].

Many hours of talks from this meeting are available online [10.155]. Here, I describe talks by two speakers at the meeting—one a government employee, and one a private sector employee—to demonstrate how business is conducted to support the U.S. government's aggressive militarism.

At LAUNCH, Mike Griffin, Under Secretary of Defense for Research and Engineering [10.158], represented one example of the government aspect of so-called government-industry initiatives. Griffin spoke of blending the "commercial space" perspective with the "national security space" perspective, thereby linking commercialization and weaponization of space together. Using the classic language of the military-industrial complex, he discussed a "synergy" between these two arenas and his view that the "commercial demand" for space is insufficient to maintain the size of the "industrial base" in space needed by government and the national security apparatus. According to DoD, space is no longer an uncontested operational domain, since China and Russia are investing heavily in space to overcome the U.S.'s current strategic advantage there. In this regard, Griffin specifically mentioned the "hypersonic glide body threats" that China [10.178] and Russia [10.108] are developing. As stated above, the U.S. is now building its own capacity in this area [10.71].

According to an article from the U.S. DoD website, the Space Development Agency (SDA) was established with the responsibility of "unifying and integrating DOD's space development efforts, monitoring the department's threat-driven future space architecture and accelerating the fielding of new military space capabilities necessary to ensure U.S. technological and military advantages in space" [10.160]. The first director of the SDA was named on October 28, 2019. The establishment of this agency is described in the "Final Report on Organizational and Management Structure for the National Security Space Components of the Department of Defense" [10.157].[14]

Griffin stated that the SDA will work outside of DoD's existing yet relatively slow acquisition system in order to move more quickly with weapons acquisitions in the context of developments occurring in private industry. Griffin emphasized the pursuit of U.S. technical dominance in space, DoD's perceived need to engage commercial sector contractors to help develop proliferated space architectures, how to improve surveillance capabilities, and how systems that fly above and below U.S. current capabilities present, in his and DoD's view, a hypersonic threat. Griffin stated that DoD's goal is to obtain necessary surveillance without creating "juicy targets" for the U.S.'s adversaries. Here, he quoted General John Hyten, commander of U.S.

---

[14]One can read more on the SDA at [10.21].

Strategic Command (STRATCOM), who stated, "I will not support buying big satellites that make juicy targets" [10.44].

Also speaking at LAUNCH but representing the private sector side of the coin, Charlie Precourt, Propulsions Systems Division Vice President and General Manager, Northrop Grumman, spoke about the OmegA space launch offering. Specifically, "the U.S. Air Force awarded Northrop Grumman a $792 million Launch Services Agreement to complete detailed design and verification of the OmegA launch vehicle and launch sites" [10.117]. In his talk, Precourt emphasized industry-government collaboration, and he contextualized his remarks in language conjoining "national security" and "space launch." Precourt discussed how the "Air Force mission" is first, that Northrop Grumman had a "national security focus," and how the OmegA program was working with a 2021 launch schedule. After launch, the stated intention is that the program increasingly will focus on commercial industry applications. Although he works in private industry now, Precourt was listed at the event as "Former Astronaut and Program Manager at NASA [National Aeronautics and Space Administration], 1991-2005" [10.155].

In what is a typical "revolving door" manner, government employees such as Precourt regularly move from agencies such as NASA to major defense contractors. Although most people think of NASA in friendly terms, certain of NASA's activities[15] have ties to U.S. military and IC "black programs." This history is well documented by J.E. David, a curator at NASA's Division of Space History, who has compiled documents on behalf of the National Security Archive at George Washington University [10.41]. This makes it all but certain that a portion of NASA's budget comes from black budgets. Indeed, NASA maintains a close relationship with defense contractors such as Northrop Grumman. For example, in August 2018, NASA's Marshall Space Flight Center awarded a Northrop Grumman employee its Outstanding Public Leadership Medal [10.170].

Civilian-military partnerships, with their long history in the U.S., are at the forefront of the development of the newest technologies intended for space. Because so much money stands to be made by private companies with launch capabilities, it comes as no surprise that certain ultrarich individuals such as Jeff Bezos, Richard Branson, and Elon Musk increasingly are developing such civilian-military partnerships by joining the ranks of defense contractors. These individuals stand to make huge profits from the development of high-tech space weapons.

---

[15]NASA's missions are described on its website [10.175].

The grey zone that characterizes civilian-military partnerships is mirrored by another ambiguous situation, namely that involving "dual-use" technology. Such technology has both civilian and also military intended applications. Systems in this category include new small-platform satellites, anti-satellite (ASAT) weapon systems, etc., both of which can be defensive and/or offensive.

Generally, because of dual-use technology and also because of the physics of space, defining what constitutes a space weapon is difficult. According to Sa'id Mosteshar:

> There is no general agreement on the definition of "space weapon" (Waldrop, 2004B, p. 329). The dynamics of space and the laws of physics render virtually any object in space a potential weapon. Also, space systems are increasingly designed or deployed with dual-use capability, and many are used for both civilian and military purposes ([10.112], citing to [10.169]).

In an episode of *Black Files Declassified*, Dan Hart, President & CEO of Virgin Orbit, describes his company as "Federal Express to orbit." Although he says he will not discuss publicly precisely what military cargo the company plans to loft—or already has lofted—into space, Mr. Hart does admit that this includes an array of new government and civilian satellites, including nanosatellites. He states, "Our mission is to put into orbit a whole new array of small satellites that are being developed commercially as well as by governments." Usage of the plural "governments" is telling, since it shows that companies such as Virgin Orbit are being contracted by governments of more than one country. To loft small satellite platforms into space, Virgin Orbit is trying to create an orbital rocket. The intention is that this small rocket be put under the wing of a 747, taken to 35,000 feet, and dropped into orbit [10.62]. However, the company's first LauncherOne flight was unsuccessful because of engine shutdown resulting from a break of a propellant line [10.49].

Microsatellites [10.26] and nanosatellites, i.e., those weighing less than 10 kilograms, allow companies of all types and sizes to access the benefits of satellites previously accessed by large companies and/or by space agencies with significant financial resources [10.7]. There are many potential uses for smaller platforms such as microsatellites and nanosatellites. These include for communications, to inspect other satellites, and for spy craft. Military interest is high. Hart states:

> ...these new satellites provide a whole different operating strategy for national security. The idea of getting to space quickly, getting to space from anywhere so that you're not just sitting in one place and launching satellites, and being able to affordably get mission capabilities up or quickly replace mission capabilities if they're disabled is of huge importance to national security [10.62].

Unfortunately, human beings are using these technologies for military purposes instead of maximizing the potential of the technologies for humanitarian and social justice objectives.

Like Richard Branson's Virgin Orbit, Jeff Bezos's Blue Origin recently joined the ranks of defense contractors. Three companies "won a $2.3 billion Air Force contract in 2018," with Blue Origin being "awarded $500 million to produce its New Glenn rocket system, a fully reusable rocket intended to carry both passengers and cargo to Earth's orbit" [10.63]. As reported by the *Financial Times*, Blue Origin was named an "essential business" by the U.S. government, which permitted it to continue to operate during the COVID-19 pandemic because of what DoD and Homeland Security state is "its future value to national security" [10.80].

Private corporations are not widely accountable, nor are they subject to congressional oversight in the same way that government agencies are—facts that undoubtedly have not been lost either on Bezos or on the powerful decision-makers within DoD. This is precisely why the U.S. has ended up with so many defense contractors, and why so many government employees utilize the "revolving door" out of the Pentagon into positions with these companies when they "retire" from government employment.

Beyond this discussion, however, the real question is, do we need weapons in space? Do our civilian companies really need to be launching military hardware into orbit? Or, has humankind reached a point at which cooperation across the globe is possible? If constructive global cooperation is possible, then what is the quickest way to re-posture society away from militarism and toward collaboration, before spending on space weapons becomes even more egregious than it already is? These important questions must be discussed by society at large, and astrobiologists and astroethicists should be important voices in these discussions.

Given the weight and potential negative impacts of weaponizing space, we need to take this second elephant of the weaponization of space down from the tightrope and maneuver it onto the astroethics bridge as quickly as possible. Scientists, ethicists, and the public at large need to come together to oppose the militarization and weaponization of space. Human cooperation rather than conflict-led depletion of global resources is the correct and the moral path forward.

10.6   How the Two Elephants Are Connected

Even down from the tightrope and brought onto the astroethics bridge of broad-based, societal discussion, the two elephants described above remain yoked together. Why is this? And, what can we do to unyoke them?

It is because the first elephant—the phenomenon—actively is being used as pretextual rationale for the second, namely the weaponization of space. The well-known aerospace engineer and space architect Dr. Wernher von Braun is reported to have warned against this unmerited situation decades ago. Von Braun apparently was concerned that the phenomenon, which is not a threat, would be used unfairly in an attempt to justify the aggressive militarization and weaponization of space.

According to Dr. Carol Rosin, von Braun opposed the weaponization of space. Rosin worked as a corporate manager at Fairchild Industries from 1974-77. There, she met von Braun in early 1974, when he was vice president of the company. At some point thereafter, Rosin became von Braun's spokesperson. Rosin states that von Braun discussed with her the efforts of other individuals to weaponize space, "to control the Earth from space, and space itself," as Rosin paraphrases von Braun. According to Rosin, von Braun thought that space-based weapons were destabilizing, too expensive, unnecessary, and undesirable. However, von Braun was concerned that those who develop weapons systems would try to manipulate the public into thinking such systems were necessary by touting so-called enemies of the U.S.—first the Russians, followed by terrorists, "nations of concerns," i.e., unstable rulers of Third World countries, asteroids, and, finally, extraterrestrials. Referring to this raising up of one so-called threat after another, Rosin states, "'And all of it,' he [i.e., von Braun] said, 'is a lie.'" Von Braun wanted Rosin to help him prevent the weaponization of space, a calling to which she agreed [10.131] [10.139] [10.154].

Coarsely portraying extraterrestrials as an ultimate other in order to create a pretextual rationale to control resources and people is undesirable and immoral [10.139] [10.154] [10.30]. Instead of following a dysfunctional military and IC dystopian playbook, we instead should proceed calmly by collecting as much data and other information as possible on UAP and USOs and by preparing human society for widespread Contact with ETI.

10.7   The Astroethics Public Policy Path Forward

Because the phenomenon is not threatening human beings, any effort to use the reality of the phenomenon to justify the weaponization of space is pretextual. Instead of being manipulated into further militarism by means of so-called national security rhetoric, human beings must engage in public policy formulation and implementation to ensure that space is preserved as a domain for peaceful, international scientific exploration. Such a moral, rational mindset also will further international cooperation in astrobiology and astroethics.

Sound public policy formulation begins with the recognition that the most significant challenges confronting human beings as a species are global in nature. Without militarism depleting economic resources, human beings will be in a better position to work cooperatively and quickly to respond to the global and multigenerational nature of the challenges it faces. A situation in which certain human beings shoot DEWs, neutral particle beams, and/or EMPs at one another is not going to solve any of the challenges facing humanity. Neither is 'awarding' Northrop Grumman, Boeing, Lockheed Martin, and/or other defense contractors hundreds of billions in government contract to develop weapons systems. Shooting at UAP also is a malformed idea. Yet, that precisely is what appears to be shown in NASA footage from the Space Shuttle Discovery's mission in September 1991, in which an unidentified aerial object dodges a beam fired from an electronic weapon fired from Earth [10.42] [10.30].[16]

Instead of aggressive militarism, I suggest that human beings pursue international cooperation and constructive activities involving the phenomenon that proceeds according to a straightforward, three-step approach.

Step 1: Re-posturing Toward Cooperation
Step 1 re-postures human beings away from conflict and towards cooperation by means of redirecting the trillions of dollars allocated to the militaries of the world and toward humanitarian projects worldwide. Weapons threaten all of humankind, which is irrational in the context of humanity's survival. To meet global challenges effectively, the world must demilitarize. This requires that the militaries of the world and defense contractsors disentangle themselves from one another, and that all concerned reposture their research, development, and manufacturing toward humanitarian ends. Society must reallocate military funding away from weapons development

---

[16]Narrator Jeremy Piven states, "This footage, captured by NASA's Discovery space shuttle on September 15, 1991, shows an extraterrestrial vehicle dodge an electronic weapon fired from Earth. Here, we see the craft slowly crossing the frame, there's a bright flash, then it makes an abrupt, right-angle turn, and flies off. Moments after the turn, we see the shot fired, across space, where the craft had just been" [10.30]. The text overlay on the video shown in the documentary film *Close Encounters of the Fifth Kind: Contact Has Begun* [10.30] and also in *The Inquisitr* article [10.42] reads "NASA Discovery space shuttle STS-48/15 September 1991/Analyzed by Jack Kasher, Ph.D. – physicist."

and toward the production of things that are useful and life-supporting rather than lethal and life-negating.

If the U.S., Russia, and China take the lead together, humankind will succeed in preserving space for peaceful purposes [10.78]. These three global leaders immediately must engage in trilateral discussions regarding how to stop the escalatory buildup of ASATs [10.76] and how to put the brakes on hypersonics. There is evidence that Russia would welcome international attempts to de-escalate conflict in space [10.137]. The Russian government has acknowledged that the implications of continued weaponization of space have dire and global ramifications [10.67]. In November 2019, a spokesperson of the Russian mission to the UN and gave a statement to the General Assembly on behalf of the Commonwealth of Independent States (CIS) regarding the need to enter international agreements to deminilitarize space [10.68].

Historically, the UN General Assembly has been an important venue for debates regarding space militarization, and in 2004 a resolution regarding the prevention of an arms race in space was accepted without any vote against it—and though four states, including the United States, withdrew from the vote [10.132]. Nevertheless, it is likely that if the U.S. and Russia were to pivot away from the weaponization of space, China would follow suit [10.179]. Even further, I believe firmly that if the U.S. were to pursue a policy of complete demilitarization, Russia, China, and the rest of the world also would demilitarize. It would be in their pragmatic interest to do so in order to make resources available for projects to benefit humankind rather than to develop and implement weapons systems.

Step 2: Dialogue
Step 2 involves pursing and strengthening international dialogue. International, academic conferences and other such forma can bridge geopolitical divides and facilitate collaborative research. Virtual conferences bringing together scientists, ethicists, and other stakeholder such as the CEOs of major corporations can create settings for dialogue that arrive at solutions and effect change. Participants can share ideas and create additional platforms for cooperation to discuss issues in astrobiology and astroethics, and also to discuss and study the phenomenon. To maximize global cooperation, participants should include researchers from all countries, even those that the governments of certain countries may want their own citizens to perceive as adversaries. Nobody is an adversary inherently, however, and real, inclusive international dialogue regarding space policy can help dismantle militarism based on political and sociocultural indoctrination. Stakeholders who profit from militarism and weapons development, such as defense contractors, politicians, and others, should be included

in such discussions, so that they can come to understand the wisdom in changing course. To facilitate this deeper level of conversation, such dialogic settings should emphasize conversation concerning higher human ideals and positive values such as morality, ethical conduct, transparency, imagination, creativity, vision, wisdom, compassion, kindness, and love.

Step 3: De-mystifying the Phenomenon
Step 3 requires educating human beings about the phenomenon and how it is yoked unfairly to the weaponization of space. It also invites human beings to develop a communicative relationship with the phenomenon so that fear and paranoia do not overcome reason and love as humanity comes of age in its widespread acknowledgement of the phenomenon's reality. Structures supporting secrecy must be dismantled, since there is no room for continued deception and manipulation.. Instead, interactivity is key, as is fostering reciprocal and mutual understanding between human beings and the ETI responsible for the operation of UAP and USOs on and around Earth.

As human beings turn away from conflict in space, and also on Earth, they also should develop and implement transnational projects that make information on the phenomenon available to the global public. For example, Italian researchers have assembled a database of 13,000 cases involving UAP over the last 70 years. As founder of Centro Ufologico Nazionale [10.40] Roberto Pinotti observes, people in Russia and China are organizing "at a government level" in order to study the phenomenon [10.125]. Such research should not occur only at the level of governments, however. It also should involve academics and the general public, so that human beings are progressively and creatively assimilated to the phenomenon itself.

Now, where to our elephants? To start, we must de-yoke them, one from another. The second—the weaponization of space—is old, tired, and should retire off the stage gracefully—with no small push by astroethicists and others. The first elephant—the phenomenon—is fresh, nimble, and something from which human beings can learn much. I suggest we watch it with interest, engage in constructive attempts to communicate with it whenever possible, and open ourselves to the wonders of the cosmos in which we live.

References

[10.1]   116th United States Congress, U.S. Government Publishing Office, https://www.congress.gov/116/bills/hr2500/BILLS-116hr2500eh.pdf, H. R. 2500: 919, lines 17-20, Washington, D.C., 2019.

[10.2]   116th United States Congress, Intelligence Authorization Act for Fiscal Year 2021, U.S. Government Publishing Office, https://www.intelligence.senate.gov/publications/intelligence-authorization-act-fiscal-year-2021, Washington, D.C., U.S. Senate Select Committee on Intelligence, 2020.

[10.3]   The 1952 Wave, [Season 4, Episode 2] in: *Unsealed Alien Files*, IMDb TV, April 24, 2015.

[10.4]   Aero Consulting Experts, *Meet the Team*, Aero Consulting Experts Meet the Team, http://www.aeroconsultingexperts.com/team-index#meet-team.

[10.5]   Aimer, R., To Catch an Alien [Season 1, Episode 3], in: *Black FilesDeclassified*, Science Channel, April 16, 2020.

[10.6]   Aimer, R., UFO Insiders, [Season 1, Episode 1], in: *Unidentified: Inside America's UFO Investigation*, History Channel, May 31, 2019.

[10.7]   Alén, *A Basic Guide to Nanosatellites*, Alén, 2020, https://alen.space/basic-guide-nanosatellites/.

[10.8]   Aliens in Alaska, [Season 1, Episodes 1 to 8], Discovery Plus (discovery+), Episode 1, Men in Black, February 15, 2021; Episode 2, Alien Flight Plan, February 22, 2021; Episode 3, Night Stalkers, March 1, 2021; Episode 4, They Walk Among Us, March 8, 2021; Episode 5, Nightmare Below Zero, March 15, 2021; Episode 6, The Mother Ship, March 22, 2021; Episode 7, Captured, March 29, 2021; Episode 8, Above and Far Beyond, April 5, 2021.

[10.9]   American UFOs [Season 1, Episode 2], in: *Black Files Declassified*, Science Channel, April 9, 2020.

[10.10]   Anderson, G., *NASA Confirms Evidence That Liquid Water Flows on Today's Mars*, NASA, November 20, 2017, https://www.nasa.gov/press-release/nasa-confirms-evidence-that-liquid-water-flows-on-today-s-mars.

[10.11]   Andresen, J., and Chon Torres, O.A., *Extraterrestrial Intelligence*, Cambridge Scholars Publishing, Newcastle upon Tyne, U.K., forthcoming.

[10.12]   The Atomic Connection [Season 1, Episode 5], in: *Unidentified: Inside America's UFO Investigation*, History Channel, June 28, 2019.

[10.13]   Baker, M., Secrets of the Space Force [Season 1, Episode 1], in: *Black Files Declassified*, Science Channel, April 2, 2020.

[10.14]   BBC, *Brazil air force to record UFO sightings*, BBC, August 12, 2010, https://www.bbc.com/news/world-latin-america-10947856.

[10.15]   Bender, B. and Desiderio, A., *Government report can't explain UFOs, but offers no evidence of aliens*, Politico, June 25, 2021, https://www.politico.com/news/2021/06/25/government-report-ufos-are-real-496319.

[10.16]   Bender, B., *Senators get classified briefing on UFO sightings*, Politico, June 19, 2019, https://www.politico.com/story/2019/06/19/warner-classified-briefing-ufos-1544273.

[10.17]   Benson, E., *Harry Reid on What the Government Knows About UFOs*, Intelligencer, March 21, 2018. https://nymag.com/intelligencer/2018/03/harry-reid-on-what-the-government-knows-about-ufos.html.

[10.18] Blumenthal, R. and Kean, L., *Do We Believe in U.F.O.s? That's the Wrong Question*, The New York Times, July 28, 2020, https://www.nytimes.com/2020/07/28/insider/UFO-reporting.html.

[10.19] Blumenthal, R. and Kean, L., *No Longer in Shadows, Pentagon's U.F.O. Unit Will Make Some Findings Public*, The New York Times, July 23, 2020, https://www.nytimes.com/2020/07/23/us/politics/pentagon-ufo-harry-reid-navy.html.

[10.20] Boyle, A., *How aliens and Apollo astronaut Edgar Mitchell got tangled up in Wiki Leaks emails*, Geek Wire, October 10, 2016, https://www.geekwire.com/2016/aliens-apollo-astronaut-edgar-mitchell-wikileaks/.

[10.21] Boyle, A., *Pentagon launches development agency seen as key to future Space Force*, CNN, March 13, 2019, https://edition.cnn.com/2019/03/13/politics/pentagon-space-force-agency/index.html.

[10.22] Brennan, J.O., *John O. Brennan on Life in the CIA: What working in intelligence has taught him about human nature*. [Ep. 111], Conversations with Tyler, T. Cowen, December 16, 2020, https://conversationswithtyler.com/episodes/john-o-brennan/. For video of this interview, see https://www.thechronicles.com/2020/12/former-cia-director-john-brennan-on.html.

[10.23] Broad, W., *Pentagon Leaves the Shuttle Program*, The New York Times, August 7, 1989, https://www.nytimes.com/1989/08/07/us/pentagon-leaves-the-shuttle-program.html.

[10.24] Browne, R., *With a signature, Trump brings Space Force in to being*, CNN, December 20, 2019, https://www.cnn.com/2019/12/20/politics/trump-creates-space-force/index.html.

[10.25] Cameron, G., *UFOs: The White House Files*, History Channel, December 6, 2019.

[10.26] Cappelletti, C., Battistini, S., Graziani, F., Small launch platforms for micro-satellites. *Adv. Space Res.*, 62, 12, 3298–3304, 2018.

[10.27] Centre National d´Études Spatiales, *Le GEIPAN ouvre ses dossiers*, CNES, 2006, https://cnes.fr/fr/web/CNES-fr/5847-le-geipan-ouvre-ses-dossiers.php.

[10.28] Chomsky, N., *Hegemony or Survival: America's Quest for Global Dominance*, pp. 220–265, Henry Holt and Company, LLC <MetropolitanBooks>, New York, NY, USA, 2003.

[10.29] Chronicle Herald, *Russian cosmonaut captures possible UFO footage*, The Chronicle Herald, August 24, 2020, https://www.pressreader.com/canada/the-chronicle-herald-metro/20200824/281625307680122.

[10.30] *Close Encounters of the Fifth Kind: Contact Has Begun*, Directed by M. Mazzola, StarContact, 2020.

[10.31] Cobb., J.D., Airline Encounters, [Season 2, Episode 5], in: *Unidentified: Inside America's UFO Investigation*, History Channel, August 8, 2020.

[10.32] Cohen, Z. et al., *Slow-moving blob' that may have been a flock of birds caused White House lockdown*, CNN, November 26, 2019, https://edition.cnn.com/2019/11/26/politics/white-house-lockdown-airspace/index.html.

[10.33] *Comité de Estudios de Fenómenos Aéreos Anómalos*, CEFAA, Committee for Studies of Anomalous Aerial Phenomena), http://www.cefaa.gob.cl/.

[10.34] Conte, M., *Pentagon officially releases UFO videos*, CNN, April 27, 2020, https://edition.cnn.com/2020/04/27/politics/pentagon-ufo-videos/index.html.

[10.35] Cooke, C., UFO Insiders, [Season 1, Episode 1], in: *Unidentified: Inside America's UFO Investigation*, History Channel, May 31, 2019.

[10.36] Cooper, G., Video clip shown in Mazzola, M., *Unacknowledged*, Auroris Media, 2017.

[10.37] Cooper, H., Blumenthal, R., Kean, L., *Glowing Auras and 'Black Money': The Pentagon's Mysterious U.F.O. Program*, The New York Times, December 16, 2017, https://www.nytimes.com/2017/12/16/us/politics/pentagon-program-ufo-harry-reid.html.

[10.38] Cowen, T.W., *Classified UFO Documents Could Be Released Within180 Days Thanks to COVID-19 Relief Bill*, Complex, December 30, 2020, https://www.complex.com/life/2020/12/classified-ufo-documents-released-180-guests-covid-19-bill.

[10.39] CTBTO Preparatory Commission, *Starfish Prime*, Outer Space, CTBTO: Comprehensive Nuclear-Test-Ban Treaty Organization, July 9, 1962, https://www.ctbto.org/specials/testing-times/9-july-1962starfish-prime-outer-space.

[10.40] CUN Italia Network, *Centro Ufologico Nazionale*, 1997, https://www.centrou fologiconazionale.net/.

[10.41] David, J.E. (Ed.), *NASA's Secret Relationships with U.S. Defense and Intelligence Agencies*, The National Security Archive Electronic Briefing Book No. 509, 2015, Washington, D.C. https://nsarchive2.gwu.edu/NSAEBB/NSAEBB509/.

[10.42] Didymus, J., *Unsolved Mystery Of NASA Footage Showing 'Intelligently Controlled' UFOs During 1991 STS-48 Mission Revisited*, The Inquisitr, August 8, 2017, https://www.inquisitr.com/2644559/unsolved-mystery-of-nasa-foot-age-showing-intelligently-controlled-ufos-during-1991-sts-48-mission-revisited/.

[10.43] Duncan, C., *What does the military know about UFOs? Senate committee wants public report*, Impact 2020, June 25, 2020, https://www.mcclatchydc.com/news/nation-world/national/article243768802.html.

[10.44] Erwin, S., *STRATCOM chief Hyten: 'I will not support buying big satellitesthat make juicy targets'*, Space News, November 19, 2017, https://space-news.com/strat-com-chief-hyten-i-will-not-support-buying-big-satellites-that-make-juicy-targets/.

[10.45] Europa Press, *EEUU preparaunafuerzamilitarparalainvestigaciónde-OVNIS*, 20 minutos, August 15, 2020, https://www.20minutos.es/noticia/4353445/0/ee-uu-prepara-fuerza-militar-investigacion-ovnis/. 11 May 2019, Montevideo, Uruguay

[10.46] Exoplanet, *Catalog*, Exoplanet, http://exoplanet.eu/catalog/.

[10.47] Extraterrestrial Encounters [Season 2, Episode 8], in: *Unidentified: Inside America's UFO Investigation*, History Channel, August 8, 2020.

[10.48] Fingas, J., *Australia aims to launch locally-made hybrid rockets by 2022*, Engadget, June 14, 2020, https://www.engadget.com/australia-hybrid-rockets-gilmour-space-191245091.html.

[10.49] Foust, J., *Virgin Orbit identifies cause of engine shutdown on first LauncherOne flight*, Space News, July 22, 2020, https://spacenews.com/virgin-or- bit-identifies-cause-of-engine-shutdown-on-first-launcherone-flight/.

[10.50] Friend, A.H. and Johnson, K., *The Sixth Service: What the Reorganization of Special Operations Forces Can Teach Us About Space Force*, War on the Rocks, September 17, 2018, https://warontherocks.com/2018/09/the-sixth-service-what-the-reorganization-of-special-operations-forces-can-teach-us-about-space-force/.

[10.51] Fuerza Aérea Uruguaya, *Comisión Receptorae Investigadora de Denuncias de Objetos Voladores No Identificados (CRIDOVNI)*, Fuerza Aérea Uruguaya, May 11, 2019, Montevideo, Uruguay, https://www.fau.mil.uy/es/artic ulos/182-comision-re-ceptora-e-investigadora-de-denuncias-de-objetos-vola-dores-no-identi-ficados- cridovni.html.

[10.52] The Galileo Project: "Daring to Look Through New Telescopes," Harvard University, https://projects.iq.harvard.edu/galileo. For a full statement of the Project Goal, see https://projects.iq.harvard.edu/galileo/project-goal

[10.53] Garamone, J., *Trump Signs Law Establishing U.S. Space Force*, U.S. Department of Defense, December 20, 2019, https://www.defense.gov/Explore/News/Article/Article/2046035/trump-signs-law-establishing-us-space-force/.

[10.54] GEIPAN, *Grouped´ Étudesetd´ Informations surles Phénomènes Aérospatiaux Non Identifiés*, http://www.cnes-geipan.fr/.

[10.55] Ghoshroy, S., The X-37B: Backdoor weaponization of space? *Bull. At. Sci.*, 71, 3, 19–29, 2015.

[10.56] Given Place Media, *The Mysterious Battle of Los Angeles, 1942*, Los Angeles Almanac, http://www.laalmanac.com/history/hi07s.php.

[10.57] Gould, J., *Congress adopts defense bill that creates Space Force*, Congress, December 17, 2019, https://www.defensenews.com/congress/2019/12/17/congress-adopts-defense-bill-that-creates-space-force/.

[10.58] Gravel, M., To Catch an Alien [Season 1, Episode 3], in: *Black Files Declassified*, Science Channel, April 16, 2020.

[10.59] Greenwood, M., *Pentagon acknowledges program to investigate UFO encounters: report*, The Hill, December 16, 2017, https://thehill.com/policy/defense/365250-pentagon-acknowledges-program-to-investi-gate-ufo-encoun ters-report.

[10.60] Gventer, C.W., *Make It So: Putting Space Force in Context*, War on the Rocks, August 15, 2018, https://warontherocks.com/2018/08/make-it-so-putting-space-force-in-context/.

[10.61] Handberg, R., War and rumours of war, do improvements in space technologies bring space conflict closer? *Def. Secur. Anal.*, 34, 176–190, 2018.

[10.62] Hart, D., Secrets of the Space Force [Season 1, Episode1], in: *Black Files Declassified*, Science Channel, April 2, 2020.

[10.63] Hartmans, A., *Jeff Bezos' Blue Origin rocket company has been deemed an essential business by the government, allowing it to continue operating amid the coronavirus outbreak*, Business Insider, March 30, 2020, https://www.

busines-sinsider.com/jeff-bezos-blue-origin-rocket-essential-busi-ness-corona-virus-outbreak-2020-3.

[10.64]   Hastings, R., *UFOs & Nukes: Extraordinary Encounters at Nuclear Weapons Sites*, Second Edition (Revised and Updated), CreateSpace Independent Publishing Platform, Scotts Valley, CA, USA, 2017.

[10.65]   Horning, C., Above and Far Beyond [Season 1, Episode 8], in: *Aliens in Alaska*, Discovery Plus (discovery+), April 5, 2021.

[10.66]   Howell, E., *Fermi Paradox: Where Are the Aliens?*, Space, April 27, 2018, https://www.space.com/25325-fermi-paradox.html.

[10.67]   Interfax America, Inc., Corridors of Power: Implications of weaponization of space to surpass those of intermediate-range missiles in Europe – Putin [(Part 1)]; Implications of weaponization of space to surpass those of intermediate-range missiles in Europe - Putin (Part 2), in: *Russia & CISGeneral Newswire*, October 11, 2019.

[10.68]   Interfax America, Inc., *SPACE INDUSTRY: CIS states call for banning space weaponization - statement*, Central Asia & Caucasus Business Weekly, November 5, 2019.

[10.69]   Isgleas, D., *Hay aún 40 casos de ovnis sin explicación*, El Pais, June 7, 2009, http://historico.elpais.com.uy/090607/pnacio-421863/nacional/Hay-aun-40-casos-de-ovnis-sin-explicacion/.

[10.70]   Jones, E.M., *"Where Is Everybody?": An Account of Fermi's Question*, Los Alamos National Laboratory, Los Alamos, NM, USA, 1985.

[10.71]   Judson, J., *How the US Army is building a hypersonic weapons business*, Defense News, October 14, 2019, https://www.defensenews.com/digital-show-dai-lies/ausa/2019/10/14/how-the-us-army-is-building-a-hypersonic-weapons-business/.

[10.72]   Justice, S., Raining UFOs [Season 1, Episode 2], in: *Unidentified: Inside America's UFO Investigation*, History Channel, June 7, 2019.

[10.73]   Justice, S., The Revelation [Season 1, Episode 6], in: *Unidentified: Inside America's UFO Investigation*, History Channel, July 15, 2019.

[10.74]   Kapnisi, C., *U-18-0001/OCC1. [Letter to John McCain and Jack Reed]*, Defense Intelligence Agency, Washington, D.C., 2018, https://fas.org/irp/dia/aatip-list.pdf.

[10.75]   Kim, H.T., Militarization and Weaponization of Outer Space in International Law. *Korean J. Air Space Law Policy*, 33, 1, 261–284, 2018, https://www.koreascience.or.kr/article/JAKO201823954940705.pdf.

[10.76]   Kopeć, R., *Broń Antysatelitarna: U Progu Drugiego Etapu Militaryzacji Kosmosu*, vol. 15, 53/2, pp. 45–72, Ksiegarnia Akademicka, 2018.

[10.77]   Korpala, E.J., SETI: Its Goals and Accomplishments, in: *Handbook of Astrobiology*, V.M. Kolb (Ed.), p. 738, 2019.

[10.78]   Larsen, P.B., Outer Space Arms Control: Can the USA, Russia and China Make this Happen. *J. Confl. Secur. Law*, 23, 1, 137–159, 2018.

[10.79]   Lee, C., *Space Force Receives Fifth SBIRS Satellite*, National Defense, January 20, 2021, https://www.nationaldefensemagazine.org/articles/2021/1/20/

space-force-receives-fifth-sbirs-satellite#:~:text=The%20U.S.%20Space%20 Force%20has,Space%2DBased%20Infrared%20System%20constellation.&- text=The%20system%20is%20expected%20to,2022%20from%20Cape%20 Canaveral%2C%20Florida.

[10.80]   Lee, D., *Jeff Bezos's rocket company gets virus lockdown exemption*, Financial Times, March 29, 2020, https://www.ft.com/content/52a1258c-4c9a- 4a6f-adb9-d3c22cac1539.

[10.81]   Lendon, B., *The US successfully tested a laser weapon that can destroy air-craft mid-flight*, CNN, May 22, 2020, https://www.cnn.com/2020/05/22/asia/ us-navy-lwsd-laser-intl-hnk-scli/index.html.

[10.82]   Lewis-Kraus, G., *How the Pentagon Started Taking U.F.O.S Seriously*, The New Yorker, May 10, 2021 hard copy issue, online April 30, 2021, https://www.newyorker.com/magazine/2021/05/10/how-the-pentagon-started-taking-ufos-seriously?utm_source=nl&utm_brand=t-ny&utm_mailing=TNY_ Magazine_050321&utm_campaign=aud-dev&utm_medium=email&bxid=- 5be9c87b24c17c6adf391be2&cn-did=53684960&hasha=309bb353e656f26160b- 41fce9ae9a83a&hash-b=9ee749fbeedf407bf048fba450d06462465dd0b8- &hashc=14d2c168037e-846148cd12103a566d8b99ccafc1a9929955a4024250 3a9e568f&es-rc=AUTO_PRINT&utm_term=TNY_Magazine

[10.83]   Li, W., *Secret UFO Files? In Canada the truth is out there–online and search-able*, The Star, July 27, 2020, https://www.thestar.com/news/can-ada/ 2020/07/27/secret-ufo-files-in-canada-the-truth-is-out-there-on-line-and-searchable.html.

[10.84]   Library and Archives Canada, *Canada's UFOs: The Search for the Unknown*, Library and Archives Canada, 2013, https://www.bac-lac.gc.ca/eng/discover/ unusual/ufo/Pages/default.aspx.

[10.85]   Lyons, K., *SpaceX launched more Starlink satellites on Falcon 9, and three Planet SkySats hitched a ride*, The Verge, June 13, 2020, https://www.theverge. com/2020/6/13/21290081/spacex-starlink-satellite-launch.

[10.86]   Mack, J.E., *Abduction: Human Encounters with Aliens*, Charles Scribner's Sons, New York, NY, USA, 1994.

[10.87]   Mack, J.E., *Passport to the Cosmos: Human Transformation and Alien Encounters*, Crown Publishers, New York, NY, USA, 1999.

[10.88]   Madrak, S., *What About That UFO Over The White House That Just Prompted A Lockdown?*, Crooks and Liars, November 27, 2019, https://crooksandliars. com/2019/11/what-about-ufo-over-white-house-prompted-0.

[10.89]   Manaugh, G., *The Growing Risk of a War in Space*, The Atlantic, June 21, 2016, https://www.theatlantic.com/technology/archive/2016/06/ weaponizing-the-sky/488024/.

[10.90]   Maranzani, B., *Dwight Eisenhower's Shocking—and Prescient—Military Warning*, History, March 23, 2017, https://www.history.com/news/dwight-eisenhowers-shocking-and-prescient-military-warning.

[10.91]   Maroni, A., *Special Access Programs and the Defense Budget: Understanding the "Black Budget"*, Congressional Research Service, The Library of Congress, Washington, D.C., 1989.

[10.92]   Martín-Torres, F.J., Zorzano, MP., Valentín-Serrano, P. *et al.* Transient liquid water and water activity at Gale crater on Mars. *Nature Geosci*, 8, 357–361, 2015. https://doi.org/10.1038/ngeo2412

[10.93]   McDonald, J.E., Science in Default: Twenty-Two Years of Inadequate UFO Investigations, presented at: *American Association for the Advancement of Science, 134th Meeting, General Symposium, Unidentified Flying Objects*, Boston, December 27, 1969, https://physics.princeton.edu//~mcdonald/JEMcDonald/mcdonald_aaas_69.pdf (text), and https://physics.princeton.edu//~mcdonald/JEMcDonald/science_in_default_aass69.mp4 (audio).

[10.94]   McMillan, T., *Inside the Pentagon's Secret UFO Program*, Popular Mechanics, February 14, 2020. https://www.popularmechanics.com/military/research/a30916275/government-secret-ufo-program-investigation/.

[10.95]   McMillan, T., UFOs vs. Nukes [Season 2, Episode 3], in: *Unidentified: Inside America's UFO Investigation*, History Channel, July 25, 2020.

[10.96]   Mellon, C., *Are UFOs a Threat to National Security? This Ex-U.S. Official Thinks They Warrant Investigation*, History, June 7, 2019, https://www.history.com/news/chris-mellon-ufo-investigations.

[10.97]   Mellon, C., Extraterrestrial Encounters [Season 2, Episode 8], in: *Unidentified: Inside America's UFO Investigation*, History Channel, August 22, 2020.

[10.98]   Mellon, C., Airline Encounters, [Season 2, Episode 5], in: *Unidentified: Inside America's UFO Investigation*, History Channel, August 8, 2020.

[10.99]   Mellon, C., "Don't Miss the Alien Hypothesis." June 24, 2021, https://www.christophermellon.net/post/don-t-dismiss-the-alien-hypothesis.

[10.100] Mellon, C., "The UAP Report and the UAP Issue:" July 6, 2021, updated July 11, 2021, https://www.christophermellon.net/post/the-uap-report-and-the-uap-issue.

[10.101] Mellon, C., Raining UFOs [Season 1, Episode 2], in: *Unidentified: Inside America's UFO Investigation*, History Channel, June 7, 2019.

[10.102] Mellon, C., The Revelation [Season1, Episode 6], in: *Unidentified: Inside America's UFO Investigation*, History Channel, July 5, 2019.

[10.103] Mellon, C., Sightings Surge [Season 2, Episode 7], in: *Unidentified: Inside America's UFO Investigation*, History Channel, August 22, 2020.

[10.104] Mellon, C., The Triangle Mystery [Season 2, Episode 2], in: *Unidentified: Inside America's UFO Investigation*, History Channel, July 18, 2020.

[10.105] Mellon, C., The UFO Cover-Up [Season 2, Episode 6], in: *Unidentified: Inside America's UFO Investigation*, History Channel, August 15, 2020.

[10.106] Mellon, C., UFOs vs. Nukes [Season 2, Episode 3], in: *Unidentified: Inside America's UFO Investigation*, History Channel, July 25, 2020.

[10.107] Mellon, C. and Fox, J., *The Phenomenon UFO Movie Interview*, Interview with J. Norton and S. Roberts for Jim Norton & Sam Roberts Show, October 5, 2020. https://www.youtube.com/watch?v=MdwIgn1RbVw.

[10.108] Missile Defense Advocacy Alliance, *Avangard (Hypersonic Glide Vehicle)*, MDAA: Missile Defense Advocacy Alliance, 2020, Alexandria, VA. https://missiledefenseadvocacy.org/missile-threat-and-proliferation/missile-proliferation/russia/avangard-hypersonic-glide-vehicle/.

[10.109] Mitchell, E., Video clip shown in *Unacknowledged*, Directed by M. Mazzola, Auroris Media, 2017.

[10.110] Mix, L.J., *Life in Space: Astrobiology for Everyone*, p. 4, Harvard University Press, Cambridge, MA, USA, 2009.

[10.111] Montesdeoca, J., *La Fuerza Aérea de Perú confirma avistamiento de ovnis cerca del aeropuerto de Lima*, CNN Latinoamérica, June 18, 2019, https://cnnespanol.cnn.com/2019/06/18/la-fuerza-aerea-de-peru-confirma-avis-tamiento-de-ovnis-cerca-del-aeropuerto-de-lima/.

[10.112] Mosteshar, S., *Space Law and Weapons in Space*, Oxford Research Encyclopedia of Planetary Science, 2019, https://doi.org/10.1093/acrefore/9780190647926.013.74.citing Elizabeth S. Waldrop, "Weaponizationof outer space: US national policy," Annals of Air & Space Law 29, 329-355, (2004).

[10.113] MUFON, *MUFON: Mutual UFO Network*, https://www.mufon.com.

[10.114] Nacouzi, G., American UFOs [Season 1, Episode 2], in: *Black Files Declassified*, Science Channel, April 9, 2020.

[10.115] The National Archives [of the United Kingdom], *UFOs: Newly released UFO files from the UK government*, National Archives, 2020, Kew, Richmond, http://www.nationalarchives.gov.uk/ufos/.

[10.116] The New York Times Archives, *Extraterrestrial Life*, The New York Times, January 10, 1963, https://www.nytimes.com/1963/01/10/archives/extraterrestrial-life.html.

[10.117] Northrop Grumman Corporation, *Northrop Grumman Signs Customer for First Flight of OmegA*, Space Ref, 2019, http://spaceref.com/news/viewpr.html?pid=55010.

[10.118] Norton-Taylor, R., *Alien nation: MoD releases final UFO files*, The Guardian, June 21, 2013, https://www.theguardian.com/uk/2013/jun/21/last-release-mod-ufo-files.

[10.119] ODNI, Office of the Director of National Intelligence. *Preliminary Assessment: Unidentified Aerial Phenomena*, pp. 3-5, McLean, VA, Office of the Director of National Intelligence, June 25, 2021, https://www.dni.gov/files/ODNI/documents/assessments/Preliminary-Assessment-UAP-20210625.pdf.

[10.120] Pasternack, A., *The Moon-Walking, Alien-Hunting, Psychic Astronaut Who Got Sued By NASA*, Vice, May 14, 2016, https://www.vice.com/en_us/article/aek7ez/astronaut-edgar-mitchell-outer-space-inner-space-and-aliens.

[10.121] Paul, D., *How angry pilots got the Navy to stop dismissing UFO sightings*, Washington Post, April 24, 2019, https://www.washingtonpost.com/national-security/2019/04/24/how-angry-pilots-got-navy-stop-dismissing-ufo-sightings/.

[10.122] Peters, T., Astroethics and Microbial Life in the Solar Ghetto, in: *Astrotheology: Science and Theology Meet Extraterrestrial Life*, M. Hewlett, J.M. Moritz, R.J. Russell (Eds.), pp. 402–403, Cascade Books, Eugene, OR, USA, 2018.

[10.123] *The Phenomenon*, 2020, https://thephenomenonfilm.com/. Directed by J. Fox, produced by Farah Films, and distributed by 1091 Media.

[10.124] Picheta, R., *Aliens definitely exist and they could be living among us on Earth, says Britain's first astronaut*, CNN, January 6, 2020, https://edition.cnn.com/2020/01/06/uk/helen-sharman-aliens-exist-scli-scn-gbr-intl/index.html.

[10.125] Pinotti, R., The Revelation [Season 1, Episode 6], in: *Unidentified: Inside America's UFO Investigation*, History Channel, July 5, 2019.

[10.126] Podcast UFO Live, *Show 398 Notes: Travis Taylor & Leslie Kean*, Podcast UFO Live, March 29, 2020, https://podcastufo.com/show-notes/travis-taylor-leslie-kean/.

[10.127] Ramírez Muro, V., *Porqué Perú reabrió la oficina debúsqueda de ovnis*, BBC, February 5, 2014, https://www.bbc.com/mundo/noticias/2014/02/140205_peru_ovni_gobierno_busqueda_vh.

[10.128] Reid, H., The Revelation [Season 1, Episode 6], in: *Unidentified: Inside America's UFO Investigation*, History Channel, July 5, 2019.

[10.129] Reid, H., UFOs vs. Nukes [Season 2, Episode 3], in: *Unidentified: Inside America's UFO Investigation*, History Channel, July 25, 2020.

[10.130] Reif, K., *Current U.S. Missile Defense Programs at a Glance*, Arms Control Association, 2019, https://www.armscontrol.org/factsheets/usmissiledefense#space.

[10.131] Riddlept, *VonBraun's Legacy - Dr. Carol Rosin*, YouTube, October 12, 2013, https://youtu.be/h9KEcuEe1U4.

[10.132] Rogan, T., *Birds, balloons, or a UFO: What violated DC airspace?*, Washington Examiner, November 27, 2019, https://www.washington-examiner.com/opinion/birds-balloons-or-a-ufo-what-violated-dc-airspace.

[10.133] Rogan, T., *Why the UFO story is far more interesting than you think*, Washington Examiner, May 28, 2019, https://www.washingtonexaminer.com/opinion/why-the-ufo-story-is-far-more-interesting-than-you-think.

[10.134] Rubio, M., *1-on-1 with Marco Rubio on Coronavirus Response*, J. De Fede, CBS Miami, July 16, 2020, https://twitter.com/defede/status/1283840918521552897.

[10.135] Schlenker, R., Airline Encounters, [Season 2, Episode 5], in: *Unidentified: Inside America's UFO Investigation*, History Channel, August 8, 2020.

[10.136] Secrets of the Space Force [Season 1, Episode 1], in: *Black Files Declassified*, Science Channel, April 2, 2020.

[10.137] Sheer, A. and Li, S., Emergence of the International Threat of Space Weaponization and Militarization: Harmonizing International Community for Safety and Security of Space. *Front. Manage. Res.*, 3, 3,100–14, 2019.

[10.138] Sightings Surge [Season 2, Episode 7], in: *Unidentified: Inside America's UFO Investigation, History Channel*, August 22, 2020.

[10.139] *Sirius*, Directed by A. Kaleka. Never ending Light Productions, 2013.

[10.140] Skuse, Benjamin. *Spacecraft in a 'warp bubble' could travel faster than light, claims physicist*, astronomy and space, March 19, 2021, https://physicsworld.com/a/spacecraft-in-a-warp-bubble-could-travel-faster-than-light-claims-physicist/.

[10.141] Smith, M.S. (Ed.), *Space Policy Online*, https://spacepolicyonline.com/.

[10.142] Smith, M., *U.S. Space Force is Now a Reality*, Space Policy Online, December 20, 2019, spacepolicyonline.com/news/u-s-space-force-is-now-a-reality/.

[10.143] Sturrock, P., Report on a Survey of the Membership of the American Astronomical Society Concerning the UFO Problem: Part 1. *J. Sci. Explor.*, 8, 1, 1–45, 1994; Part 2. *J. Sci. Explor.*, 8, 2, 153-95, 1994; Part 3. *J. Sci. Explor.*, 8, 3, 309-46, 1994.

[10.144] Sturrock, P.A., *The UFO Enigma: A New Review of the Physical Evidence*, Warner Books, New York, 1999.

[10.145] Taylor, T., The Alien Protocols [Season 13, Episode 3], in: *Ancient Aliens*, Prometheus Entertainment and A&E Television. History Channel, 2018.

[10.146] Taylor, T., Revelations [Season 1, Episode 8], in: *The Secret of Skinwalker Ranch*, History Channel, 2020.

[10.147] Truman, H., *Quotes From Government and Military On The Record About UFOs*, MUFON, 1950, https://mufonct.com/quotes-from-government-and-military-on-the-record-about-ufos/.

[10.148] Tucker, P., *Pentagon Wants to Test A Space-Based Weapon in 2023*, Defense One, March 14, 2019, https://cdn.defenseone.com/b/defense-one/interstitial.html?v=9.21.3&rf=https%3A%2F%2Fwww.defenseone.com%2Ftechnol-ogy%2F2019%2F03%2Fpentagon-wants-test-space-based-weap-on-2023%2F155581%2F.

[10.149] Tucker, P., Secrets of the Space Force [Season 1, Episode 1], in: *Black Files Declassified*, Science Channel, April 2, 2020.

[10.150] The UFO Insiders [Season 1, Episode 1], in: *Unidentified: Inside America's UFO Investigation, History Channel*, May 31, 2019.

[10.151] UFO, [Season 1, Episodes 1 to 4], Showtime Networks Inc., Episode 1, 101, August 8, 2021; Episode 2, 102, August 15, 2021; Episode 3, 103, August 22, 2021; Episode 4, 104, August 29, 2021.

[10.152] *UFOs and Nukes: The Secret Link Revealed*, Directed by R. Hastings. Veritable Pictures, 2016.

[10.153] UFOs vs. Nukes [Season 2, Episode 3], in: *Unidentified: Inside America's UFO Investigation*, History Channel, July 25, 2020.

[10.154] *Unacknowledged*, Directed by M. Mazzola, Auroris Media, 2017.

[10.155] United States Chamber of Commerce, LAUNCH: *The Space Economy*, U.S. Chamber of Commerce, 2019, https://www.uschamber.com/event/launch-thespace-economy.

[10.156] United States Department of Defense, *Establishment of Unidentified Aerial Phenomena Task Force*, U.S. Department of Defense, Washington, D.C, 2020, https://www.defense.gov/Newsroom/Releases/Release/Article/2314065/establishment-of-unidentified-aerial-phenomena-task-force/.

[10.157] United States Department of Defense, *Final Report on Organizational and Management Structure for the National Security Space Components of the Department of Defense*, United States Department of Defense, Washington, D.C., USA, 2018, https://media.defense.gov/2018/Aug/09/2001952764/-1/-1/1/ORGANIZATIONAL-MANAGEMENT-STRUCTURE-DOD-NATIONAL-SECURITY-SPACE-COMPONENTS.PDF.

[10.158] United States Department of Defense, *Michael D. Griffin: Under Secretary of Defense for Research and Engineering*, U.S. Department of Defense, Washington, D.C., https://www.defense.gov/Our-Story/Biographies/Biography/Article/1489249/michael-d-griffin/.

[10.159] United States Department of Defense, *Office of the Under Secretary of Defense (Comptroller)/Chief Financial Officer: February 2020*, Office of the Secretary of Defense, Washington, D.C., 2020, https://comptroller.defense.gov/Portals/45/Documents/defbudget/fy2021/fy2021_Budget_Request_Overview_Book.pdf.

[10.160] United States Department of Defense, *Space Development Agency Gets First Permanent Director*, U.S. Department of Defense, Washington, D.C., 2019, https://www.defense.gov/Explore/News/Article/Article/2000839/space-development-agency-gets-first-permanent-director/.

[10.161] United States National Security Agency Central Security Service, *UFO Documents Index*, National Security Agency Central Security Service, https://www.nsa.gov/news-features/declassified-documents/ufo/.

[10.162] United States Senate Committee on Armed Services, *FY 2020 National Defense Authorization Act*, 2020, https://www.armed-services.senate.gov/imo/media/doc/FY%202020%20NDAA%20Executive%20Summary.pdf, p. 5.

[10.163] UNOOSA, *2222 (XXI). Treaty on Principles Governing the Activities of States in the Exploration and Use of Outer Space, including the Moon and Other Celestial Bodies*, United Nations: Office for Outer Space Affairs, 1966, article 4, https://www.unoosa.org/oosa/en/ourwork/spacelaw/treaties/outerspacetreaty.html.

[10.164] UNOOSA, *Treaty on Principles Governing the Activities of States in the Exploration and Use of Outer Space, including the Moon and Other Celestial Bodies*, United Nations: Office for Outer Space Affairs, 1966. Available at https://www.unoosa.org/oosa/en/ourwork/spacelaw/treaties/introouterspacetreaty.html.

[10.165] US National Archives, *Eisenhower's "Military-Industrial Complex" Speech Origins and Significance*, YouTube, 2011, https://www.youtube.com/watch?v=Gg-jvHynP9Y.

[10.166] Vallee, J., *Anatomy of a Phenomenon: The detailed and unbiased report of UFOs*, Ace Books, Inc., New York, NY, USA, 1965.

[10.167] Vallee, J. and Aubeck, C., *Wonders in the Sky: Unexplained Aerial Objects from Antiquity to Modern Times and Their Impact on Human Culture, History, and Beliefs*, J.P. Tarcher (Ed.), pp. 29–449, Penguin, New York, NY, USA, 2009.

[10.168] Verissimo, T., *Conheça o fundo sobre OVNIs do Arquivo Nacional*, Arquivo Nacional, September 13, 2018, https://www.gov.br/arqui-vonacional/ pt-br/canais_atendimento/imprensa/copy_of_noticias/conheca-o-fundosobre-ovnis-do-arquivo-nacional.

[10.169] Waldrop, E., Weaponization of outer space: US national policy. *Ann. Air Space Law*, 29, 329–355, 2004.

[10.170] Walker, H., *Northrop Grumman Innovation Systems Employee Honored with NASA's Outstanding Public Leadership Medal*, Northrop Grumman, 2018, https://news.northropgrumman.com/news/features/northrop-grumman-inno-vations-systems-employee-honored-with-na-sas-out-standing-public-leader-ship-medal.

[10.171] Wall, M., *X-37B: The Air Force's Mysterious Space Plane*, Space, May 15, 2020, https://www.space.com/25275-x37b-space-plane.html.

[10.172] Warner, M., UFOs vs. Nukes [Season 2, Episode 3], in: *Unidentified: Inside America's UFO Investigation*, History Channel, July 25, 2020.

[10.173] Wikipedia, *<2008> Christopher Mellon*, Wikipedia: The Free Encyclopedia, 2020, https://en.wikipedia.org/wiki/Christopher_Mellon.

[10.174] Wikipedia, *<2014> Travis S. Taylor*, Wikipedia: The Free Encyclopedia, 2020, https://en.wikipedia.org/wiki/Travis_S._Taylor.

[10.175] Wilson, J.R. (Ed.), *NASA Missions*, National Aeronautics and Space Administration [NASA], 2020, https://www.nasa.gov/missions.

[10.176] Wilson, J.R., *The new era of high-power electromagnetic weapons*, Military & Aerospace Electronics, November 19, 2019, https://www.militaryaerospace.com/power/article/14072339/emp-high-power-electromagnetic-weapons-railguns-microwaves.

[10.177] Woody, C., *The US now has a Space Force and a Space Command. It's not clear what each will do, but commanders are excited*, Business - Insider, December 27, 2019, https://www.businessinsider.com/us-establishes-space-force-to-support-space-command-operations-2019-12.

[10.178] Yeo, M., *China unveils drones, missiles and hypersonic glide vehicle at military parade*, Defense News, October 1, 2019, https://www.defense-news.com/global/asia-pacific/2019/10/01/china-unveils-drones-missiles-and-hypersonic-glide-vehicle-at-military-parade/.

[10.179] Zhang, H., Action/Reaction: U.S. Space Weaponization and China. *Arms Control Today*, 35, 10, 6–11, 2005.

[10.180] Zhao, Y., *Space Commercialization and the Development of Space Law*, Oxford Research Encyclopedias, Oxford University Press, July 30, 2018, https://oxfordre.com/planetaryscience/oso/viewentry/10.1093$002facre-fore$002f9780190647926.001.0001$002facrefore-9780190647926-e-42.

11
Microbial Life, Ethics and the Exploration of Space Revisited

Charles S. Cockell

School of Physics and Astronomy, University of Edinburgh, Edinburgh, UK

Abstract

What might motivate us to not wantonly destroy microbial life on Earth, and if we found it elsewhere, beyond Earth? Critiques against the idea that microbial life has some form of non-instrumental value have ranged from the fact that we destroy microbes on a day-to-day basis to arguments based on perceived opposition to the notion that they have intrinsic value. I challenge these perceptions and suggest that if there is a weakness in the intrinsic value argument, it lies elsewhere, not least in the linguistic problems caused by the term itself. The idea of intrinsic value for microbes may be strongly bound up with the sense of our own virtue, the notion that we might protect microbes because to destroy them reflects badly on ourselves. Regardless of where we locate the argument for intrinsic value, it does not detract from the fundamental observation that we can consider microbial life to have a value beyond its mere use as a resource.

Keywords: Microbes, ethics, intrinsic value, instrumental value, space

11.1 Introduction

Do microbes deserve any protection? This apparently simple question turns out to be more complex than it appears. The answer to it has implications for a number of situations that include but are not limited to: 1) the protection of natural environmental microbial communities [11.4] [11.18] [11.19]; 2) the conservation of specific microbes enriched or

Email: c.s.cockell@ed.ac.uk

Charles S. Cockell: https://www.ph.ed.ac.uk/people/charles-cockell, www.astrobiology.ac.uk

Octavio Alfonso Chon Torres, Ted Peters, Joseph Seckbach and Richard Gordon (eds.) Astrobiology: Science, Ethics, and Public Policy, (233–254) © 2021 Scrivener Publishing LLC

isolated from the environment [11.1] [11.5]; 3) the question of whether it is ethically acceptable to deliberately drive a microbe to extinction (for example a medically relevant microbe) [11.21], which also applies to viruses, although their status as living entities is a much-disputed matter; and 4) planetary protection and the contamination of hypothetical biospheres and ecosystems on other planetary bodies [11.13] [11.34] [11.40] [11.43].

There is no controversy in the observation that microbes have practical uses: they have instrumental value to human beings. Their vast diversity of uses in artificial settings, such as beer, yogurt and drug production, and their importance in natural environments, such as their role in biogeochemical cycles, give us plenty of reasons to protect or conserve them so that these uses and benefits to humanity and the biosphere on which we depend can be maintained in the future [11.1] [11.4] [11.5] [11.18] [11.19]. These motivations influence many of the situations discussed in the first paragraph. For each of the four categories listed, we would protect microbes, based on an instrumental concern, for the following reasons: 1) natural microbial communities carry out biogeochemical transformations ultimately of importance to humans, such as the cycling of carbon in grasslands, rainforests and the oceans; 2) microbial isolates have uses in drug, food and drink production and must be maintained so these uses can be realized; 3) even if we were to perceive no ethical quandary with driving a dangerous microbe to extinction, we might want to retain stocks for research and experimental purposes; and 4) the contamination of hypothetical extraterrestrial microbial communities could compromise important scientific information.

A more controversial question is whether microbes deserve any protection beyond these purely instrumental concerns. For some people, such as those who might hold that only rational beings can have any non-instrumental value, then the answer is in the negative. There are other arguments deployed to support the supposition that microbes do not have non-instrumental value. The notion that microorganisms have a value beyond their purely instrumental use is often referred to as their intrinsic value. While this term is open to critique because it begs the question of how anything can have a value beyond the value ascribed to it by a valuer, which seems to bring us back to the matter of instrumental value, the concept is broadly used to capture the value that we ascribe to something beyond its mere usefulness. Callicott [11.6] defined intrinsic value as follows: "it is valuable in and for itself—if its value is not derived from its utility, but is independent of any use or function it may have in relation to something or someone else. In classical philosophical terminology, an intrinsically valuable entity is said to be an 'end-in-itself,' not just a 'means' to another's end."

To ascribe intrinsic value requires an adequate theory of from where this value might emanate. For the microbial world, a number of authors have implicitly and explicitly addressed this question. For example, Callicott [11.6], pursuing his definition above, investigated the intrinsic value of "nonhuman" species. Varner [11.52] provided a concept for a sliding scale of intrinsic value in which he explicitly included microbial life ("non-conscious organisms" as he puts it, which he explains include single-celled organisms). Attfield [11.2] considered trees, but his arguments seem applicable to microbes since he discusses the inability to protect these entities most of the time and the nature of their, quite simple, biological interests. Cockell [11.12] [11.15] examined intrinsic value for microbes and discussed a variety of theories of intrinsic value and their applicability to microbes, such as the classic argument of Feinberg [11.23].

In this chapter, I consider a diversity of more recent critiques of the concept of intrinsic value applied to microbes. I explore ways to break through this impasse in understanding why we might protect microbes beyond their instrumental value.

11.2 Critiques of Intrinsic Value

In the terrestrial and space exploration ethical literature there have been a number of critiques of the concept of intrinsic value when applied to microorganisms. In this section, I discuss some of these arguments and examine them. I follow with a section in which I propose that there may well be a problem with the concept of intrinsic value for microbes, but not for the reasons alluded to in this section.

11.2.1 The Argument from Existing Destruction

A prominent argument against intrinsic value for microbes derives from the fact that we destroy microbes on a day-to-day basis. For example, in cleaning ourselves, our homes, our food production factories and endless other localities, we lay waste to an unknown but large number of the organisms. Thus, the thinking goes, this is sufficient to render microbes without intrinsic value.

In the extraterrestrial case, this argument has been stated by McKay [11.35]: "It is important to note that the basis for the ethical issues does not come from assigning intrinsic value to microbes *per se,* although this has been suggested [11.12] [11.15]. On Earth, we freely kill microorganisms."

It is true that many microorganisms are unavoidably killed, particularly if we wish to maintain a basic level of hygiene. Patrouch [11.38] described a fictional state of affairs where microbial rights were recognized and bleaching our houses was banned. His story was followed by another *reductio ad absurdum* in the same magazine in which the Legal Rights for Germs Movement takes up a court case on behalf of a pneumonia-causing bacterium against a university medical department, which is charged with having killed the bacterium inside a patient using antibiotics [11.20]. These amusing think pieces aside, unless we wish to return to a medieval state of medical affairs (and even then people killed microbes while cooking their food), then we have to accept unavoidable destruction of a large number of microbes. It is also clear that this destruction of individual organisms generally places microbes outside any credible application of extreme versions of biocentric individualism that demand that we consider the interests of individual organisms.

However, the mere unavoidable destruction of organisms is not an argument on its own for the denial of intrinsic value. That would invite the exclusion of any organisms from any consideration of intrinsic value if some of those members were unavoidably destroyed. There are some organisms we can try not to destroy, particularly large ones such as whales and elephants, but it would seem to be a highly subjective ethical boundary to declare that intrinsic value should be denied to any organism for which human activities and interests result in unavoidable destruction of some of their representatives. Indeed, that would be the very antithesis of the reason for proposing intrinsic value in the first place as being something separate from instrumental value.

The argument for the denial of intrinsic value on the basis of unavoidable destruction would have greater strength if there were no instances where we could protect microbes for reasons other than instrumental uses, since it would invite the obvious implication that although the argument for intrinsic value might be attractive, it would have no practical relevance. However, as I shall illustrate later with the protected stromatolites of Shark Bay, Australia, some microbial communities can be protected, which brings such communities into the purview of an argument about whether they have value beyond their instrumental uses. Nevertheless, my conclusion here is that unavoidable destruction of many microbes does not *a priori* negate intrinsic value for all microbes.

11.2.2   The Argument from Sheer Numbers

The argument from sheer numbers is related to the argument from existing destruction in that both derive from the small size of microorganisms and the immensity of their numbers, but its difference lies in its view that microbes cannot have intrinsic value purely on the basis that there are so many of them. This argument is independent of whether we can or cannot avoid destroying them, although clearly in practical terms these two problems are linked: the vast number of microbes in many environments makes it difficult for human interests to be implemented without killing some, maybe many, microbes.

The argument from sheer numbers has a strong relationship to the idea that organisms have greater value the rarer they are. For instance, we might not value a whale which belongs to an abundant species as much as a species on the brink of extinction. There has long been a public ethical perception that things that are rare have greater value and therefore merit our protection [11.26]. In a more informal way, one can see evidence of this perception in the way that people value some rare antiquity, such as a rare postage stamp, and perceive less value in a mass-produced article, such as a modern postage stamp for everyday use.

The weakness in this argument lies in the linking of the intrinsic value to the number of objects that are being considered. If an organism has intrinsic value on account of its biological or cognitive capacities, for example, that we might think put it beyond mere instrumental value, then this value either exists in the object or not, irrespective of how many objects there are. Nevertheless, the problem remains. Sober [11.48] provided, in my view, perhaps the most elegant and simple elaboration of this essential problem: "But for an environmentalist, this holistic property—membership in an endangered species—makes all the difference in the world: a world with n sperm and m blue whales is far better than a world with $n + m$ sperm and 0 blue whales. Here we have a stark contrast between an ethic in which it is the life situation of individuals that matters, and an ethic in which the stability and diversity of populations of individuals are what matter."

On a more prosaic level, the problem with rarity as an intensifier of value invites the potentially egregious idea that by making a life form rarer, we would increase the intrinsic value of the remaining representatives. Thus, to enhance the value of organisms, we should destroy as many of them as possible and their subsequent rarity would finally render them worthy of intrinsic value.

It may well be the case that large numbers of organisms—microbes, as is the case in this discussion—mean that on a very practical level, destruction of large numbers is less likely to have consequences for the biosphere and its functioning, but large numbers in themselves, seem not to provide a basis for the denial of intrinsic value.

In a paper on environmental ethics and size, I explored the way in which microbial numbers affect the view of their value in more detail [11.15] by imagining a world where polar bears are one micron in size, invisible to the naked eye, and fill our soils and lakes, but in all other respects are physiologically identical to the polar bears we know (this is physically implausible, but this is a thought experiment). In contrast, the microbes, big ugly bags of water, are the size of dogs and are visible at our scale (leaving aside the linguistic problem that they are no longer microscopic). I then invited the reader to consider what would happen if an area was to be cleared, in the process killing many billions of polar bears and a few dog-sized microbes. Which would get the benefit of protection? These types of thought experiments help elucidate the role of the size and number as problems in environmental ethics, these two factors being riven through these discussions.

The small physical size of microbes, which is linked to their large number since the world can host large numbers of these microscopic entities, may well be linked to other capacities that do traditionally impinge on intrinsic value. For example, generally, the smaller the organisms, the less cognitive capacity [11.15]. There are many exceptions, such as large nets of fungal hyphae that are larger than a blue whale yet possess less cognitive capacity, but in general, if we were to claim that sentiency—the ability to feel pleasure and pain—was the basis of intrinsic value [11.3], for example, then one could claim that the large number of microbes is correlated with their small size and their lack of possession of the cognitive capacities to ascribe intrinsic value. Thus, one can see how certain intrinsic value arguments might indirectly link to the large numbers and small sizes of microorganisms.

In summary, regardless of what we think about which creatures merit intrinsic value, if at all, the large number of microbes does not seem directly defensible, as a reason in itself, to deny intrinsic value.

11.2.3   The Argument from Impracticality

An argument against intrinsic value for microbes is the argument from impracticality. This argument is a relative of the argument from destruction and argument from sheer numbers, and it might be said to derive from a combination of these two arguments. It does not explicitly draw on the fact that we destroy large numbers of microbes, nor does it explicitly suggest that a large number, in itself, whether we kill them or not, denies intrinsic value. The argument from impracticality simply says that we cannot protect microbes, regardless of our moral proclivities, because they are so small and are in such vast numbers that it is just not practical to do so. The problem lies with how one would even protect them in any way that makes sense. What is one to do, build a tiny fence around some microbes in a patch of soil? This argument posits that the inability to protect microbes means that no meaningful ethic can be constructed for microbes because the field of ethics relates to our real interactions with the world. If there is no way to protect microbes, then there is no intersection with ethics in the first place.

From a conceptual point of view, the argument is logical. However, although it is true that the destruction of many microbes is unavoidable, there are microbial communities we can protect, thus placing microbes within the realm of ethics.

In the Shark Bay region of Australia are large mounds of microbes, predominantly cyanobacteria, that form structures of about one meter in diameter. Thought to be representative of the sorts of macroscopic microbial structures that could have existed on Archean Earth about 3.5 billion years ago, these stromatolites are protected under UNESCO World Heritage Site status. Indeed, in the UNESCO citation for this site, the microbes loom large: "However it is for its stromatolites (colonies of microbial mats that form hard, dome-shaped deposits which are said to be the oldest life forms on earth), that the property is most renowned" [11.51]. In this case, we have an example of World Heritage status directed toward the unequivocal protection of microorganisms. What deliberations were behind this text is irrelevant to the essential point that despite the everyday impracticality of protecting many microbes, here are some natural microbial communities that have fallen within the remit of an ethos of protection. The mounds of microbes also provide an example that stands in contrast to the argument from existing destruction and the argument from sheer numbers. The UNESCO treatment of these microbes is the paradigmatic example of the fact that regardless of what our theoretical arguments might be, an environmental ethic directed toward microbes in the natural environment generally operates at the level of species or communities of organisms, although in protecting communities we are, by default, protecting the individual microbes that comprise them. I use the phrase "natural environment" because in the laboratory we might well protect relatively small numbers of individual organisms in a frozen vial, and in the extreme case, we can even use single-cell sorting technologies to isolate and preserve the progeny of a single microorganism. Thus, nowadays, a concern for microbial preservation can in practical terms be directed toward a single organism, although this is generally not the case in natural environments. These laboratory methods for preserving and conserving microbes are now standard procedure [11.1].

The conflict between the realities that some representatives of organisms cannot be protected, thus potentially denying them the possibility of being subject to any ethical code we develop, and the implementable protection of others is not a new problem. Attfield [11.2] tackled the same conundrum with trees. He recognized that trees might be organisms we could wish to protect, but that it is not always possible: "There are, of course, in practice ample grounds for disregarding the interests of trees at most junctures. Human and/or animal interests are almost always at stake, and mere vegetation can be forgotten where those interests would be imperiled. The good of trees might outweigh some of our whims: but it does not outweigh our interests except where our interests depend on it. But this is not to make trees of no ethical relevance in themselves...and there is always the residual possibility of their interests being of greater significance than any other which are at stake."

Attfield's argument is not a self-contained argument for microbial ethics, but he does illustrate the same conundrum that applies to microbes, and it raises the question of whether there can be a meaningful ethic for anything that is unprotectable.

Goodpaster [11.25] described an ethical position that we would like to implement as regulative, rather than a practically implementable one which he described as operative. Cockell [11.12] suggested that this was the case with individual microbes that usually cannot be protected. We might want to protect individual microbes as a principle, but the practical needs of our lives usually prevent us from doing so, as Attfield would have it, "at most junctures." As a rejoinder to Cockell's observation, Persson [11.39] points out that if an ethic cannot be applied in practice, then it cannot be an ethic at all, since ethics is about the very matter of how we deal with objects and organisms in the real world. This is surely an important normative point, but it only applies if there are *never* situations where the type of life under consideration can be protected.

However, similarly to Attfield's argument for trees, Cockell [11.12] pointed out that for microbes there are situations where we can protect collections of individual microbes. Cockell stated: "If we have the option of walking on a microbial community that grows at the edge of a lake or walking around it to protect it, we walk around it." Although this behavior, and whether it is merited, in itself is a point of argument, the principle is that microbes are not completely unprotectable, even natural collections of individual microbes. The cyanobacterial stromatolites of Shark Bay are a practical manifestation of this idea. Thus, the argument from impracticality cannot be a reason for the denial of microbial intrinsic value.

11.2.4   The Argument from Prevailing View

Another argument against intrinsic value is the notion that the idea of intrinsic value for microbes is just intuitively absurd. In the application to the extraterrestrial case, the claim for microbial intrinsic value is perceived to have led to a "mania" about the protection of microbial life [11.47]. There are statements that have been made that when extracted and repeated give force to this view. For example, Smith quotes Cockell's statement [11.11] that "microbes have intrinsic worth equal to, if not greater than, that of any other species," and then presents this to a variety of professional philosophers to see what they have to say about this view. We might point out that in the same paper, Cockell elaborates the many ways in which individual microbes must inevitably be killed and examines arguments for the protection of communities or microbial functions, instead of individuals. Thus, despite the bold assertion, no maniacal preservation of microbes was proposed. However, ignoring this, we might proceed with an examination of the response to this assertion.

Similarly, Sagan's comment [11.44] that "If there is life on Mars, I believe we should do nothing with Mars. Mars then belongs to the Martians, even if the Martians are only microbes" is subjected to the same analysis through a poll.

The concern with this method of examining the veracity of intrinsic arguments is two-fold. One is circularity and the other a more fundamental objection.

The circularity problem emerges from the situation that just because a large number of professionals think that something is absurd, it is by definition absurd. On this basis, civilization cannot make any intellectual progress, since we can never break from an existing paradigm that later transmogrifies into a prejudice. I have little doubt that if any group of people had been told 500 years ago that women are equal to men, should be paid the same and have the vote, it would have been ranked as absurd. Indeed, the very job of philosophers and academics in general is to question ideas and construct hypotheses that challenge our existing views and lead to advances beyond them. There is certainly no harm, and it is academically interesting, to discover what professional philosophers think about the intrinsic value of microorganisms, but the notion of intrinsic value being a mania cannot be upheld because a certain number of philosophers say it is.

But more interestingly, Smith's data itself does not seem to support the notion of a mania. A surprising ~26% of professional philosophers fall into the categories of thinking that Cockell's rather bold statement is either patently true (~4%), probably true (~3%) or plausible (~19%) [11.47]. That about 4% of professional philosophers think that Cockell's statement that microbes have intrinsic value "equal to if not greater than other species" is "patently true," implying that no further discussion is required, is astonishing, particularly given that Cockell himself went on to elaborate the different ways in which microbes might not merit protection, such as those with medically significant negative impact [11.11].

Similarly for Sagan's statement, the same three categories of individuals yielded ~28% (plausible or stronger) support [11.47]. Instead of thinking that strong support for the intrinsic value of microbes constitutes a mania, the polling data rather suggests we should think more about it.

That view aside, there is potentially a second objection, a more fundamental one, which is the argument from authority. Of course, it is true to say that professional philosophers have an understanding of many of the issues at stake here. However, the veracity of an argument is not demonstrated merely by the authority of the individuals(s) professing the point of view. If many professional philosophers think that a statement on microbial intrinsic value is absurd, we would certainly want to ask them in more detail why they hold this view and to discover the reasoning behind it, but data from polling cannot in and of itself constitute the argument. Instead, we must know those arguments before the burden is placed on others to provide the countervailing argument. More constructively, we might consider that the polling data could suggest that discussions about microbial value have not had much of an airing in the professional ethics community and that there is a great deal of discussion to be developed on this subject.

In summary, discovering the views of professionals or laypersons about the intrinsic value of microbes is important [11.47], but not an *a priori* argument for or against such value. Nevertheless, the polling that has been accomplished suggests an intriguing percentage of professionals willing to endorse even quite extreme statements about microbial intrinsic value. The view that microbial intrinsic value ranges from being plausible to patently true cannot be dismissed unless we assume that ~25% of professional philosophers are gripped by irrational manias about microbes. Conversely, it is equally important to realize that these percentages provide no a *priori* support to the notion of intrinsic value either.

11.2.5 The Argument from Respect

In this section, rather than critiquing a critique of intrinsic value, I critique an argument for intrinsic value. Over several papers, Cockell examined the value of microorganisms, both in the context of Earth-based problems and their application to life elsewhere, either (currently) speculative extraterrestrial life or the more practical and real concerns of planetary protection. A consistent thread in these papers has been a notion that intrinsic value is merited from the point of view of some form of respect or recognition of the independent trajectory of living things. For example, Cockell states [11.12]: "Part of our reverence for the microbial world must surely reside in the awe we feel for the sheer scale of their [microbes] biogeochemical processes and their longevity on Earth." Later, Cockell [11.17] states that "a form of intrinsic value, rooted in a respect for the trajectories and potential of other life forms and biospheres, allows us to take a more cautious approach to the exploration of worlds." In attempting to understand from where this notion of intrinsic value might emanate, he attempts to root it in the basic biological "interests" that microbes have [11.12], particularly following the classic argument of Feinberg [11.23]. However, with Callicott [11.6], Cockell does not propose that intrinsic value is some objective thing that exists in an organism independent of people, but holds that it is a value projected onto things by humans, a "truncated" intrinsic value in Callicott's terminology.

Cockell's view seems to be that intrinsic value is somewhere instantiated in our sense that microbes are "doing their thing" independent of humans and that it encourages in us a sense that these organisms should be left alone on their trajectories. A weakness in Cockell's analysis is that he is wed to the term intrinsic value, probably because the term has long been used and it is therefore helpful to elaborate a position using terms that are widely understood by the community. This is indeed the direction that others take. Randolph and McKay [11.42] state that "intrinsic value provides an important check against humans intentionally or unintentionally skewing their analysis so that promoting human interests is always the most ethical option."

The major conceptual problem with intrinsic value, one recognized previously [11.6], is that without a valuer, the concept of an object or living thing having some value that objectively inheres within it that can exist independent of any valuer is physically dubious. Surely without a valuer, objects are just objects in the Universe, living or not? Under such circumstances, the notion of "value" makes little sense.

There have been attempts to find an objective basis of intrinsic value. For example, some see it manifested in "order" or "integrity" [11.28]. Moore

[11.37] famously considered that a world could be objectively beautiful, and Varner [11.53] proposed that intrinsic value could objectively exist in an object as a variant of this concept. Wilks [11.54] rooted intrinsic value in the fact that living things show autonomously ordered complexity that is manifested in the capacity for self-organization. The problem with all these "objective" accounts is that they suffer from a potentially spurious physical basis. What exactly is Leopold's "order" and "integrity." We might observe the same thing of a cubic lattice of sodium chloride. The existence of an objective "beauty" in the Universe without an observer to have this concept can be questioned. Over the past decade or so, astonishingly complex and unpredictable autonomously ordered complexity is observed in non-biological objects, or parts of biological objects that exhibit patterns as complex as the simplest bacterium and that *in vitro* represent the same complexity observed by things deemed to be living [11.45] [11.46].

A problem with attempting to find an objective basis for intrinsic value that reaches down to microbes is that any definition that circumscribes microbes is essentially a definition of life, which so far has proven intractable [11.10] [11.36]. The difficulty of finding a definition of life suggests that any attempt to define intrinsic value in some physical objective property that can robustly include microbes and all other things we consider to be alive is likely to meet identical hurdles.

An additional problem with intrinsic value is undoubtedly provoked by words within it. The word intrinsic encourages the need to find something objective within an object, although its intention is usually to capture the idea that the value of the object lies outside and independent of human uses, even if that value is projected by humans [11.6]. The word value conjures up many reasonable questions about how that value can possibly exist without a valuer, and if it is a value projected by a valuer, then in some sense we could argue that it is an artificial human construct and therefore a subspecies of instrumental value; in the sense that intrinsic value is merely a tool to guide our interactions with an object or living thing, it therefore serves us. But the encouragement to think like this rather misses the point that intrinsic value is commonly merely meant to convey that the value in an object is different from its use as a resource of some kind, regardless of from where the value originates.

11.3 What of Intrinsic Value?

The foregoing discussion shows that there are many criticisms of the concept of intrinsic value for microbes. I argue that people are right to critique both the term and the concept, but that often the repudiation of intrinsic value for microbes is done for the wrong reasons, and where attempts have been made to defend it, such as in the case of Cockell [11.12] [11.17], the link between the argument being proposed and the traditional use of the term intrinsic value is tenuous.

Perhaps one problem is that we have become too attached to the very term intrinsic value with all of its linguistic and philosophical problems, a matter long foreshadowed by Callicott's cogent discussion of this very challenge [11.6]. The arguments distract from what the notion of intrinsic value, in my view, is often trying to achieve. One might, in a very prosaic way, say that instrumental value encourages us to ponder how we might interact with an object or living thing as a resource of some kind (whether that is a physical resource or even just something to admire visually), whereas intrinsic value, at least for living things in the wild, is about the opposite sentiment: capturing our sense that the object should generally be left alone, in the case of living things to engage in some genetically determined trajectory that might lead to some future biological state. This is, of course, unless we could provide very satisfactory proof that human intervention would be beneficial for those organisms, which often, apparently paradoxically, includes improving their chances of being left alone (for example in Shark Bay).

The sense that we should leave things alone may be strongly rooted in virtue ethics [11.8]. We might genuinely have a regard for the other entity and a sense that it should be unmolested to carry out its own biological functions, but a strong element of that view for microbial life is likely derived from our own sense of self-worth, where many people think that it is a bad human characteristic to wantonly destroy or use things as mere ends in themselves for the fulfillment of human wishes. The virtue ethical account of this view of microbes resides in the idea that an intelligence that merely uses biological resources for its own ends, disrupting other biota in the process, offends our sense of responsibility to the biosphere and our inwrought sense of appreciation for other life forms that have shared the same biological trials and challenges that our own species has engaged in, however unscientific that sentiment may on close inspection appear. This account of intrinsic value cannot work for all organisms. We do not shy away from murder just because we feel that killing people makes us look bad, although it certainly does that as well. We really do, in Callicott's definition [11.6], think that humans have value "in and for themselves." The extent to which the notion of intrinsic value for different complexities of organisms with different cognitive capacities is merely rooted in our own sense of virtue is worthy of further exploration. It may be the case that the virtue ethical account of intrinsic value becomes more strongly mixed with more detached notions of organisms having value "in and for themselves" the more cognitively complex organisms become.

Do we need to figure out whether intrinsic value is something deeper than virtue ethics, something truly intrinsic to an object or something projected by us, and even then only because of our own sense of self-worth? I propose that we actually do not. The discussion is an intellectually interesting one. However, in practical terms, it is sufficient to recognize that we have this view of other biota possessing some value beyond its mere use and it should mollify and soften a purely instrumental view of the microbial world and our interactions with it.

This view encourages a further question: should we just drop the phrase intrinsic value altogether, particularly when we speak of entities such as microbes? In my view, that would be a good way to avoid much confusion. However, intrinsic value has for a long time been used to capture the notion of non-instrumental value, and although traditions should not dictate the way we operate, its widespread use, like any scientific or common term, can be even more confused by introducing yet new ones. The crucial matter is that we are clear about its weaknesses and its linguistic problems. Callicott [11.6], while recognizing many of the problems elaborated here and endorsing the idea that value can only be derived from a valuer, nonetheless launched a spirited defense of intrinsic value for nonhuman species under the umbrella of the definition provided earlier.

Having said that, there might be terms that do a better job of capturing what is expressed in this chapter. Wilson [11.55] famously described our affinity with the biosphere and its component organisms as biophilia. We might talk of the "biophilic virtue" of caring for microbes and all other organisms instead of their intrinsic value. Cockell [11.16] used the term "teloempathy" to capture the idea of having an empathy with living things as fulfilling some sort of biological trajectory. It should be emphasized that Cockell was very explicit in underscoring that this is *not* derived from the restricted Aristotelian sense of *telos* as some *pre-defined* direction or goal. Evolution has no goal. Cockell used the word *telos* (τέλος) in its acceptably non-restricted sense [11.29] to mean that the object has some trajectory or eventual end that is a result of the expression of its genetic material (even if that end is not predictable or pre-ordained). This is the same broader sense as the use of the word telogenesis to mean the end state of erosional and oxidation processes in geology or the final growth state of a hair or feather follicle in biology, the latter reflecting the widespread use of the prefixes tel- or telo- in biology to mean an end state (noting of course that evolutionary trajectories may not ever reach a definitive "end" state as such).

Despite this, it was interpreted as Aristotelian teleology anyway [11.54], which goes to show how poisonous the word *telos* can be. Cockell was motivated to follow Callicott's use of the term bioempathy [11.6], but because "bio" might be construed as Earth-centric and related to the biology we all know, the word teloempathy was proposed as a term that would capture the same sentiment directed toward any form of life in the Universe that has some trajectory, but the "biology" may be structured very differently from Earth life. Another possible term could be "biorespect," suggested by Taylor [11.50]. Ehrenfeld [11.22] uses the term "non-economic value," but this term may not be helpful since instrumental value is not always economically motivated.

The paragraph above is probably leaving the reader feeling in a state of linguistic dizziness, and that is the point I raised prior to that. We probably could find a term other than intrinsic value that better elaborates what we are talking about without the conceptual baggage, but it would seem that we are likely to introduce further confusion with new terms, each one of which might be subjected to its own detailed critique.

Otiose discussions of definitions are not very useful, but it behooves us to understand that there is a problem with the term intrinsic value. However, as I argue here, those problems should not draw us into the idea that the whole notion of some value projected onto objects other than their instrumental uses is therefore suspect. We just may not have done the best job of elaborating and expressing our sense of this other value we wish to ascribe to these living things. When this value involves organisms that are small, numerically abundant and subject to daily unavoidable destruction, it becomes an even more difficult task to define the locus of this value and how it should be squared with instrumental value. I suggest that this has likely been the root of much of the skepticism of microbial intrinsic value. I further suggest that the debate about the place of microbes in our ethical frameworks is one that merits ongoing discussion.

11.4   Adjudicating Other Interests

This notion that microbes might have some sort of intrinsic value, or whatever we wish to call it, does not block the adjudication of competing interests provided we think that intrinsic value comes in degrees.

The idea that intrinsic value can be elaborated in degrees is an old one. If we root intrinsic value in the sense that living things are on some trajectory and they should be left alone [11.6] [11.12], then something akin to Varner's concept [11.53] might be useful and would deal with the problem of microbes. Indeed, Varner, unusually for environmental ethics papers, explicitly includes microbial life in his formulation. He proposed that the intrinsic value of organisms can vary according to their possession of "biological interests," "non-categorical desires" and "ground projects," each one of these representing successfully more complex manifestations of behavior.

Varner's version of a continuum of intrinsic value would be congruent with our sense that microbes, which in his terminology only possess "biological interests," would generally be subordinate to a dog, for example, that possesses some "non-categorical desires" [11.15], when we come to adjudicating a competing situation between them. Both of these organisms' interests would, in most cases, be subordinate to a human that has future "ground projects" that are frustrated by its deliberate demise. Varner's concept of a continuum of intrinsic value is motivated by the same objective to ascertain how to deal with competing interests as Sturba's account [11.49] of biocentric pluralism and concepts discussed by VanDeVeer [11.52] on "interspecific justice."

Another attraction of Varner's scheme is that it accords well with our physical view of life. If a living thing is merely a lump of matter in the Universe that happens to have a DNA code and it replicates and evolves, facts that do not in themselves seem to provide much basis for a concept of intrinsic value, his scheme, which roots our non-instrumental concerns about living things in their potentialities and trajectories, accords well with our view about the primacy of humans over dogs over microbes, for instance, and why we even think that any of these objects have non-instrumental values to humans.

More recently, Wilks [11.54] proposed a continuum of intrinsic value based on the ability of organisms to respect the intrinsic value of others, a "Respect for Intrinsic Value Test." Although Wilks suggests that organisms exhibiting sentience (examples given are dolphins and dogs) can respect in some rudimentary way the intrinsic value of at least some other members of their species, it is not clear that they have any capacity to respect intrinsic value in other species, particularly if we were to adopt Wilks's own definition of intrinsic value, which rests on an organism having autonomously ordered complexity, a concept that seems too difficult for any organism other than humans to apprehend.

For example, when a dog starts to dig and scratch at a microbial mat at the edge of a lake, I do not find myself in an ethical quandary. The dog does not respect the intrinsic value of the microbes and the microbes do not respect the intrinsic value of the dog. Should I allow the dog to continue killing the microbes or should I kill the dog on behalf of the microbes? This is not usually an ethical dilemma.

Varna's formulation, however, does provide a foundation for the adjudication of these interests. These gradations do not rely on the requirement that organisms themselves respect the intrinsic value of others [11.53], merely that we recognize that the complexity of the behaviors that we associate with the non-instrumental value of living things vary and that allows us to provide some adjudication on competing interests. Nor does this gradation require that we define in some objectively physical way what intrinsic value is, other than that it refers to the property that living things (for which we do not need to provide a definitive definition) can be valued more than as objects of use and that this value might be modified by our sense that they have different potentialities and cognitive capacities.

These formulations do not make things clear-cut, ethics is rarely that simple, but they do have sufficient granularity to allow us to adjudicate many situations that are practically important, such as whether the interests of humans or even a dog should trump the interests of microbes in given situations. In the hypothetical scenario of the discovery of microbes on other planets, they allow us to see how the objectives of humans might trump the interest of microbes in the course of human exploration [11.14].

11.5   Do We Need a Cosmocentric Ethic for Microbial-Type Life?

Finally, I address the matter directly of the consequences of these deliberations for the (currently speculative) discovery of an alien life form.

The notion that the robotic and human expansion into space requires an ethical framework that is up to the challenge of this new frontier is an important one, whether it deals with extraterrestrial life specifically or not [11.7] [11.9] [11.24] [11.30] [11.31] [11.32] [11.33] [11.54]. One phrase used to capture this idea is a "cosmocentric ethic" [11.31].

Would the discovery of microbe-like life elsewhere require a new "cosmocentric" ethic to deal with it? Although the discovery of such life would certainly motivate a protocol on how to deal and interact with it [11.40], which might involve procedures that are novel, it is not clear that microbial life *per se* would require some new ethical basis other than the one we develop for microbial life on Earth. We would presume that extraterrestrial microbe-like life grows, reproduces and evolves; indeed, that is in essence the behavior that brings physical material in the Universe into the purview of matter that we loosely call "life," so in some sense the argument is circular. However, assuming that all material that we wish to include in the class of material we call life exhibits these behaviors (that is not to say we could not be persuaded to think otherwise by the discovery of something entirely radical and new "out there," but this remains within the realms of science fiction at the current time), then the ethics we develop for the simplest creatures on Earth—microbes—seems suitable for this other alien life, whatever its biochemistry.

There is the matter of whether, if this other life is on an independently evolved tree of life (an ethical quandary that we might refer to as a problem of "originism" [11.16]), it would need a new type of cosmocentric ethic. It we were to hold a highly restricted view that ethics only applies to life that is related to Earth's biotic communities, an implicit (although I am sure not deliberate) consequence of views such as those held by Leopold [11.28] as Callicott pointed out [11.7], then the alien life would be different in that it would not lie within our ethical sphere at all. In contrast, others have suggested that life on a different tree of life might merit special treatment [11.27] [11.30] [11.41]. It would certainly be the case that living things on a different tree of life could have special instrumental value, as this life would give us profound new biological and biochemical insights about modes of growth, reproduction and evolution in an entirely alien biochemistry. Furthermore, we would gain new fundamental and industrial knowledge from a new biochemistry. However, the matter of whether it is on a different tree of life is merely a statement about biochemistry and origin. If we locate intrinsic value in almost any of the arguments previously proposed (because we should leave things that we define as life alone [11.6], because living things have integrity [11.28], beauty [11.37], or basic interests [11.23] [11.53], or they are autonomously ordered in complexity [11.54], etc.), then these properties seem independent of the phylogenetic tree if these living things share the property of being replicating, evolving entities of some kind. Thus, at first blush, there seems no *a priori* reason for needing some new and unique cosmocentric ethic. Whatever argument we derive for intrinsic value for terrestrial life would seem applicable to alien life.

If alien life were eventually to be found to come in different complexities, from a single cell-like entity that merely divides in two every now and then to creatures that build spaceships, then we may find that a large number of other ethical debates apply to these life forms with little modification. Varner's work [11.53] on biocentric individualism, which explores a sliding scale of intrinsic value from basic biological interests to creatures like humans that have "ground projects," whether you agree with it or not, seems applicable to alien life with this range of complexity, regardless of its biochemistry. The only situation in which we might find ourselves confronted by a moral situation that is truly alien to us would be our ethical response to an entity vastly intellectually superior to us.

In conclusion, we should consider how our cosmic adventures affect ethics, including the ethics we apply to the simplest of organisms, but I contend that the ethical positions we adopt for terrestrial life are likely to be largely useful for considering our interaction with alien life and may not need modification at all. Instrumental values might well change our behavior toward, and our wish to preserve, a novel and alien microbial biochemistry discovered on another planet, but the core decision on whether we think that such single-celled entities have any intrinsic value can be decided from deliberations on terrestrial microbes and the way in which we value this type of life.

11.6   Conclusions

Understanding the ethics we apply to microorganisms is a fascinating but surprisingly complex endeavor. Their large numbers, small size and for some fraction, unavoidable destruction in the course of human activities, makes them highly enigmatic with respect to many of the traditional arguments of environmental ethics. In this chapter, I have tried to separate these quandaries and some of their corresponding arguments and discuss where I think their weaknesses might lie. The term intrinsic value, used to capture the notion of a non-instrumental value in microbes, is problematic, mainly for linguistic and epistemological reasons, but it does a useful job in capturing this "other" value that we project onto the microbial world. It may act to mollify our worst environmental excesses. This so-called intrinsic value, at least in the case of microbes, may ultimately derive from virtue ethics and a sense of what standards our own behavior demands, but that does not change the practical consequences. I argue here that it is an important consideration in our treatment of the microbial world on Earth and, if we ever find it, extraterrestrial life. Traditions die hard, but other terms might be conceivable that do a better job of capturing what traditionally has been defended under the rubric of intrinsic value. The resolution of these arguments on Earth will likely lead to ethics that can be applied to extraterrestrial life with little or no modification, if we eventually find it.

References

[11.1]   Arora, D.K., Saika, R., Dwievdi, R., Smith, D., Current status, strategy and future prospects of microbial resource collections. *Curr. Sci.*, 89, 488–495, 2005.

[11.2]   Attfield, R., The good of trees. *J. Value Inq.*, 15, 35–54, 1981.

[11.3]   Bentham, J., *Introduction to the Principles of Morals and Legislation*, Clarendon Press, Oxford, 1823.

[11.4]   Bodelier, P.L.E., Toward understanding, managing, and protecting microbial ecosystems. *Front. Microbiol.*, 2, 80, 2011.

[11.5]   Budhiraja, R., Basu, A., Jain, R., Microbial diversity: significance, conservation and application. *Natl. Acad. Sci. Lett. (India)*, 25, 189–201, 2002.

[11.6]   Callicott, B., On the intrinsic value of non-human species, in: *The Preservation of Species*, B. Norton (Ed.), pp. 138–172, Princeton University Press, Princeton, 1986.

[11.7]   Callicott, B., Moral considerability and extraterrestrial life, in: *Defense of the Land Ethic: Essays in Environmental Philosophy*, B. Callicott (Ed.), pp. 263–266, State University of New York Press, Albany, 1989.

[11.8]   Capper, D., The search for microbial Martian life and American Buddhist ethics. *Int. J. Astrobiology*, 19, 244–252, 2020.

[11.9]   Chon Torres, O., Astrobiology and its influence on the renewal of the way we see the world from the teloempathic, educational and astrotheological perspective. *Int. J. Astrobiology*, 19, 330–334, 2020.

[11.10] Cleland, C.E., Chyba, C.F., Defining 'life'. *Orig. Life Evol. Biosph.*, 32, 387–393, 2002.

[11.11] Cockell, C.S., The rights of microbes. *Interdiscip. Sci. Rev.*, 29, 141–150, 2004.

[11.12] Cockell, C.S., The value of microorganisms. *Environ. Ethics*, 27, 375–390, 2005.

[11.13] Cockell, C.S., Planetary protection—A microbial ethics approach. *Space Policy*, 21, 287–292, 2005.

[11.14] Cockell, C.S., Duties to extraterrestrial microscopic organisms. *J. Br. Interplanet. Soc*, 58, 367–373, 2005.

[11.15] Cockell, C.S., Environmental ethics and size. *Ethics Environ.*, 13, 23–39, 2008.

[11.16] Cockell, C.S., Ethics and extraterrestrial life, in: *Humans in outer space—Interdisciplinary perspectives*, U. Landfester, N.-L. Remuss, K.-U. Schrogl, J.-C. Worms (Eds.), pp. 80–101, Springer, Berlin, 2011.

[11.17] Cockell, C.S., The ethical status of microbial life on Earth and elsewhere: In defence of intrinsic value, in: *The Ethics of Space Exploration*, J.S.J. Schwartz and T. Milligan (Eds.), pp. 167–179, Springer, Berlin, 2016.

[11.18] Cockell, C.S. and Jones, H.L., Advancing the case for microbial conservation. *Oryx*, 43, 520–526, 2009.

[11.19] Colwell, R.R., Biodiversity amongst microorganisms and its relevance. *Biodivers. Conserv.*, 1, 342–345, 1992.

[11.20] Cox, D.L., University medical versus *Diplococcus pneumoniae*. *Analog Sci. Fiction Fact*, 97, 153–156, 1978.

[11.21] Dixon, B., Smallpox, imminent extinction and an unresolved dilemma. *New Sci.*, 26, 430–432, 1976.

[11.22] Ehrenfeld, D., The conservation of non-resources. *Am. Sci.*, 64, 648–656, 1976.

[11.23] Feinberg, J., The rights of animals and unborn generations, in: *Philosophy and Environmental Crisis*, W.T. Blackstone (Ed.), pp. 43–68, University of Georgia Press, Athens, 1974.

[11.24] Fogg, M.J., The ethical dimensions of space settlement. *Space Policy*, 16, 205–211, 2000.

[11.25] Goodpaster, K., On being morally considerable. *J. Philos.*, 22, 308–325, 1978.

[11.26] Gunn, A., Preserving rare species, in: *Earthbound: New Introductory Essays in Environmental Ethics*, T. Regan (Ed.), pp. 289–335, Waveland Press, Illinois, 1984.

[11.27] Hargrove, E., *Foundations of Environmental Ethics*, Prentice-Hall, New Jersey, 1989.

[11.28] Leopold, A., *A Sand County Almanac*, Oxford University Press, Oxford, 1949.

[11.29] Liddell, H.G., Scott, R., *A Greek-English Lexicon*, Perseus Digital Library, https://www.perseus.tufts.edu/hopper/.

[11.30] Lupisella, M., The rights of Martians. *Space Policy*, 13, 89–94, 1997.

[11.31] Lupisella, M. and Logsdon, M., Do we need a cosmocentric ethic? *Int. Astronaut. Sci.*, 1997. Paper IAA-97-IAA.9.2.09.

[11.32] Marshall, A., Ethics and the extraterrestrial environment. *J. Appl. Philos.*, 10, 227–236, 1993.

[11.33] McKay, C.P., Does Mars have rights?, in: *Moral Expertise*, D. MacNiven (Ed.), pp. 184–197, Routledge, London, 1990.

[11.34] McKay, C.P., Davis, W., Planetary protection issues in advance of human exploration of Mars. *Adv. Space Res.*, 9, 197–202, 1989.

[11.35] McKay, C.P., The search on Mars for a second genesis of life in the Solar System and the need for biologically reversible exploration. *Biol. Theory*, 13, 103–110, 2018.

[11.36] Mix, L.J., Defending definitions of life. *Astrobiology*, 15, 15–19, 2015.

[11.37] Moore, G.E., *Principia Ethica*, Cambridge University Press, Cambridge, 1903.

[11.38] Patrouch, J., Legal rights for germs. *Analog Sci. Fiction Fact*, 97, 167–169, 1977.

[11.39] Persson, E., The moral status of extraterrestrial life. *Astrobiology*, 12, 976–984, 2012.

[11.40] Race, M.S. and Randolph, R.O., The need for operating guidelines and a decision-making framework applicable to the discovery of non-intelligent extraterrestrial life. *Adv. Space Res.*, 30, 1583–1591, 2002.

[11.41] Randolph, R.O., Race, M.S., McKay, C.P., Reconsidering the theological and ethical implications of extraterrestrial life. *CTNS (Center Theology Natural Sciences, Berkeley) Bull.*, 17, 1–8, 1997.

[11.42] Randolph, R.O. and McKay, C.P., Protecting and expanding the richness and diversity of life: An ethic for astrobiology research and space exploration. *Int. J. Astrobiology*, 13, 28–34, 2014.

[11.43] Rummel, J.D., Planetary protection in the time of astrobiology: Protecting against biological contamination. *Proc. Natl. Acad. Sci.*, 98, 2128–2131, 2001.

[11.44] Sagan, C., *Cosmos*, Ballentine Books, New York, 1985.

[11.45] Sanchez, T., Chen, D.T.N., DeCamp, S.J., Heymann, M., Dogic, Z., Spontaneous motion in hierarchically assembled active matter. *Nature*, 491, 431–435, 2012.

[11.46] Schaller, V., Wieber, C., Semmrich, C., Frey, E., Bausch, A.R., Polar patterns of driven filaments. *Nature*, 467, 73–77, 2010.

[11.47] Smith, K.C., The curious case of the Martian microbes: Mariomania, intrinsic value and the prime directive, in: *The Ethics of Space Exploration*, J.S.J. Schwartz and T. Milligan (Eds.), pp. 195–208, Springer, Berlin, 2016.

[11.48] Sober, E., Philosophical problems for environmentalism, in: *The Preservation of Species: The Value of Biological Diversity*, B.G., Norton (Ed.), pp. 173–194, Princeton University Press, Princeton, N.J, 1986.

[11.49] Sturba, J., From biocentric individualism to biocentric pluralism. *Environ. Ethics*, 17, 191–207, 1995.

[11.50] Taylor, P., The ethics of respect for nature. *Environ. Ethics*, 3, 197–218, 1981.

[11.51] No Author, *Shark Bay*, no date, https://whc.unesco.org/en/list/578/.

[11.52] VanDeVeer, D., Interspecific justice. *Inquiry*, 22, 55–70, 1979.

[11.53] Varner, G., Biocentric individualism, in: *Environmental Ethics: What Really Matters, What Really Works*, D. Schmitz and E. Willott (Eds.), Oxford University Press, Oxford, 2002108–120.

[11.54] Wilks, A.F., Kantian foundations for a cosmocentric ethics, in: *The Ethics of Space Exploration*, J.S.J. Schwartz and T. Milligan (Eds.), pp. 181–192, Springer, Berlin, 2016.

[11.55] Wilson, E.O., *Biophilia*, Harvard University Press, Cambridge, MA, 2009.

12

Astrobiology, the United Nations, and Geopolitics

Linda Billings

NASA's Astrobiology and Planetary Defense Programs, Sarasota, Florida, USA

Abstract

In the context of the UN Outer Space Treaty of 1967, this chapter focuses on astrobiology's primary astroethical consideration, to ensure that: 1) astrobiology missions to other planetary bodies are in compliance with planetary protection requirements. That is, astrobiologists must ensure that their experiments and instruments pose no risk of contaminating other planetary bodies with terrestrial biology; and, 2) extraterrestrial biology, should it exist, does not contaminate the terrestrial environment. Formulation of pertinent public policy is becoming urgent because of recent plans by both the private and public sectors to land on off-Earth sites and even to colonize Mars.

Keywords: Astrobiology, space law, space ethics, space policy, planetary protection

12.1 Introduction

In the context of the UN Outer Space Treaty of 1967, this chapter focuses on astrobiology's primary ethical consideration to ensure that: 1) astrobiology missions to other planetary bodies are in compliance with planetary protection requirements. That is, astrobiologists must ensure that their experiments and instruments pose no risk of contaminating other planetary bodies with terrestrial biology; and, 2) extraterrestrial biology, should it exist, does not contaminate the terrestrial environment.

Email: billingslinda1@gmail.com

Octavio Alfonso Chon Torres, Ted Peters, Joseph Seckbach and Richard Gordon (eds.) Astrobiology: Science, Ethics, and Public Policy, (255–270) © 2021 Scrivener Publishing LLC

While law and ethics are not interchangeable, law tends to reflect the ethical principles of the society that produced it. The 1967 United Nations Treaty on Principles Governing the Activities of States in the Exploration and Use of Outer Space, Including the Moon and Other Celestial Bodies, commonly known as the Outer Space Treaty (OST), "provides the basic framework on international space law," according to the United Nations Office of Outer Space Affairs.[1] The OST reflects the common values and ethical principles of the member states of the UN. As of June 2019, 109 nations were states parties to the treaty [12.7].

In the United States, which ratified the treaty in 1967, the treaty has the status of "the law of the land," according to the U.S. Constitution. The OST establishes that states parties to the treaty—that is, national governments—are responsible for the actions of both public- and private-sector actors in space.

The treaty establishes the following principles:

- The exploration and use of outer space shall be carried out for the benefit and in the interests of all countries and shall be the province of all mankind;
- Outer space shall be free for exploration and use by all States;
- Outer space is not subject to national appropriation by claim of sovereignty, by means of use or occupation, or by any other means;
- States shall not place nuclear weapons or other weapons of mass destruction in orbit or on celestial bodies or station them in outer space in any other manner;
- The Moon and other celestial bodies shall be used exclusively for peaceful purposes;
- Astronauts shall be regarded as the envoys of mankind;
- States shall be responsible for national space activities whether carried out by governmental or non-governmental entities;
- States shall be liable for damage caused by their space objects; and
- States shall avoid harmful contamination of space and celestial bodies.

OST articles VI and IX are especially relevant to astrobiology ethics.

---

[1]https://www.unoosa.org/oosa/en/ourwork/spacelaw/treaties/introouterspacetreaty.html

Article VI establishes that: "States Parties to the Treaty shall bear international responsibility for national activities in outer space, including the Moon and other celestial bodies, whether such activities are carried on by governmental agencies or by non-governmental entities, and for assuring that national activities are carried out in conformity with the provisions set forth in the present Treaty. The activities of non-governmental entities in outer space, including the Moon and other celestial bodies, shall require authorization and continuing supervision by the appropriate State Party to the Treaty. When activities are carried on in outer space, including the Moon and other celestial bodies, by an international organization, responsibility for compliance with this Treaty shall be borne both by the international organization and by the States Parties to the Treaty participating in such organization."[2]

Article IX establishes that, "In the exploration and use of outer space, including the Moon and other celestial bodies, States Parties to the Treaty shall be guided by the principle of co-operation and mutual assistance and shall conduct all their activities in outer space, including the Moon and other celestial bodies, with due regard to the corresponding interests of all other States Parties to the Treaty. States Parties to the Treaty shall pursue studies of outer space, including the Moon and other celestial bodies, and conduct exploration of them so as to avoid their harmful contamination and also adverse changes in the environment of the Earth resulting from the introduction of extraterrestrial matter and, where necessary, shall adopt appropriate measures for this purpose. If a State Party to the Treaty has reason to believe that an activity or experiment planned by it or its nationals in outer space, including the Moon and other celestial bodies, would cause potentially harmful interference with activities of other States Parties in the peaceful exploration and use of outer space, including the Moon and other celestial bodies, it shall undertake appropriate international consultations before proceeding with any such activity or experiment. A State Party to the Treaty which has reason to believe that an activity or experiment planned by another State Party in outer space, including the Moon and other celestial bodies, would cause potentially harmful interference with activities in the peaceful exploration and use of outer space, including the Moon and other celestial bodies, may request consultation concerning the activity or experiment."[3]

Thus, existing space law—that is, the OST—is adequate to govern public- and private-sector space activities. What is needed, to back up the law, is a

[2]https://2009-2017.state.gov/t/isn/5181.htm#treaty
[3]Ibid.

regulatory regime to govern public- and private-sector space activities, which are expanding in the 21st century. Different states parties to the OST may have different approaches to regulating public- and private-sector activities in space, and political ideology plays a role in establishing such approaches. In the United States, for example, neoliberal political ideology has established a pattern of "light-touch" regulation, and private-sector actors in the space sector have been lobbying for an even lighter-touch approach.

A scientific and ethical dilemma facing astrobiology in the 21st century is the possibility that public- and private-sector actors might land humans on potentially habitable planetary bodies—such as Mars—before astrobiological exploration of those bodies is complete. Further adding to this dilemma is that consideration of ethical issues in astrobiology remains the domain of the natural and social sciences and the humanities, not with decision makers.

12.2   What is Astrobiology?

Astrobiology research focuses on three basic questions: How does life begin and evolve? Does life exist elsewhere in the Universe? What is the future of life on Earth and beyond? In recent years, astrobiology research has focused more and more on the link between the "astro" and the "bio" in astrobiology—that is, what makes a planetary body habitable. "Habitability" has become a major buzzword in astrobiology as researchers have learned more about potentially habitable extraterrestrial environments in our solar system and beyond and deepened their understanding of how and when the early Earth became habitable.

Major topics of research in astrobiology today include identifying abiotic sources of organic compounds, the synthesis and function of macromolecules in the origin of life, early life and the development of increasing complexity, the co-evolution of life and environment, and identifying, exploring, and characterizing environments for habitability and biosignatures.

12.3   Ethical Issues in Astrobiology

In the U.S. National Aeronautics and Space Administration's (NASA's) 2015 Astrobiology Strategy,[4] Lucas Mix and Connie Bertka addressed this question: What is the role for ethics in astrobiology? They wrote:

---

[4]https://astrobiology.nasa.gov/nai/media/medialibrary/2016/04/NASA_Astrobiology_Strategy_2015_FINAL_041216.pdf

"Interest in the history, character, and extent of life is intimately tied to questions of value. Astrobiological questions impact public thinking about our ethical obligations toward living entities. Legal and ethical limitations on experiments, particularly those traveling to other planets, shape the science we do. What role do definitions of life play in ethical typologies? What role do definitions of life play in cultural and religious cosmologies? What resources are available within various cultural and religious traditions for the incorporation of non-Terran life into worldviews? To what extent is human exceptionalism and/or Terran exceptionalism necessary or desirable? Do humans have non-Terran ethical obligations? Can they be agreed upon in a socially plural fashion? Does Astrobiology have implications for Terran environmental ethics?"

As mentioned above, this chapter will focus on astrobiology's primary astroethical consideration, to ensure that: 1) astrobiology missions to other planetary bodies are in compliance with planetary protection requirements. That is, astrobiologists must ensure that their experiments and instruments pose no risk of contaminating other planetary bodies with terrestrial biology; and, 2) extraterrestrial biology, should it exist, does not contaminate the terrestrial environment. The latter consideration applies to samples of extraterrestrial bodies returned to Earth. That is, astrobiology's primary ethical consideration is to ensure compliance with planetary protection policy and guidelines.

12.4   Astrobiology and Planetary Protection

The international Committee on Space Research (COSPAR) has a planetary protection policy[5] in place, compliance with which is voluntary, as COSPAR has no enforcement authority. The U.S. National Aeronautics and Space Administration has a planetary protection policy in place, compliance with which is mandatory.[6] COSPAR policy forms the basis for NASA policy. COSPAR's Panel on Planetary Protection (PPP) is responsible for reviewing and updating COSPAR guidelines. COSPAR PPP members include representatives from China, France, Germany, India, Japan, Russia,

---

[5]https://cosparhq.cnes.fr/assets/uploads/2019/12/PPPolicyDecember-2017.pdf
[6]NASA Policy Directive (NPD) 8020.7G, "Biological Contamination Control for Outbound and Inbound Planetary Spacecraft" (https://nodis3.gsfc.nasa.gov/displayDir.cfm?Internal_ID=N_PD_8020_007G_&page_name=main&search_term=8020%2E7), and NASA Procedural Requirements (NPR) 8020.12D, "Planetary Protection Provisions for Robotic Extraterrestrial Missions" (https://nodis3.gsfc.nasa.gov/displayDir.cfm?Internal_ID=N_PR_8020_012D_&page_name=AppendixB&search_term=8020%2E12).

Spain, and the U.S. Through the COSPAR PPP, COSPAR, NASA, and other interested space agencies work to ensure that their policies and practices are in concurrence. COSPAR and NASA have had concurrent planetary protection (in their earlier years, planetary quarantine) policies in place since the 1960s [12.10].

According to the COSPAR PPP, "It is the State that ultimately will be held responsible for wrongful acts committed by its jurisdictional subjects. The Outer Space Treaty does not require States Parties to use the COSPAR guidance framework on planetary protection in fulfilling Article IX obligations. However, States Parties have for fifty years implemented Article IX by using COSPAR and following its planetary protection guidance framework." [12.7].

The purpose of planetary protection policy is to preserve pristine extraterrestrial environments for scientific exploration—that is, the search for evidence of extraterrestrial life. This purpose is in keeping with Article IX of the OST. Planetary protection policy does not explicitly call for preserving pristine extraterrestrial environments for their own sake. As the COSPAR PPP points out, "protecting solar system bodies for their own sake, protecting unique solar system environments or historical sites are specifically not included in the COSPAR Planetary Protection Policy."[7]

In 2010, COSPAR held a workshop to address ethical considerations for planetary protection in space exploration [12.11]. One of the goals of this workshop[8] was to determine whether planetary protection policy should be extended to protect other aspects of planetary environments within an ethical and practical framework intended to preserve pristine extraterrestrial environments for scientific exploration. Workshop participants considered the value of protecting pristine environments for their own sake and the intrinsic value of all life, wherever it may be.

Workshop recommendations were as follows [12.11]:

1. An expanded framework for COSPAR planetary protection policy/policies is needed to address other forms of "harmful contamination" beyond what is currently addressed (i.e., biological and organic constituent contamination). Such a policy framework should be developed within the scope of the UN Outer Space Treaty (including Article IX on harmful contamination).

---

[7]Ibid.
[8]The author was a participant in this workshop.

2. COSPAR should maintain the existing, effective planetary protection policy virtually intact while examining in parallel how to address issues of ethical considerations related to life, non-life, environmental management and multiple uses.

3. COSPAR should add a separate and parallel policy to provide guidance on requirements/best practices for protection of non-living/non-life-related aspects of Outer Space and celestial bodies.

4. COSPAR should consider that the appropriate protection of potential indigenous extraterrestrial life shall include avoiding the harmful contamination of any habitable environment—whether extant or foreseeable—within the maximum potential time of viability of any terrestrial organisms (including microbial spores) that may be introduced into that environment by human or robotic activity.

5. To begin the process of integrating ethical considerations into COSPAR Planetary Protection Policy, the group recommended that specific wording be added to the COSPAR preamble and policy on planetary protection related to life and non-life, biological planetary protection, and environmental disturbances.

With the possible exception of recommendation #4, as of this writing, these recommendations have not been implemented.

Workshop participants also formulated the following recommendations[9] [12.11] relating to "the ethics of making [planetary protection] policy: broadening and sustaining dialogue":

1. COSPAR should encourage its members and associated states to initiate and sustain a broad-reaching public dialogue about the ethical aspects of space exploration and planetary protection and to conduct public engagement and public consultation efforts at national and/or regional levels concerning ethics in space exploration.

2. COSPAR policy regarding space exploration and the preservation of outer space environments should take into account and reflect the international trend toward sincere consultation with a broad range of publics about the ethical and policy issues associated with space exploration, as has been

---

[9]The author was a co-author of these recommendations along with Susanna Priest.

put into practice for consultation about developments in biotechnology, nanotechnology, neuroscience, and so on, in both Europe and the United States and Canada.

3. Toward addressing the challenges of assessing and incorporating public opinion in policy and planning, COSPAR should ask its Panels on Planetary Protection and Space Exploration to hold a workshop on public engagement, consultation, and participation in public policy making, involving relevant experts. The purpose of this workshop is to inform members about the premises, principles, and purposes of public engagement activities, and to disseminate best practices.

At the time of this writing, these recommendations have not been implemented.

12.5   Conflicting Ideologies

At the same time that astrobiologists are intent upon exploring multiple environments in our solar system that they deem potentially habitable— Mars, Jupiter's moon Europa, Saturn's moons Enceladus and Titan, and now even Neptune's moon Triton—human-exploration advocates, inside and outside NASA, continue to argue, as they have for years, that planetary protection requirements are unnecessarily restrictive. In the United States, this critique gained traction in the Obama and Trump administrations.

The conflict between advocates of the need for planetary protection and advocates of human exploration and settlement of other planetary bodies is rooted in ideology [12.2].

In the 21st century, some advocates of the human exploration and settlement of space have been arguing that astrobiology and planetary protection should not hinder the advance of humans into space. Other advocates, and critics, of human exploration—including astrobiologists, philosophers of science, and environmental ethicists—have been arguing for a more measured and careful approach to human exploration, including measures for planetary protection.[10]

---

[10]See, for example, *Futures* 110 (June 2019), a special issue of the journal on human colonization of other worlds, edited by Kelly Smith and Keith Abney. The papers in this issue were a product of a roundtable discussion of the topic at the 2018 meeting of the Society for Social and Conceptual Issues in Astrobiology. The author was a contributor.

The former position is driven by neoliberal/libertarian ideology—let individuals and businesses proceed largely unfettered by government. Among critics of neoliberalism, Noam Chomsky has described this theory of political economy as the "Washington consensus," designed by the U.S. government and international financial institutions to "liberalize trade and finance, let markets set prices...end inflation...privatize" [12.6].

The latter position is driven by humanistic ideology and the principles of environmental ethics. The American Humanist Association describes humanism as "a progressive philosophy of life that, without theism or other supernatural beliefs, affirms our ability and responsibility to lead ethical lives of personal fulfillment that aspire to the greater good."[11] The principles of environmental ethics are addressed elsewhere in this volume.

Jeff Bezos, the billionaire founder and leader of the aerospace company Blue Origin, and Elon Musk, the billionaire founder and leader of the aerospace company SpaceX, have both expressed their desire to develop human colonies in space [12.8] [12.9]. Bezos is a self-described libertarian and Musk tends to express libertarian views about the roles of government and business and the rights of individuals.

NASA has a policy in place on planetary protection for human missions to Mars.[12] As this policy notes, "There is presently insufficient scientific and technological knowledge to establish detailed requirements and specifications to enable NASA to incorporate planetary protection into the development of crewed spacecraft and missions. Thus, this NASA Policy Instruction (NPI) establishes policy guidelines and describes the approach for obtaining the scientific information and developing the technologies and procedures over the next few years that are needed to draft an NPR for crewed planetary missions." However, the policy also states, "The intent of this planetary protection policy is the same whether a mission to Mars is conducted robotically or with human explorers. Accordingly, planetary protection goals should not be relaxed to accommodate a human mission to Mars."

Nonetheless, in 2018, under the Trump administration, NASA established a new Regulatory and Policy Committee (RPC) of its NASA Advisory Council (NAC). The RPC was tasked with advising NASA on civil space regulation and policy "to catalyze America's civil space economy

---

[11]https://americanhumanist.org/what-is-humanism/definition-of-humanism/
[12]NSA Policy instruction NPI 8020.7, https://nodis3.gsfc.nasa.gov/OPD_docs/NPI_8020_7.pdf

and advance other policy objectives."[13] The primary topic for its first meeting was planetary protection policy. The RPC's first chair, Mike Gold, a corporate attorney and a critic of planetary protection requirements, joined NASA in 2019 as a special advisor to the administrator for international and legal affairs, tasked with promoting a set of eight principles for the extraction and utilization of extraterrestrial resources, developed by the RPC, to a wider audience domestically and internationally.[14]

At the RPC's first meeting on November 18, 2018, Gabriel Swiney, lead international space lawyer for the U.S. Department of State, gave a briefing on "harmful contamination: what's legally required, and what's not." In a December 10, 2018, report to the NAC, the RPC recommended, responding to Swiney's briefing, that "NASA should establish a multi-disciplinary team of experts from industry, the scientific community, and relevant government agencies, to develop U.S. policies that properly balance the legitimate need to protect against the harmful contamination of the Earth or other celestial bodies with the scientific, social, and economic benefits of public and private space missions. The recommended multi-disciplinary team should be tasked with producing a detailed policy, provided to a joint session of the Regulatory and Policy Committee, the Science Committee, and the HEO Committee, that will describe best practices for the Administration, the science and research community, and private sector, to protect against harmful contamination and adverse changes in the environment of the Earth. The multi-disciplinary team should also explore the use of the term 'Planetary Protection' relative to other terms utilized in the Outer Space Treaty."

The NAC concurred with this recommendation, stating: "COSPAR [planetary protection] regulations[15] are becoming obsolete and do not properly account for the possibilities of human spaceflight and private sector missions. Creating a multi-disciplinary team to craft a balanced policy that can be implemented by NASA (and eventually COSPAR itself) will help to encourage new, innovative, human spaceflight, robotic, and private sector missions to Mars and other celestial bodies. The more of these missions that take place the more science, exploration, and commerce can be conducted.... If NASA adopts the COSPAR guidelines[16] without any

---

[13]NASA Advisory Council Regulation and Policy Committee Terms of Reference, November 15, 2018.
[14]When he joined NASA, Gold stepped down as RPC chair.
[15]COSPAR has no regulatory authority. Compliance with COSPAR planetary protection guidelines is voluntary.
[16]NASA and COSPAR have worked together for decades to ensure that NASA policy and COSPAR guidelines are in agreement.

review or revisions they will have a chilling effect on robotic, human space-flight, and private sector missions. The costs and complexity of conducting space missions will not be moderated and could become problematic. The result will be less science, exploration, and commercial activities, harming both national and global interests."

In response, NASA created a Planetary Protection Independent Review Board (PPIRB). The PPIRB issued a report[17] on October 18, 2019, including approximately 80 findings and recommendations, including the following:

"The context in which PP is conducted is profoundly and rapidly changing…The PPIRB findings and recommendations presented in this report apply to the current era and generally are made with a 3–5 year horizon in mind.…

NASA should reassess its PP guidelines at least twice per decade with an IRB-like body.…

NASA should establish a standing forum for the discussion and res-olution of emergent PP issues that includes input from government, private sector, and perhaps even non-US private sector enterprises.…"

It is unclear whether and how NASA will respond to the PPIRB's recom-mendations. However, it appears that one step has been taken to respond to the latter recommendation for a standing forum. In September 2020, the U.S. National Academies' Space Studies Board established a new Committee on Planetary Protection, requested and funded by NASA, intended to provide the space agency with external expert advice "on mea-sures undertaken to protect the biological and environmental integrity of extraterrestrial bodies and to preserve the integrity of Earth's biosphere."[18] At a December 2019 meeting of the COSPAR PPP,[19] Blue Origin chief scientist Steve Squyres said his company "takes planetary protection very seriously and plans to be a good steward." SpaceX principal Mars develop-ment engineer Paul Wooster said "planetary protection "has been worked through" for his company's Red Dragon missions to Mars. However, it remains to be seen how these and other companies will comply with planetary protection policies. Compliance is necessary to ensure that

---

[17]NASA Planetary Protection Independent Review Board report to NASA/SMD, October 18, 2019. https://www.nasa.gov/sites/default/files/atoms/files/planetary_protection_board_report_20191018.pdf
[18]https://www.nationalacademies.org/our-work/committee-on-planetary-protection
[19]Meeting minutes, https://cosparhq.cnes.fr/assets/uploads/2020/04/PPP_2019-meeting-December-COSPAR-PPP-Minutes-Open-Session.pdf

astrobiological exploration—the search for habitability and life in the solar system—will be able to continue.

12.6   International Cooperation—or Not?

Historian Andrew Bacevich has observed: "For the United States, the passing of the Cold War yielded neither a 'peace dividend' nor anything remotely resembling peace." Bacevich has documented the construction and maintenance of a post-Cold War United States as "the world's only superpower": "Whatever means were employed, the management of empire assumed the existence of bountiful reserves of power—economic, political, cultural, but above all military. In the immediate aftermath of the Cold War, few questioned that assumption. The status of the United States as 'sole superpower' appeared unassailable.... This was not hypernationalistic chest-thumping; it was the conventional wisdom" [12.1].

While some U.S. politicians continue to claim that the United States is the world's only superpower and the leader of the world, it is not. From its inception in 1958, the U.S. space program has been an instrument of foreign policy [12.3]. While the United States spends more money on space exploration than any other nation, it is by no means the only nation that is "leading" in the space exploration arena. The United States, Russia, and China have the capability of building space stations and launching people into space. In addition to these nations, India and Japan have launch vehicles capable of sending robotic spacecraft into the solar system. These five nations, in addition to several private-sector launch vehicle providers[20], are sending satellites and other robotic spacecraft into the solar system for many other nations.

At a time when space exploration is becoming increasingly international, multinational, and transnational, U.S. space policy and actions continue to declare that the United States is uniquely suited to lead the world in space exploration, while others may follow, as lesser partners. Successive administrations have insisted on the need to sustain U.S. dominance in the global space arena.

Historically and presently, the U.S. national rhetoric of space exploration advances a conception of outer space as a place of wide-open spaces and limitless resources—a space frontier. The metaphor of the frontier, with its associated images of pioneering, homesteading, claim-staking, and taming, has been persistent in American history, including the 60-year history of U.S. space flight [12.2].

---

[20]Including Arianespace, SpaceX, and Rocket Lab.

The Obama administration's space policy[21] called for "preserving the space environment and the responsible use of space." However, the policy refers only to space situational awareness and the minimization of space debris as means of preserving the space environment. The policy's commercial space guidelines call for "Minimiz[ing], as much as possible, the regulatory burden for commercial space activities and ensure that the regulatory environment for licensing space activities is timely and responsive." Its civil space guidelines call for "Maintain[ing] a sustained robotic presence in the solar system to: conduct scientific investigations of other planetary bodies; demonstrate new technologies; and scout locations for future human missions." The Trump administration has updated national space policy with several space policy directives, including SPD-2, "Streamlining regulations on commercial use of space."[22]

12.7   Conclusions

In this contribution, I have reminded us of the decisive role in formulating public policy played by the UN Outer Space Treaty of 1967. I have then focused on astrobiology's primary astroethical consideration, *Planetary Protection*. I want to ensure that: 1) astrobiology missions to other planetary bodies are in compliance with planetary protection requirements. That is, astrobiologists must ensure that their experiments and instruments pose no risk of contaminating other planetary bodies with terrestrial biology; and, 2) extraterrestrial biology, should it exist, does not contaminate the terrestrial environment. I believe that new formulation of pertinent public policy is becoming urgent, because of recent plans by both the private and public sectors to land on off-Earth sites and even to colonize Mars.

It is unclear how COSPAR and NASA planetary protection policies will adapt to changes in the global environment for space exploration. Though public opinion polling[23] does not show great support for it, both the Obama and Trump administrations made human exploration a top priority and

---

[21]Barack Obama: Advancing the Frontiers of Space Exploration. Barack Obama 2008, August 16.

[22]https://www.whitehouse.gov/presidential-actions/space-policy-directive-2-streamlining-regulations-commercial-use-space/

[23]See, for example, "Majority of Americans believe it is essential that the U.S. remain a global leader in space" (2018). Pew Research Center, June 6. This poll asked respondents to rank nine priorities for NASA, and human missions to Mars and the Moon ranked eighth and ninth. https://www.pewresearch.org/science/2018/06/06/majority-of-americans-believe-it-is-essential-that-the-u-s-remain-a-global-leader-in-space/

promoted greater private-sector activity in space, without regulations that might hinder that activity. The rationale for astrobiology is scientifically sound, thanks to decades of work in the science community [12.4]. The rationale for human exploration is ideological, not scientific. The goals and objectives of astrobiology and the goals and objectives of human exploration are in conflict.

The U.S. Congress continues to forbid NASA to collaborate with China in space activities. Under the Trump administration, U.S.-China relations grew especially tense [12.5]. U.S. president Trump's 2020 executive order promoting mining of the Moon and other bodies in the solar system[24] drew criticism from the Russian space agency Roscosmos: "Attempts to expropriate outer space and aggressive plans to actually seize territories of other planets hardly set the countries (on course for) fruitful cooperation,' said Roscosmos Deputy Director General on International Cooperation Sergey Saveliev."[25]

Meanwhile, other nations are collaborating with China and Russia on space activities.[26] China and Russia are represented on COSPAR's Panel on Planetary Protection. The National Space Science Center of the Chinese Academy of Sciences[27] and the Russian Academy of Sciences, along with research organizations from many other nations, space-faring and not, are institutional members of COSPAR.[28]

Since the beginning of the Space Age, COSPAR has provided a forum for advancing international cooperation in the scientific exploration of space. It remains to be seen how COSPAR will be able to continue this service in the changing global political environment of the 21st century.

---

[24]Executive Order on Encouraging International Support for the Recovery and Use of Space Resources, https://www.whitehouse.gov/presidential-actions/executive-order-encouraging-international-support-recovery-use-space-resources/.

[25]Plans to seize other planets' territories damage cooperation (2020). Roscosmos, April 7. http://en.roscosmos.ru/21369/

[26]See, for example, https://www.esa.int/Enabling_Support/Space_Transportation/International_cooperation: "Russia is ESA's first partner in its efforts to ensure long-term access to space. There is a framework agreement between ESA and the government of the Russian Federation on cooperation and partnership in the exploration and use of outer space for peaceful purposes, and cooperation is already underway in two different areas of launcher activity that will bring benefits to both partners."

[27]http://english.cssar.cas.cn

[28]https://cosparhq.cnes.fr/about/members/national-scientific-institutions/

References

[12.1]   Bacevich, A.J., *The limits of power: the end of American exceptionalism*, Henry Holt, New York, 2008, Also see Bacevich, A.J. (2020). *The age of illusions: how America squandered its Cold War victory*. New York: Henry Holt.

[12.2]   Billings, L., Ideology, advocacy, and space flight – evolution of a cultural narrative, in: *Societal Impact of Space Flight (SP-200-4801)*, S.J. Dick and R.D. Launius (Eds.), NASA History Division, Washington, D.C., 2007.

[12.3]   Billings, L., Fifty years of NASA and the public: What NASA? What publics?, in: *NASA's First 50 Years: Historical Perspectives (NASA SP-2010-4704)*, S.J. Dick (Ed.), pp. 151–182, NASA History Division, Washington, D.C., 2010.

[12.4]   Billings, L., Astrobiology in culture: the search for extraterrestrial life as "science". *Astrobiology*, 12, 10, 966–975, 2012.

[12.5]   Bradsher, K. and Myers, S.L., *Beijing hardens resolve to defy U.S., even while calling for cooperation*, New York Times, 2020, May 28.

[12.6]   Chomsky., N., *Profit Over People: Neoliberalism and Global Order* (with an introduction by R. McChesney), p. 20, Seven Stories Press, New York, 1999.

[12.7]   Coustenis, A., Kminek, G., Hedman, N. *et al.*, The COSPAR Panel on Planetary Protection Role, Structure, and Activities. *Space Res. Today*, 205: August, 2019. https://cosparhq.cnes.fr/assets/uploads/2019/07/PPP_SRT-Article_Role-Structure_Aug-2019.pdf.

[12.8]   Ferrell, P., *Elon Musk, Jeff Bezos and 'The Martian' dream: expect a crash landing*, Marketwatch, New York City, New York, USA, October 6, 2015.

[12.9]   Fuchs, M., *Earthlings be scared of these billionaire space cadets*, Marketwatch, New York City, New York, USA, September 24, 2015.

[12.10] Meltzer, M., *When Biospheres Collide: A history of NASA's Planetary Protection Programs*, pp. 81–82, NASA SP-2011-4234, NASA History Division, Washington, D.C., 2011.

[12.11] Rummel, J.D., Race, M.S., Horneck, G. (Eds.), *Report: COSPAR Workshop on Ethical Considerations for Planetary Protection in Space Exploration*, Princeton University, Princeton, N.J., USA, 2012, 8-10 June 2010.

13

An Ethical Assessment of SETI, METI, and the Value of Our Planetary Home

Chelsea Haramia[1]* and Julia DeMarines[2,3†]

Philosophy Department, Spring Hill College, Mobile, Alabama, USA
Astronomy Department, Breakthrough Listen Lab, University of California, Berkeley, California, USA
Blue Marble Space Institute of Science, Seattle, Washington, USA

Abstract

Ethically responsible searches for extraterrestrial intelligence require careful reflection. Different projects and goals raise different ethical concerns. At the same time, there are overarching ethical considerations that apply to any search for intelligent life in the cosmos. In this chapter, we explain both specific and general concerns and address questions of inherent risk, opaque outcomes, intended effects, value, representation, oversight, and the ethical implications of available alternatives. We also highlight justifications for an expansive and inclusive understanding of moral considerability, drawing on ecological and ethical reasoning and arguing for a planetocentric and ecosystemic approach to SETI and METI debates.

Keywords: SETI, METI, ethics, risk, value

13.1 A Brief History of SETI and METI

The search for extraterrestrial intelligence (SETI) has been an ongoing human effort over the last 60 years, initiated by Project Ozma in 1960. The first endeavors were humble and observed just two stars, Tau Ceti and Epsilon Eridani, over a narrow radio band (around 1.42 GHz) for 150 hours over a four-month period [13.26]. There were no positive

Corresponding author: charamia@shc.edu
†*Corresponding author:* Julia.DeMarines@berkeley.edu

Octavio Alfonso Chon Torres, Ted Peters, Joseph Seckbach and Richard Gordon (eds.) Astrobiology: Science, Ethics, and Public Policy, (271–292) © 2021 Scrivener Publishing LLC

detections, but these seminal observations, led by Dr. Frank Drake, planted what would grow and evolve into the robust SETI searches that we have today. Currently, SETI researchers are observing across the radio frequencies with some of the world's largest telescopes, such as the Green Bank Observatory and the Parkes Radio Telescope, and even performing optical searches with telescopes such as the Automated Planet Finder, the VERITAS Cherenkov telescope, and space-based observatories such as the TESS exoplanet finder [13.13].

Our desire to search the cosmos for signs of an "other" raises the question: How observable are we? Our planet has been communicating its presence of life through an atmospheric chemical disequilibrium since the rise of photosynthetic oxygen over the last 2.5 billion years. While astrobiologists aim to observe chemical disequilibrium in exoplanetary atmospheres in the hope of detecting biological life, these observations will fail to unequivocally determine if life indeed resides on these worlds unless we are able to make *in situ* observations [13.25]. Furthermore, evidence of biological life is not necessarily evidence of an advanced civilization. Because of this, SETI researchers are on the lookout for extraterrestrial technology. Detecting a technosignature, such as a repeated laser pulse or a powerful modulated narrowband signal that cannot be explained by natural phenomena, could be a sign of intelligent life. Carl Sagan writes, "such modulation patterns are never observed for naturally occurring radio emissions and implies [sic] the transmission of information [13.24]." Humans on our planet have been broadcasting the presence of a technologically advanced civilization since the use of high-powered radio transmitters over the last 100 years. Many of our powerful technological emissions are used for military purposes (such as ballistic missile radar) and for planetary safety (such as near-Earth asteroid radar scanning). These strong signals are observable at intergalactic distances to an observer with an Arecibo-like telescope. It should be mentioned that these signals are *unintentional* leakage generated by everyday human activities, such as television signals and satellite communications.

Unlike this unintentional leakage, there have also been a number of intentional attempts to send observable messages. These projects are commonly known as "active-SETI," or "METI" (Messaging ExtraTerrestrial Intelligence). They have taken the form of physical messages, such as the Golden Records on the two Voyager spacecrafts sent by NASA, and of high-powered narrowband radio beacons, such as the famous Arecibo message. A handful of radio astronomers have also sent radio messages. The latter include messages sent by Douglas Vakoch of METI International and Alexander Zaitsev of Russia's Institute of Radio-engineering and

Electronics [13.17]. Intentional, directed messages have also been sent by several private entities, such as Lone Signal, which ran a crowdfunding campaign to send a message to the red dwarf star, Gliese 526; and Spain's Sónar music festival, which broadcast a message to known exoplanet, GJ273b. The goal of some of these messages is to initiate contact with extraterrestrial intelligence, while others, like the Arecibo message aimed at a globular cluster 25,000 light years away, are more ceremonial in nature. The search for intelligent life in the cosmos is evolving, and these recent developments show that a cohort of METI proponents are diverging from the more traditional SETI approaches. We turn now to an ethical analysis of this divergence.

13.2   Ethical Analyses of SETI and METI

The key distinction between the diverging projects of SETI and METI · involves the mode of the search. SETI projects, including the Berkeley SETI Research Center, which is the world's most comprehensive and sensitive SETI search to date, do not engage in active SETI [13.14]. Instead, their searches are considered passive searches in that they are not aiming to elicit contact, but instead are merely listening for the presence of others. METI projects, however, intentionally aim to elicit a response to our own signals and are thereby considered active searches. Active and passive searches are not ethically identical. Each has unique features, and the appropriate ethical analysis of the project of searching for intelligent life depends in part on *how* each search is carried out. The most obvious difference between active and passive searches is a difference in risk. While no one claims that passive searches are risk-free, it seems inherently riskier to announce our presence than it is to merely attempt to detect the presence of others. However, not all who contribute to this debate agree that risk is inherent to active messaging. Adam Korbitz states that

> "we have no way of knowing if ETI's position toward humans or other extant civilizations would be one of egoism (selfishness or even predation), ethical altruism, or something in between resembling a recognition of equality between races or civilizations. Given this vacuum of knowledge, we do not currently have reason to believe that Active SETI is inherently risky, presenting a clear and present danger to the human race" [13.12].

One problem with this reasoning is that it conflates inherent risk with clear and present danger. If something is inherent to an activity, it is fundamental to, and ever present in, the activity itself. But risk may be inherent without the manifestation of clear and present danger. Driving in a car is inherently risky. Hurtling around at varying speeds while trying to expedite travel and avoid obstacles risks physical harm, and the high numbers of automobile-related injuries and death bear this out. But there is not a clear and present danger of injury or death every time you get behind the wheel. Clear and present danger is not inherent to driving. With respect to the activity itself, the risk is inherent, and the danger is potential. The specific conditions in a given instance of driving determine whether and when the danger moves from potential to realized (e.g., inclement weather, intoxicated drivers, etc.). These conditions are contingent on different driving contexts and not present in every act of driving and thereby not fundamental to the activity itself. Therefore, we can distinguish whether an activity is inherently risky from whether it presents a clear and present danger. The conditions under which we encounter an extraterrestrial civilization would determine whether and when humanity faces a clear and present danger. And we are indeed ignorant of the conditions under which we will (if ever) encounter ETI, so it is true that active searches do not at the moment present a clear and present danger to humanity. But that by itself does not entail that they are not inherently risky.

To understand the inherent risk of active searches and how it compares to passive searches, we can distinguish between considerations of what we hope to discover versus considerations of what we hope to interact with. Presumably, when searching for intelligent life in the universe, we hope to discover such life, whatever it may be. But we do not necessarily wish to interact with any intelligent life or civilization, whatever it may be. Some intelligent life might not be worth interacting with. In general, we presumably want to interact with beneficial or benign intelligent life, and we want to avoid interacting with hostile or damaging intelligent life. Granted, these presumptions gloss over the possibility of mixed features of hostility, benevolence, and neutrality; and of course it's not clear that we would ever be in a good position to understand whether detected intelligent life had any given set of features. Nonetheless, we will want to *evaluate* these possibilities to the best of our abilities. Active searches, by their very nature, inhibit our ability to preemptively evaluate the features of detected intelligent life. The aim is to "elicit a reply," according to Doug Vakoch [13.31]. This means that such messages set out to force an interaction with whatever we may find, and that could compress the time allowed for evaluation—or it could remove the possibility of evaluation altogether.

Passive searches, by their very nature, significantly increase the chances that we may evaluate any intelligent life we find before committing to an interaction with it. This is not a guarantee of course. Our passive searches may still be noticeable, and our passive presence may still be detectable, given our unintentional radio leakage, among other things. But successful passive searches do not guarantee the detection of our presence the way that successful active searches do. By analogy, if you have a video doorbell and someone rings it, you may see who they are and evaluate whether they are a threat without their knowing you're home. Contrast this with a scenario where your door is wide open, and you're singing your favorite song loudly as someone approaches. If that person is a threat, then you're in a much more vulnerable position. You're calling attention to yourself and rendering yourself open to potential harm or attack.

Proponents of METI and of active searches might reply by extending this analogy a little further. If your door is always closed, you might miss opportunities for, say, pleasant or useful interactions with your neighbors. This loss is a harm insofar as it makes you worse off than you would have been had you opened your door and announced your presence. By extension, failing to announce our presence might cause us to miss out on valuable benefits and opportunities that may arise only if others notice that we're here. And even if we never make contact, simply engaging in active searches might benefit the SETI community and scientific investigation more generally regardless of detection [13.12].[1] The potential beneficial outcomes of active searches should not be overlooked in our deliberations, and, ideally, we should consider them alongside the potential risks of harmful outcomes. Many who express concerns about METI focus on the risk of harm [13.4] [13.6] [13.11] [13.19] [13.27].[2-6] But METI proponents

---

[1] This point is argued for in more detail by Korbitz. See: Korbitz, A. (2014) "The Precautionary Principle: Egoism, Altruism, and the Active SETI Debate", in *Extraterrestrial Altruism: Evolution and Ethics in the Cosmos*, ed. Douglas A. Vakoch. Springer-Verlag, Berlin, #112.

[2] See, for example: Smith, K. (2020) "METI or REGRETTI: Ethics, Risk, and Alien Contact", in *Social and Conceptual Issues in Astrobiology*, eds. Kelly C. Smith and Carlos Mariscal. Oxford University Press, New York, USA.

[3] See, for example: Race, M. S. (2013) "Space Exploration and Searches for Extraterrestrial Life: Decision Making and Societal Issues", in *Encountering Life in the Universe*, eds. Chris Impey, Anna H. Spitz, and William Stoeger. The University of Arizona Press, Tucson, USA.

[4] See, for example: Haramia, C. and DeMarines, J. (2019) The imperative to develop an ethically-informed METI analysis. *Theology and Science* 17(1).

[5] See, for example: Buchanan, M. (2016) Searching for trouble? *Nature Physics* 12.

[6] Another example involves the Dark Forest Theory, which posits that, out of a desire for self-preservation, ETI may be proceeding as silently as possible, ready to destroy any other intelligent life it comes across because there's no way to know whether other intelligent life will destroy your civilization if given the chance. This theory stems from the second novel in the Remembrance of Earth's Past series. See: Cixin, L. (2008) *The Dark Forest*. Tom Doherty Associates, LLC, New York, NY.

often focus on the risk of missed opportunities for benefit [13.12] [13.30] [13.31].[7-9]

Considerations of harm and benefit in this debate target outcomes, or future states of affairs. Moral dilemmas involving risk often require conjecture about available outcomes. Both active and passive searches raise ethical questions regarding the outcomes we ought to aim for, and both METI critics and METI proponents tend to highlight the potential outcomes they find most salient. In the philosophical world, ethicists have had a lot to say about the role that outcomes per se play in making the appropriate moral determinations. But the scope of the application of this work usually comprises typical predictive scenarios. In typical predictive scenarios, we're often in a decent or even excellent epistemic position to consider and then choose between competing outcomes. A new medication might save a lot of lives, or it might fail and cause a lot of harm. We can run tests that give us confidence in one outcome over another, and then choose the ethically and scientifically appropriate course of action.

We are not similarly epistemically positioned when it comes to deliberating about the outcomes of METI projects. There are important atypical features in the active search for intelligent life. These atypical features are notably parallel to another burgeoning technological and biological project, and here we can draw on the work that others have done in the risk evaluation of synthetic biology and the creation of artificial living cells and apply it to the project of searching for intelligent life.[10] Mark A. Bedau and Mark Triant [13.2] focus on the field of synthetic biology, and they apply relevant concerns in that field to similar concerns with space exploration. Success in both of these endeavors will entail grappling with a new understanding of life in this world, and until the details of these understandings manifest alongside our successes, we are ill-equipped to make predictions about the risks involved. Many others have noted that active messaging involves engaging in new activities the results of which are largely if not

---

[7]See, for example: Vakoch, D. (2011) Asymmetry in active SETI: a case for transmissions from earth. *Acta Astronautica* 68.

[8]See, for example: Vakoch, D. (2016) In defense of METI. *Nature Physics* 12.

[9]See, for example: Korbitz, A. (2014) "The Precautionary Principle: Egoism, Altruism, and the Active SETI Debate", in *Extraterrestrial Altruism: Evolution and Ethics in the Cosmos*, ed. Douglas A. Vakoch. Springer-Verlag, Berlin.

[10]Thanks to Dr. Janella Baxter for providing consultation on synthetic biology projects.

entirely opaque to us [13.8] [13.12].[11,12] We are, as some say, "deciding in the dark" [13.2].

Because the outcomes of synthetic biology projects and of active searches for ETI are excessively difficult to determine, there is no way to avoid the claim that either could possibly lead to catastrophe. When analyzing the risks of synthetic biology or of active messaging, one might initially assume that certain common risk principles will thereby be relevant to those debates. Bedau and Triant challenge that assumption [13.2]. In particular, they challenge the commonly cited Precautionary Principle. They characterize the principle in the following way: "Do not pursue a course of action that might cause significant harm, even if it is uncertain whether the risk is genuine" [13.2]. If this is the correct principle to follow, then active messaging is permissible only when METI proponents can provide convincing reason that messaging will not cause significant harm.

But this is not as simple as testing new medication, and the problem with this principle is that it may not be well-equipped to deal with the complexities that arise when we are deciding in the dark [13.2]. When it comes to both the creation of new forms of life in the laboratory or the discovery of intelligent life out in the cosmos, claims about the potential results of these projects are founded on nearly pure speculation. The Precautionary Principle tells us to avoid projects until they are proven safe enough. But the discovery or creation of new life cannot be proven safe enough without actual work on precisely those projects that the principle requires us to postpone. The same goes for any new technologies that one must first engage with in order to successfully make predictions about. Any time that engaging in the project itself is a necessary part of gauging the risk of the project, the principle cannot permit the activity. The Precautionary Principle, then, requires the endless postponing of many projects involving new technologies and unknown outcomes. This seems like an undesirable implication. Sometimes we may be justified in moving forward, even when we are deciding in the dark. So, what could justify this move?

While we arguably cannot successfully assign probabilities to the proposed outcomes of contact, perhaps we can successfully assign a probability to the proposal that we will make contact with an extraterrestrial civilization through messaging. One might argue that there is good reason

---

[11]See, for example: Denning, K. (2010) Unpacking the great transmission debate. *Acta Astronautica* 67.

[12]See, for example: Korbitz, A. (2014) "The Precautionary Principle: Egoism, Altruism, and the Active SETI Debate" in *Extraterrestrial Altruism: Evolution and Ethics in the Cosmos*, ed. Douglas A. Vakoch. Springer-Verlag, Berlin, #112.

to engage in active messaging in the face of the inherent risks involved because the likelihood of our making contact is actually extremely low. Indeed, the Fermi Paradox famously highlights the fact that our searches have found no verifiable signs of life in a galaxy that should arguably contain a number of advanced civilizations [13.23]. If this silence is evidence of how unlikely it is that we will make contact with any intelligent life, then why not try our hand at messaging? One reasonable consideration when answering this question is time itself. Interstellar exploration is in practice an intergenerational phenomenon. And the potential harms to humanity could increase the further into humanity's future we extend our focus. Of course, so could the potential benefits, given our ignorance about alternative outcomes. The point is that we may be legitimately concerned for future generations of humans even if we don't think any human beings are currently at risk. Considering the moral status of future humans is an important part of the ethical analysis of searching, in part because they matter morally and are in no position to speak for themselves. Another important consideration when answering such questions emphasizes that the significance of even an extremely low probability of contact from cosmic announcements still requires a serious conversation about how to proceed. While assuming for the sake of argument that the probability of contact is quite low, we can nonetheless recognize that active searches raise the following ethical concerns.

Let's assume it is extremely unlikely that we will contact ETI through current messaging practices. Notably, some potential harmful effects of active messaging are nonetheless extremely serious in that they are bound to be (at least) global in nature, if realized. Kelly Smith notes a potential existential threat to Earth arising from active messaging, stating that, "we simply can't rule out an existential risk to all of humanity—indeed, to all life on Earth. In other words, there is some chance METI could result in what would literally be, from a terrestrial perspective anyway, the *worst possible* outcome" [13.27]. How should we think through activities that risk such serious consequences, even if only minimally? While we may not know how probable such harmful consequences are, we know that there is the potential for such harm. Alastair Norcross demonstrates that even small risks of very great harm justify constraints and proscriptions on behaviors that allow such risks [13.16]. This, for example, is why we endorse regulations on child car seat usage and on airline safety checks. Chances are very good that you are not going to get into a crash the next time you drive with your kid or take a flight. But that does not justify forgoing the car seat, and you'd likely be horrified to discover that your airline decided to skip the safety checks for your flight. These reactions are evidence that risk-minimizing behaviors are required precisely when the potential harm is serious enough, even when it is extremely unlikely that the conditions that lead to harm will manifest.

So even if we assume for the sake of argument that there is a very low probability of contact, we may still be morally required to exercise caution. And, because the potential harm we're risking is so serious, intergenerational, and global in nature, perhaps we are justified in prescribing certain behaviors and protocols to those whose actions make our presence more noticeable. These concerns can be and have been directly applied to the METI debate. On the one hand, Smith draws a distinction between objective risks and acceptable risks. He maintains that the actual probability of a risk is not all that goes into determining whether the risk is acceptable, noting "[e]ven with a risk that is objectively quite low, it might be perfectly rational to refuse to accept it, especially if the negative consequences are sufficiently dire [13.27]." On the other hand, Korbitz points out that the potential benefits of active messaging may end up saving humans from some other existential threat, and that we cannot count this outcome as any more or less probable, due to our wholesale ignorance regarding the factors that would allow us to assign probabilities to different outcomes of messaging [13.12].

Furthermore, as we have seen, active messaging is not the only way that information travels beyond the boundaries of Earth. Not only have humans' relatively weak signals been leaking into space for decades, other, stronger transmissions have entered the scene as well [13.10].[13] Notably, organizations such as the Arecibo Observatory currently send high-powered, targeted signals into space in an attempt to detect near-Earth asteroid trajectories. If we've already been unintentionally broadcasting our presence in these other ways, perhaps active messaging itself adds no new risks to the situation. There is an unsettled debate about whether we have already unintentionally announced our presence in a loud enough manner to attract the attention of ETI through commonplace radio leakage, and that debate is beyond the scope of this chapter. One may nonetheless argue that the powerful broadcasting involved in other projects, such as the project of tracking asteroids, is not ethically analogous to the powerful broadcasting of active messaging; and, because of this, we may draw different ethical conclusions regarding such activities.

---

[13]See Table 13.1 for the relative detectability of different forms of radio leakage. Source: Haqq-Misra, J. *et al.* (2013). The benefits and harms of transmitting into space. *Space Policy*. 29:1, #40-48.

Table 13.1 Relative detectability of sources of radio leakage [13.10].[14]

Transmitter	Total time transmitted (years)	Power (W)	Gain	Frequency (MHz)	Radius (distance) in light years a signal is detectable by a 1 km² array
Broadband (Cell phones and TV)	1	10^8	10	100 – 2700	0.03
TV Carrier Wave (one station)	1	3×10^5	10	100 – 2700	50
Arecibo Radar (Continuous Wave)	10^{-2}	8×10^5	10^8	2380	200,000
USAF PAVE PAWS Military Radar	0.1	10^5	3×10^4	430	60,000
Goldstone (Continuous Wave)	10^{-2}	4×10^5	10^8	8650	90,000
Evpatoria (Continuous Wave)	10^{-3}	10^5	3×10^7	6000	20,000
Lone Signal at Madley 49	0.1	10^4	10^6	6000	1.5

[14]Adapted from: Haqq-Misra, J. *et al.* (2013). The benefits and harms of transmitting into space. *Space policy*. 29:1, #40-48.

Let's consider an important difference between the two kinds of projects we have in mind. On the one hand, the aim of the Arecibo Observatory here was to identify another type of potential harm—the harm posed by an asteroid on a collision course with Earth—and to identify that threat early enough to do something about it. In this case, abstaining from increasing our cosmic presence means that we fail to address a specific potential threat—the threat of an impending asteroid. METI proponents, on the other hand, are not aiming to mitigate any other particular threat. Instead, they hope that we will benefit from contact with extraterrestrial intelligence. Risking unrealized potential benefit to humanity from contact with ETI is not morally on par with risking the direct, known harm of a humanity-ending asteroid impact. This means that there is an ethical asymmetry to the available alternatives in these two activities. Arecibo Observatory was attempting to mitigate the clear threat of asteroid impact while incidentally announcing our presence and risking the potential harm of announcing our presence to ETI without first evaluating them, whereas METI proponents target no particular threat while also intentionally risking the potential harm of announcing our presence to ETI without first evaluating them.

The presence of available alternatives is arguably crucial when performing ethical analyses or assessing moral responsibilities. Such considerations lead to the ethical asymmetry above, and they also lead to the following ethical observation: what you are permitted to do when there are no good alternatives can be far different from what you are permitted to do when there are multiple similar options to consider. For example, performing life-saving surgery without anesthetic before the invention of anesthetic was, all else equal, clearly morally permissible, despite any pain the patient experienced during the procedure. But performing life-saving surgery here and now without anesthetic, when anesthetic is readily available, is, all else equal, clearly morally *im*permissible because of the gratuitous pain it would cause. Options matter morally, and here we draw attention to an important divergence between the aforementioned activities that require us to decide in the dark: synthetic biology and active messaging. The alternative to engaging in synthetic biology projects that aim to synthesize new life is to abandon these synthetic biology projects. Yet the alternative to METI and to current active searches for extraterrestrial intelligence is not to abandon the search for extraterrestrial intelligence. The alternative is passive searches. Given that passive searches (i) are inherently less risky than active searches, (ii) allow us to proceed with a search for intelligent life, and (iii) allow us to continue to pursue potential benefits of interaction with ETI; there is ethical reason to favor passive searches over active ones under our current circumstances.

This is not to claim that the inherent riskiness of active messaging must render the activity morally forbidden, because it is not the case that risky behavior must always be avoided. We still drive our children around, fly in planes, use new medications, and engage in scientific projects the effects of which have not been fully determined. And we may collectively choose to take on the inherent risks of active messaging. But how we arrive at this decision matters morally. There is at present no oversight to METI activities, which compounds concerns about risk and representation. Alongside these concerns, one might claim that we ought to gain some sort of consent or consensus from humanity regarding messaging activities.

Issues of oversight, consent, consensus, and representation are key to the debate about how to solve the ethical issues raised by searching, messaging, and contact. We will turn to these concerns and a discussion of proposed solutions in a moment. First, however, we want to mention one more important ethical issue. Absent from our discussion so far have been concerns about risks to alien civilizations and their environments. Indeed, a significant amount of the established debate about the ethical responsibilities of researchers is focused primarily on risks to humans. While there is less focus on the risks that we might pose to extraterrestrial others, this too is a crucial ethical question, and we are operating in the dark here as well. Ethical responsibility and moral obligations to others depend in part on the details and context of our relation to those others. We simply do not know how we are situated in relation to extraterrestrial others, but we risk positioning ourselves poorly if we act recklessly, and there is good moral reason to evaluate this risk to other intelligent beings as carefully as we evaluate the risks to humanity itself. Intelligent beings are the most common targets of moral consideration, though they are not the only targets in ethical debates. Any successful ethical solution to the concerns surrounding the search for ETI will take seriously all that is worthy of moral consideration, so we turn now to a discussion of how to approach ethical questions and issues of moral consideration raised by active and passive searches.

13.3   Ethical Proposals for the Road Ahead

Here, we offer some guiding ideals and proposed solutions that we hope will help to cultivate a solid ethical foundation for current and future searches. Ethicists commonly maintain that humanity has, in general, made significant moral progress over the course of human history. One of the hallmarks of moral progress is a move away from imposed hierarchies of superiority and inferiority. The historical evils of colonialism, slavery, and genocide, for example, all relied on this framework, and the ethical solutions to these problems have required that humans regularly reject these hierarchies and replace them with ethical values such as inclusion, equity, increasing welfare, and copious respect. In this way, ethical progress is arguably expansive, and the arc of human history bends toward an ever-more-inclusive moral community. Ethical debate has thoroughly challenged many of the presumed boundaries of moral consideration, and good reasoning has shown us that features such as race, gender, borders, and capabilities cannot justify excluding others from moral consideration. Beyond this, environmental and animal ethicists have shown that the land, the environment, non-human animals, species, and ecosystems are all worthwhile targets of moral consideration as well. Finally, building on this progress, we would like to consider certain ethical approaches that explicitly focus on our planetary and cosmic situation. This type of approach is compatible with multiple normative theories, so we will leave open what the appropriate normative correlation with these proposed ethical values should be while at the same time highlighting some of the important cosmic features of our ethical situation.[15] In this spirit, we consider the proposal for a radically inclusive planetocentric ethics as a starting point for addressing the ethical issues raised by active and passive searches.

Holmes Rolston calls for an Earth ethic that positions all members of humanity as Earthlings first, unified by our planetary home, and he maintains that we have a moral obligation to protect this home [13.21]. He claims that the natural world is our home in a morally relevant way, much more so than the material structures or imaginary boundaries we impose on sub-parts of the natural world. He states,

> "[i]t is not enough to be a good "citizen," it is not even enough to be "international," because neither of these terms have enough "nature" or "earthiness" to them. Neither is "worldly" enough…"Global" is a more holistic theme; it knows cultural systems intertwined with natural systems. The wisdom we need is not just that of politicians, generals, technocrats; it is that of ecologists in the philosophical sense…We need environmental ethics without borders" [13.22].

---

[15]Normative theories attempt to ground claims of what makes actions morally right or morally wrong. When applied to real-world activities, these theories often arrive at the same conclusions about rightness or wrongness of particular actions even as they appeal to different justifications for those conclusions, and that is the type of compatibility we allude to here.

Under Rolston's view, protecting our Earth from human and non-human threats is vital because the Earth is our home, but this does not mean that Earth itself is more valuable than any other planet [13.20]. Each planet manifests some unique potential in the natural world and thereby has a good of its own independent of its effects on the world around it.

This good possessed by Earth and other planetary bodies is, according to Rolston, objectively valuable. If humans recognize it at all, it is merely because we have discovered it. We had no hand in creating it. An alternative account of value maintains that all value is imposed on the world by human valuers, and none of it is objective. While decisively ruling out this alternative is well beyond the scope of this chapter, we will note here the following strange implication of the opposing claim that nothing in nature is objectively valuable. This position entails that, before human subjects arrived on the scene to value nature, nothing in nature possessed true value [13.5].[16] Indeed, all of the life that preceded human life, all of the ecosystems that we never encountered, and all of the natural world that we have yet to encounter in the cosmos—including other forms of life in the cosmos—are, under this view, valueless until we decide to value them. The strong anthropocentrism involved in positioning ourselves as determiners of value rather than discoverers of value should be carefully scrutinized. The assumption of objective value in the universe facilitates a far more unbiased approach, and it also requires our intellectual humility as we take seriously the possibility that we humans may or may not get it right when making claims of value. Assuming, then, that objective value exists, that which is objectively valuable deserves our moral consideration. The proper targets of moral consideration inform our appropriate moral deliberations, and moral deliberations are crucial to solving the ethical problems raised by active and passive searches.

We tend to pick out and isolate the targets of moral consideration when assessing our moral responsibilities. What we owe to wild animals is typically presented as a separate question from what we owe to, say, our children. But one noteworthy and morally relevant consideration in cosmic debates is the interconnectedness of the natural world. Nothing that physically exists is independent of that which surrounds it. As John Muir said, "[w]hen we try to pick out anything by itself, we find it hitched to everything else in the universe [13.15]." For example, the workings of our early solar system were enormously influential on the current state of life on

---

[16]See: Callicott, J.B. (1984) Non-anthropocentric value theory and environmental ethics. *American Philosophical Quarterly* 21.

Earth, such as the giant asteroid impact 65 million years ago that caused a mass extinction and set the stage for life on Earth as we know it today. Life on Earth today contains much of what we think deserves our moral consideration. These causal connections suggest that our ethical and scientific targets of consideration cannot be appropriately considered as isolated entities and must be integrated into more holistic analyses. As researchers or as moral agents, we often become narrowly focused. While this is typically useful, we should not lose sight of the ways in which the causal network, or systemic processes, of the natural world contribute to our considerations. We are causally, temporally, and systemically connected to all of terrestrial nature, to our Solar System, to our galaxy, and beyond. This placement, along with our moral agency, is why we may legitimately feel called on to act responsibly when we engage in practices that may disrupt our home—whether we consider "home" in terms of our planet, our Solar System, our galaxy, or even our entire Universe.

Rolston points out that ecology is the study of living creatures' homes [13.22]. Woodruff T. Sullivan builds on this to propose what he calls a "planetocentric ethics" [13.28]. Whether the living creatures in question are microbial organisms on Mars or human societies on Earth, Sullivan aims to

> "extend the concept of eco-logy (home-study, etymologically) such that home refers to more than just our neighborhood, or our watershed, or even our biosphere, but indeed refers to the entire Solar System. In planetocentric ethics each planetary body has intrinsic value as part of an ecosystem writ large, and can be modified only with strong justification" [13.28].

This is indeed a broader account of moral consideration than is put forth in typical, circumscribed ethical contexts. However, if we really are "hitched to everything else in the universe," as Muir states [13.15], and if the ecosystem of living creatures extends beyond our Solar System, as detection of ETI could confirm, then perhaps we should extend our ethics beyond our planet and beyond even our Solar System and into interstellar space. This expansive and radically inclusive approach would require that we enlarge the scope of our moral considerations far more than we are used to. But that is something that we are equipped to do, as we have seen in our interplanetary and interstellar deliberations to date [13.1],[17] as well as in our moral progress as human beings. The broader our reach and awareness has become, the more willing and able we have been to expand our consideration and deliberation about the ethical features of our activities. We are still in the beginning stages of learning the details of other solar systems, other galaxies, and the potential for intelligent life elsewhere in the cosmos. As our knowledge and awareness grows, so too does the scope of our moral consideration and ethical deliberation. This calls for a commitment to careful ethical analyses alongside our scientific analyses as we grow in our knowledge of the natural world and the universe. It does not, however, call for a set of detailed ethical principles that we can currently presume will always apply regardless of what we discover and learn. What we propose here is an ethically-informed method rather than a set of immutable moral judgments.

Methodologically, then, we ought to care about *how* we collaborate on and perform our interstellar activities. Many have called for some sort of oversight or regulation of active messaging [13.3] [13.18] [13.27] [13.31],[18-20] and some have called for a much more inclusive and representative debate to take place *before* active messaging occurs [13.11].[21,22] One important reason to engage in methodology before messaging is that the products of our methodology—such as regulations or codes of conduct or degrees of consensus—are revisable and changeable in light of new evidence or information, whereas the products of messaging—the interstellar messages themselves—cannot be revised, recalled, or erased in light of new

---

[17]For example, Seth Baum extends typical consequentialist reasoning to suggest that we should be location-egalitarian about objective value, acknowledging that its existence is not contingent on any particular spatio-temporal location on Earth or elsewhere in the cosmos. See: Baum, S.D. (2016) "The Ethics of Outer Space: A Consequentialist Perspective", in *The Ethics of Space Exploration*, eds. James S.J. Schwartz and Tony Milligan. Springer International Publishing, Switzerland.

[18]See, for example: Smith, K. (2020) "METI or REGRETTI: Ethics, Risk, and Alien Contact", in *Social and Conceptual Issues in Astrobiology*, eds. Kelly C. Smith and Carlos Mariscal. Oxford University Press, New York, USA.

[19]See, for example: Peters, Ted. "Astroethics: Engaging Extraterrestrial Intelligent Life-Forms", in *Encountering Life in the Universe*, eds. Chris Impey, Anna H. Spitz, and William Stoeger. The University of Arizona Press, Tucson, USA: 172.

[20]See, for example: Wright, J.T. (2018) Is it ethical to transmit powerful radio signals? https://rockethics.psu.edu/everyday-ethics/is-it-ethical-to-transmit-powerful-radio-signals-1

[21]See, for example: Haramia, C. and DeMarines, J. (2019) The imperative to develop an ethically-informed METI analysis. *Theology and Science* 17(1).

[22]See, for example: Smith, K. (2020) "METI or REGRETTI: Ethics, Risk, and Alien Contact", in *Social and Conceptual Issues in Astrobiology*, eds. Kelly C. Smith and Carlos Mariscal. Oxford University Press, New York, USA.

evidence or information. As Jason T. Wright notes, regulating active messaging "preserves our options in the future. Regulations on METI can be reversed, but signals sent into space cannot be taken back" [13.32].[23] Again, options matter morally. And figuring out the methods that will produce the most ethically sound guidelines for active messaging is an important part of the debate about how to properly oversee active messaging on behalf of humanity. Thomas Cortellesi, who argues in favor of active searches as merely part of a longstanding continuum of astrobiological signaling, nonetheless comments on the state of current messaging activities, noting that

> "METI reduces to a commons problem wherein the actions of individuals may decide the course of history for all concerned, even if the risk to humankind is slight. As such, those who act on their own may be 'engaged in unauthorized... diplomacy'; it follows therefore that we ought to at least be careful about what we transmit, and discourage individuals from doing so by themselves. A more specific and workable set of guidelines are needed to better enshrine, establish, and standardize METI methodology, and here is where well-intentioned passivists and metiists agree" [13.7] [13.9].

In practical ethical debates, we ought to seize on agreement and cohesion wherever it is possible and use that to achieve ethically responsible progress on our projects. The methodological approach we propose is not without challenges, however, in particular, challenges to the prospects of agreement and cohesion, as Jill Tarter and others have noted [13.29]. After SETI researchers[24] held a workshop in the early 1990s to address concerns surrounding messaging, they determined, in part, that "[t]ransmission is a diplomatic act, an activity that should be undertaken on behalf of all humans [13.29]." Interestingly, they concluded from this that "[w]e lack the cultural maturity to accomplish such a cohesive action" [13.29]. While SETI researchers took

---

[23]Wright, J.T. (2018) Is it ethical to transmit powerful radio signals? https://rockethics.psu.edu/everyday-ethics/is-it-ethical-to-transmit-powerful-radio-signals-1

[24]As Tarter also notes, this workshop was attended primarily by white men working in industrialized countries. As with many other disciplines, STEM has a history of disproportionately representing members of dominant groups, and its demographics are not representative of humanity itself. If we care about proper representation and inclusion in these debates, we need to take seriously our history of incidental and systematic exclusion of non-dominant peoples from positions of authority and respect. Some feminist epistemologists have argued that we may even prioritize knowledge from members of underrepresented groups, insofar as their social and global positions allow them access to special forms of knowledge not typically available to members of dominant groups. This knowledge may be crucial to engendering a truly well-informed public debate about messaging.

this as good reason to postpone messaging until we are mature enough to reach an appropriate consensus, others may appeal to the assumption that we will not reach consensus to justify the conclusion that we are permitted to send messages absent consensus or appropriate representation. It is understandable to worry about reaching a sufficient degree of agreement among humans. However, perhaps we ought not be as pessimistic or fatalist about a general consensus as these worries indicate. As Smith notes,

> "Consensus is not an all or nothing phenomenon: a signed consent document is at one end of the continuum, while letting anyone with access to a radio telescope do whatever they wish is at the other. It's a false dichotomy to force a choice between these two options, as there are plenty of alternatives between these extremes that reasonable people can pursue" [13.27].

One alternative that he proposes is to engage experts in various fields in the project of developing some oversight for active messaging projects. Along with Smith, we find it quite reasonable to call for the explicit involvement of experts in the fields of public policy, philosophy, anthropology, history, social science, and many other scientific disciplines—alongside the expert astronomers and astrobiologists already engaged in this activity— as a means for developing, at minimum, some best practices regarding active messaging specifically. And the general public has a role to play as well, and we should avoid jumping quickly to the conclusion that humans could not possibly reach a reasonable level of agreement on this issue. We have argued elsewhere that there is reason to be more optimistic about cross-cultural agreement than many assume, especially if we commit to science communication and public philosophy projects that aim to assess degrees of consensus and generate legitimate cohesion [13.11]. In line with this, Smith concludes that we should be

> "collecting better information about public attitudes towards METI. If it turns out that the public is actually *in favor* of METI, or even indifferent, then many of the concerns expressed here would become moot. Since this is an empirically tractable question, there is no justification for continuing to either assume or ignore public opinion" [13.27].

The most ethically appropriate course of action may then be to engage in methodologically sound activities such as careful deliberation, representative debate and oversight, and genuine work toward assessing consensus on messaging among members of humanity. Notably, this course of action is quite compatible with passive searches and current SETI activities. It also preserves common ethical and ecological values and allows for a planetocentric, ecosystemic approach to guide our debates and deliberations. It leaves open what the ethically sound conclusions will be for METI and active searches, and it redirects our focus onto the moral responsibilities we have here and now to proceed carefully and inclusively. These are responsibilities we have to our home and its inhabitants, as well as to the remarkable natural system that is our Universe, about which we still have much to learn.

References

[13.1]   Baum, S.D., The Ethics of Outer Space: A Consequentialist Perspective, in: *The Ethics of Space Exploration*, J.S.J. Schwartz and T. Milligan (Eds.), Springer International Publishing, Switzerland, 2016.

[13.2]   Bedau, M.A. and Triant, M., "Social and Ethical Implications of Creating Artificial Cells", in: *Encountering Life in the Universe*, C. Impey, A.H. Spitz, W. Stoeger (Eds.), The University of Arizona Press, Tucson, USA, 2013.

[13.3]   Billingham, J. and Benford, J., *Costs and difficulties of large-scale messaging, and the need for international debate on potential risks*, 2011, https://arxiv.org/abs/1102.1938.

[13.4]   Buchanan, M., Searching for trouble? *Nat. Phys.*, 12, 720, 2016.

[13.5]   Callicott, J.B., Non-anthropocentric value theory and environmental ethics. *Am. Philos. Q.*, 21, 299–309, 1984.

[13.6]   Cixin, L., *The Dark Forest*, Tom Doherty Associates, LLC, New York, NY, 2008.

[13.7]   Cortellesi, T., Reworking the SETI paradox: METI's place on the continuum of astrobiological signaling. *JBIS*, 73, 7, 2020.

[13.8]   Denning, K., Unpacking the great transmission debate. *Acta Astronaut.*, 67, 1399–1405, 2010.

[13.9]   Gertz, J., Post-detection SETI protocols and METI: the time has come to regulate them both. *JBIS*, 69, 2017, https://arxiv.org/abs/1701.08422.

[13.10] Haqq-Misra, J. *et al.*, The benefits and harms of transmitting into space. *Space Policy*, 29, 1, #40–48, 2013.

[13.11] Haramia, C. and DeMarines, J., The imperative to develop an ethically-informed METI analysis. *Theol. Sci.*, 17, 1, 38–48, 2019.

[13.12] Korbitz, A., The Precautionary Principle: Egoism, Altruism, and the Active SETI Debate, in: *Extraterrestrial Altruism: Evolution and Ethics in the Cosmos*, D.A. Vakoch (Ed.), Springer-Verlag, Berlin, 2014.

[13.13] Lacki, B. *et al.*, *One of everything: The Breakthrough Listen exotica catalog*, 2020, [Preprint], Submitted to The Astrophysical Journal Supplement. Available from: arXiv:2006.11304v1 [astro-ph.IM].

[13.14.] Lebofsky, M. *et al.*, *The Breakthrough Listen search for intelligent life: Public data, formats, reduction and archiving*, Publications of the Astronomical Society of the Pacific, 2019, 131:1006, #124505.

[13.15] Muir, J., *My First Summer in the Sierra*, The Riverside Press, Cambridge, 1911.

[13.16] Norcross, A., Puppies, pigs, and people: eating meat and marginal cases. *Philos. Perspect.*, 18, 229–245, 2004.

[13.17] Patton, P., *Language in the Cosmos II: Hello There GJ273b*, Universe Today, 2018.

[13.18] Peters, T., Astroethics: Engaging Extraterrestrial Intelligent Life-Forms, in: *Encountering Life in the Universe*, C. Impey, A.H. Spitz, W. Stoeger (Eds.), The University of Arizona Press, Tucson, USA, 2013.

[13.19] Race, M.S., Space Exploration and Searches for Extraterrestrial Life: Decision Making and Societal Issues, in: *Encountering Life in the Universe*, C. Impey, A.H. Spitz, W. Stoeger (Eds.), The University of Arizona Press, Tucson, USA, 2013.

[13.20] Rolston, H., The Preservation of Natural Value in the Solar System, in: *Beyond Spaceship Earth: Environmental Ethics and the Solar System*, E.C. Hargrove (Ed.), Sierra Club Books, San Francisco, USA, 1986.

[13.21] Rolston, H., Global Environmental Ethics: A Valuable Earth, in: *A New Century for Natural Resource Management*, R.L. Knight and S.F. Bates (Eds.), Island Press, Washington D.C., USA, 1995.

[13.22] Rolston, H., *A New Environmental Ethics: The Next Millennium for Life on Earth*, Routledge, New York and London, 2012.

[13.23] Sagan, C. and Drake, F., The search for extraterrestrial intelligence. *Sci. Am.*, 232, 5, 1975.

[13.24] Sagan, C. *et al.*, A search for life on Earth from the Galileo spacecraft. *Nature*, 365, 6, 1993.

[13.25] Schwieterman, E.W. *et al.*, Exoplanet biosignatures: a review of remotely detectable signs of life. *Astrobiology*, 18, 6, 2018.

[13.26] SETI Institute, *Early SETI: Project Ozma, Arecibo Message*, The SETI Institute, Mountain View, CA, 2020, https://www.seti.org/seti-institute/project/details/early-seti-project-ozma-arecibo-message.

[13.27] Smith, K., METI or REGRETTI: Ethics, Risk, and Alien Contact, in: *Social and Conceptual Issues in Astrobiology*, K.C. Smith and C. Mariscal (Eds.), Oxford University Press, New York, USA, 2020.

[13.28] Sullivan, W.T., Planetocentric Ethics: Principles for Exploring a Solar System That May Contain Extraterrestrial Microbial Life, in: *Encountering Life in the Universe*, C. Impey, A.H. Spitz, W. Stoeger (Eds.), The University of Arizona Press, Tucson, USA, 2013.

[13.29] Tarter, J.C., Contact: Who Will Speak for Earth and Should They?, in: *Encountering Life in the Universe*, C. Impey, A.H. Spitz, W. Stoeger (Eds.), The University of Arizona Press, Tucson, USA, 2013.

[13.30] Vakoch, D., Asymmetry in active SETI: a case for transmissions from earth. *Acta Astronaut.*, 68, 476–488, 2011.

[13.31] Vakoch, D., In defense of METI. *Nat. Phys.*, 12, 890, 2016.

[13.32] Wright, J.T., *Is it ethical to transmit powerful radio signals?*, The Rock Ethics Institute, University Park, PA, 2018, https://rockethics.psu.edu/everyday-ethics/is-it-ethical-to-transmit-powerful-radio-signals-1.

14

The Axiological Dimension of Planetary Protection

Erik Persson

Department of Philosophy, Lund University, Lund, Sweden

Abstract

Planetary protection is not just a matter of science. It is also a matter of value. This is so independently of whether we only include the protection of science or if we also include other goals. Excluding other values than the protection of science is thus a value statement, not a scientific statement and it does not make planetary protection value neutral. It just makes the axiological basis (that is, the value basis) for planetary protection more limited in a way that is inconsistent with the axiological grounds for back contamination, ethically questionable and strategically unwise. However we look at it, we cannot get away from the conclusion that the axiological dimension of planetary protection is a task that needs to involve experts on value theory as well as experts from a range of different sciences and also include opinions from outside the academic community.

Keywords: Planetary protection, Outer Space Treaty, environmental space ethics, epistemic values, axiology

14.1 Introduction

Rules, standards and practical decisions about planetary protection need to be based in good, solid science. Reliable information about the survivability of Earth life on the target bodies and in interplanetary space is obviously very important. So is knowledge about trajectory biasing and decontamination of spacecraft and payloads as well as a good account of the remaining bioload after decontamination. I will call this the "epistemic" (that is, knowledge)

Email: erik.persson@fil.lu.se

Octavio Alfonso Chon Torres, Ted Peters, Joseph Seckbach and Richard Gordon (eds.) *Astrobiology: Science, Ethics, and Public Policy*, (293–312) © 2021 Scrivener Publishing LLC

dimension of planetary protection. As is the case when I write this chapter, the epistemic basis for planetary protection is far from complete. There is still a large amount of uncertainty and science constantly makes new discoveries of microbes surviving in extreme environments, including in circumstances similar to those in clean room facilities, interplanetary space and on Mars. These things aside, even with a much better epistemic basis, in fact, even if we (per impossibility) had complete knowledge about all the things listed above, that would not be enough to make rational decisions about planetary protection. That a decision is rational, typically means, or at least implies, purposefulness. That is, in order to make rational decisions about planetary protection, we need to know its purpose. What is the purpose of planetary protection? What is it that we want to protect and why? If we do not know what it is we want to protect, why we want to protect it, for how long, to what extent and its importance in relation to other values, rational decisions are impossible, no matter how much and how good the knowledge we have.

We therefore need answers to both the epistemic questions and the value, or "axiological" questions. The axiological questions about planetary protection also need to be considered in a wider perspective: What other values are at stake and how does planetary protection relate to them? These questions in turn lead to yet other questions: Why should we spend resources on planetary protection, and what amount of resources are justified? The axiological dimension of planetary protection is thus no less complex than the epistemic dimension. It is also central to all decisions regarding planetary protection and needs to be considered thoroughly, carefully and transparently.

In this chapter, I am going to explore the axiological dimension of planetary protection. I will try to identify the stated and unstated axiological basis for planetary protection the way it looks today and I will try to put it in a historical context. In addition, I will say a little bit about what I see as the future of the axiological dimension of planetary protection.

In his chapter, I am going to stick to planetary protection in our own solar system. If and when we start sending spacecraft aimed at particular bodies beyond our solar system, we will not just face unprecedented levels of uncertainty but also a different set of axiological questions. This situation, though interesting in its own right, will not be covered in this chapter.

14.2   The Relation Between the Epistemic and the Axiological Dimensions of Planetary Protection

Planetary protection is about avoiding contamination of other worlds with invasive Earth microbes in connection with missions to these worlds, and about avoiding contamination of the Earth biota by invasive microbes form other worlds in connection with sample return missions. Complete sterilization of spacecraft and other equipment is not possible with present technology, however, and it might never be. We also do not yet know everything we need to know about our target bodies to be able to say exactly what it would take for Earth microbes to survive there [14.13] [14.57]. In addition, we keep discovering Earth microorganisms that are hardier than expected. Smith *et al.* [14.57] found for instance that several microorganisms found on the Mars Science Laboratory were resistant to multiple physiological stress factors, that many were resistant to desiccation and UVC, and that a small number also could make use of chemical energy sources available on Mars (perchlorate and sulphate). Nicholson *et al.* [14.33] conclude from their study that the regolith at the Phoenix landing site "would likely prove rather benign to potential terrestrial spacecraft contaminants such as spores of *Bacillus* spp., and indeed may even support their germination and growth."

These are only a couple of examples to illustrate the practical problems involved. We also have to consider the hardware that we want to "survive" the decontamination treatment. Some microbes seem to withstand a much harsher treatment than most of the hardware. Finally, we always have to take evolution into consideration. The fact is that cleanroom facilities exert a strong evolutionary pressure in favor of microbes that can survive the decontamination procedures we submit them to.

In other words, a probability of zero is clearly not compatible with visiting the targets in question at all. So, why not instead just aim for a level of probability for contamination that is as low as possible? The go to source of information for non-experts in any area, that is, Wikipedia, actually states that this is the aim of planetary protection. They acknowledge the impossibility of reaching a zero probability of contamination but instead they suggest that: "The aim of planetary protection is to make this probability (for contamination of the target body) as low as possible" [14.60].

If we accept this aim, we immediately have to ask, "as low as possible, given what circumstances?" What is possible depends on the context, including a whole range of value decisions. What is possible if you have unlimited funding, might be completely impossible if you have a tight budget. In the real world we do not have unlimited access to money, time or any other resource, and every resource we spend on realizing one value, often means less resources for realizing other values.

Given the present standards for planetary protection and overall mission costs, planetary protection stands for a few percent of the mission costs [14.18]. Though this does not seem unreasonable, there are those who think even this is too much [14.19] [14.52] [14.61]. There are also many other worthwhile things to use our limited resources on and the competition for money, talent and other resources is harsh. It is therefore necessary to make trade-offs with other values. In addition to monetary costs, planetary protection also complicates experiments [14.29]. It is, in other words, necessary to make trade-offs with other values, both internal and external. If planetary protection, on the other hand, can satisfy more than one value, the costs will be much easier to motivate.

Either way, saying that planetary protection should aim to keep the risk of contamination as low as possible is not particularly helpful. What we need to do instead is to determine the maximum acceptable probability of contamination, given the instrumental value of planetary protection for the goal(s) it aims to facilitate, and the end value of those goals in relation to other end values that they compete with.

We can thus conclude that in addition to a strong epistemic basis, planetary protection also needs a strong axiological basis. Without the latter, it is in fact not even possible to set out criteria for what constitutes a strong epistemic basis, since without an axiological basis we do not even know what facts we need to consider to build a strong epistemic basis for planetary protection. Let us therefore see what some more official sources have to say about the axiological basis for planetary protection.

14.3 The Axiological Dimension of Planetary Protection Today

The organization that is put in charge of deciding the global standard for planetary protection is the Committee for Space Research (COSPAR) under the International Science Council (ISC). In the 2002 version of COSPAR's Planetary Protection Policy, the organization formulates the basis for its work as follows:

"Although the existence of life elsewhere in the Solar System may be unlikely, the conduct of scientific investigations of possible extraterrestrial life forms, precursors and remnants must not be jeopardized. Moreover, the Earth must be protected from the potential hazard posed by extraterrestrial matter carried by a spacecraft returning from another planet. Therefore, for certain space mission/target planet combinations, controls on contamination shall be imposed, in accordance with issuances implementing this policy" [14.15].

What does this mean? Let us leave the formulation about the Earth environment for the moment (we will come back to that later) and focus on the so-called forward contamination, that is, contamination of other worlds by

Earth microbes. COSPAR states that "the conduct of scientific investigations...must not be jeopardized." This indicates that the main goal of planetary protection has to do with protecting the science, while protecting extraterrestrial life is important only as a means to that goal. This interpretation finds additional support in the latest published version of COSPAR's planetary protection policy, according to which: "The conduct of scientific investigations of possible extraterrestrial life forms, precursors, and remnants must not be jeopardized" [14.28]. In philosophical terms, this means that extraterrestrial life from the perspective of planetary protection, has "instrumental epistemic value."

A closer look at COSPAR's planetary protection policy also tells us that the protection of a target body from contamination is assumed to be temporary. It states, for instance, that "the probability that a planetary body will be contaminated during the period of exploration should be no more than $1*10^{-3}$" [14.28]. The instrumental epistemic value of life on a planetary body will only last for the expected time it will take to find and sufficiently understand it. After that, any need for protection vanishes. This can only be interpreted as meaning extraterrestrial life has no value beyond its value as a study object, that is, beyond its instrumental epistemic value.

We can thus summarize COSPAR's position regarding the axiological basis for planetary protection as follows:

1. Knowledge of whether life exists or has existed in other worlds has a very high end value.
2. A thorough understanding of this life, if it exists, has a very high end value. (It cannot be conclusively determined from the guidelines alone whether the value in (1) and (2) is end value or instrumental value but since no attempt is made to motivate it by referring to its use, it may be assumed that it is a matter of end value.)
3. Extraterrestrial life on target bodies have instrumental value as sources of (1) and (2).
4. Extraterrestrial life does not have any other value than that stated in (3) or if it does it is negligible in relation to all other values that can be promoted by the same resources.
5. The Earth biota has a very high value that cannot be compromised.

Someone might be tempted to object that the present axiological basis for planetary protection does not explicitly deny that extraterrestrial life has value beyond its instrumental epistemic value, but this objection does not work. Ignoring a value, explicitly or implicitly, is in fact the same as saying that it is non-existent or at least negligible. A description can afford to be silent about certain matters but a decision that clearly affects a certain phenomenon has to be either-or. Either we take measures to protect extra-terrestrial life beyond the projected study time or not. If we decide not to, we have de facto denied that it has (other than negligible) value beyond their value as study objects, whether we do it explicitly or just implicitly.

Can the above list of values be taken to be the consensus view regarding the axiological basis for planetary protection? It seems so. Most scientists and all space agencies seem to accept and adhere to COSPAR's planetary protection guidelines (also noted by [14.16]). This also seems to be true regarding the axiological basis for those guidelines. Most research papers on planetary protection only mention the protection of science as moti-vation for planetary protection [14.34] [14.43], and the same seems to be true of space agencies and other influential organizations in this area. In 1992, the Space Studies Board of the U.S. National Academy of Sciences (NSA) recommended, for instance, that only Mars landers with life detec-tion instruments have to be sterilized to the same level as the Viking land-ers. Other Mars landers may have a lower degree of sterilization, which do not include heat sterilization. These recommendations have been followed by all NASA missions to Mars since then [14.59], which indicates that NSA and NASA also accept that Mars life has instrumental value in relation to life detection and no or very low value beyond that.

14.4 The Nature of Epistemic Values

It is probably easy to confuse a decision to only consider epistemic value (values related to knowledge) with a decision to only consider epistemic aspects (knowledge). This is also in my experience a very common mistake. The truth is, however, that even if the two look alike, they are not. To only consider epistemic aspects is in fact not even an option, since any decision (contrary to a mere description) is per necessity a value statement. It is a statement that certain things should be pursued, protected, etcetera, while others should not. A decision that is only based on what is of use to the pro-motion of knowledge is a decision that only knowledge or the promotion of knowledge has value in itself. It is not a statement of a scientific fact. To put it in more technical terms, that something is epistemic means it has to do with knowledge. That a statement is branded as knowledge implies that it is supposedly true, that is, that it tells us something of how the world actually is. It is therefore easy to be trapped into thinking that epistemic values must be objectively true, so that if something has epistemic value it is objectively true that it is valuable and if it does not have epistemic value, it is objectively true that it does not have value, or at least that if it has value, it is of no concern for science. The value of knowledge is, however, no different from any other value in this regard. I am, as a knowledge seeker, not saying this in order to diminish the value of knowledge, far from it. I am saying this to clear up a possible source of confusion about the axiological basis for planetary protection and to make this basis more transparent in accordance with the spirit of philosophy and science. Epistemic value is a value among other values and it is not possible to refer to or assume a set of epistemic values while also claiming that one does not concern oneself with values.

So far, we have established that there is a rather strong consensus in the scientific community that the axiological basis for planetary protection is to protect science, or more specifically, the axiological basis for planetary protection is the search for and eventual understanding of extraterrestrial life, through protection of the instrumental epistemic value of extraterrestrial life. The next thing we need to do is to try to understand its background and its implications. Why does the axiological dimension of planetary protection look the way it does and where does it come from? Is it based in international law or is it an internal construction within the research community? Is it binding or is it possible to be more or less strict, and is there room for adding values just as the epistemic basis is sometimes upgraded with increased knowledge?

14.5   The Outer Space Treaty and the Axiological Dimension of Planetary Protection

According to COSPAR's planetary protection policy, their guidelines are based on the Treaty on Principles Governing the Activities of States in the Exploration and Use of Outer Space, Including the Moon and Other Celestial Bodies (also known as the Outer Space Treaty, or for short, OST) [14.28]. Also, other reports as well as academic papers about planetary protection almost unequivocally describe COSPAR's work as following directly from OST [14.12] [14.21] [14.42] [14.43]. Cockell and Horneck [14.10] even go so far as claiming that COSPAR's planetary protection policy "is merely based on a scientific interpretation of the Outer Space Treaty."

The article in the Outer Space Treaty that deals with planetary protection and that is commonly stated as making up the legal basis for COSPAR's planetary protection policy is Article IX. The relevant part reads:

> "In the exploration and use of outer space, including the Moon and other celestial bodies, States Parties to the Treaty shall be guided by

the principle of cooperation and mutual assistance and shall conduct all their activities in outer space, including the Moon and other celestial bodies, with due regard to the corresponding interests of all other States Parties to the Treaty. States Parties to the Treaty shall pursue studies of outer space, including the Moon and other celestial bodies, and conduct exploration of them so as to avoid their harmful contamination and also adverse changes in the environment of the Earth resulting from the introduction of extraterrestrial matter and, where necessary, shall adopt appropriate measures for this purpose..." [14.58].

An obvious problem when trying to interpret what this means for our question is that OST, including Article IX, is notoriously vague. Some have interpreted the fuzziness as evidence that one cannot really draw any conclusions about the axiological basis for planetary protection from OST. The U.S. National Academy of Sciences, in fact, claims that "Article IX of the Outer Space Treaty, however, is ambiguous with respect to whether its focus is on protecting celestial bodies themselves or the scientific interests of those countries exploring them" [14.32].

This is not an assessment that is shared by everyone, however. Cypser [14.17] claims that "While these policies [COSPAR's planetary protection policies] have been criticized by some as inadequate, they are consistent with current terrestrial international space law which recognizes no absolute protection for alien life-forms or alien environments."

That COSPAR's policies are consistent with OST as well as international space law in general does not seem surprising considering the vagueness of the latter, but more interesting are the questions whether COSPAR's policy originates from OST and international space law or if the axiological basis, completely or partly, comes from somewhere else and, if so, how much freedom does OST leave COSPAR to formulate its policy including its axiological dimension.

Cypser [14.17] argues that the focus of OST (and of international space law in general) is to respect the interests of the states that are party to the treaty. This interpretation is supported by the following quote from Article IX: "Parties to the Treaty shall...conduct all their activities in outer space,... with due regard to the corresponding interests of all other States Parties to the Treaty" [14.58]. It is also supported by the frequent mentioning of the importance of respecting the interests of other states also in other parts of OST, as well as in international law in general [14.17] [14.58].

Where does this leave us regarding the axiological basis for planetary protection? Cypser [14.17] takes her reasoning a couple of steps further. The only interest of the states parties to the treaty that she mentions, is to protect their ongoing space programs, and this also seems to be the only interest that according to her is relevant for planetary protection. Why does she draw this conclusion? An obvious answer could be: "What other national interests could there be?" This answer is not completely satisfying, however. Why can the states that signed and ratified the treaty only be interested in their ongoing space programs? For starters, the term "ongoing" seems a bit inappropriate in relation to planetary protection that to a large extent is about protecting future missions against unwanted effects of present missions. I also cannot find any clear evidence in OST or anywhere else for why the frequent urges to respect the interests of other states would exclusively refer to the interest in space programs. One thing to consider here is that the OST is signed and ratified by several nations without an active space program. Why would they sign and ratify the treaty if it is only about protecting ongoing space programs? It may be because they hope to start space programs later, but I do not find anything that excludes the possibility that they could have other interests in space; for instance, that the people of these countries find other values worth protecting in space.

Cypser in fact takes her reasoning one more step. She states: "The history of international planetary protection standards and the policies and practices of the space-faring nations make it clear that the stipulation to 'avoid harmful contamination' contained in the Outer Space Treaty was intended to protect the integrity of future scientific experiments and not to guarantee the preservation of alien environments" [14.17].

It is true that scientific experiments are an important motivation for space missions, but it is far from the only one. In the beginning of the space race, political and military motives played a very important role (though military use of other planets is ruled out in the treaty). For instance, in the future, the motivations for space missions will probably include more business ventures.

Contrary to this, one might say that biocontamination is only a threat to missions aimed at finding and understanding extraterrestrial life, which means it still makes sense to claim that planetary protection only aims at protecting scientific experiments. This is not necessarily true, however. There may, for instance, be commercial interests that are affected by planetary protection in both positive and negative directions. Positive because extraterrestrial life may have commercial value [14.9] [14.39], negative because planetary protection may impose limitations that get in the way of commercial exploitation. It also seems entirely plausible that to many people extraterrestrial life could have value in its own right, independently of its instrumental value for science or business [14.39].

There is also another important point that needs to be made here. The OST, just like any other international treaty, can be seen as a contract between the parties of the treaty regulating the parties' behavior towards each other. In that sense, it is obvious that respect for the interests of the other parties is key to understanding the treaty. However, this also means that it has to be acknowledged that OST does not cover all interests that may be relevant to consider and can thus not be taken as the only thing determining the actual guidelines. In particular, there is no contract or treaty that can put itself above ethics, especially ethical considerations regarding the parties' relation to non-signatories, and especially in relation to entities that cannot be signatories, including for obvious reasons, extra-terrestrial organisms. In other words, even though it is probably correct that the axiological basis for COSPAR's guidelines is consistent with OST, there is nothing, and there can be nothing in this or any other treaty that excludes an inclusion of other values in the axiological basis for planetary protection, especially not values that are morally required. Ignoring the ethical aspects of any issue is just not an option, no matter what it is about, why one is doing it, what kind of instructions one got, or what the law happens to say about it.

Before we go on to investigate the ethical aspects, however, let me point out another problem with the assumption that the axiological basis for planetary protection used by COSPAR is directly provided by OST, namely that COSPAR existed and had already issued recommendations before the formulation of OST. In fact, these recommendations did not even change substantially after the establishment of OST. What sense can we make of these curious facts?

14.6   The Axiological Dimension of Planetary Protection – Historical Background

A look at some of the earliest documents, including for instance, early documents from COSPAR and the NAS, as well as articles in space science journals, might help us understand how the axiological basis for planetary protection has developed.

The scientific community started to be concerned about potential contamination of other worlds in the mid-fifties. The International Astronautical Federation started looking into the question in 1956 [14.17] [14.30], and the NAS started their investigations in 1957 [14.17]. It is thus safe to say that concerns about contamination was present in the history of space flight from the start.

Based on these early initiatives, it also seems clear that concern about future search for life in the solar system was the driving motive. An early official statement about planetary protection that includes a motivation came in 1958 when the NAS urged scientists to "... plan lunar and planetary studies with great care and deep concern so that initial operations do not compromise and make impossible forever after critical scientific experiments" [14.31]. The initiative behind NAS's engagement in contamination issues came from the scientific community through a letter from Joshua Lederberg sent to NAS in 1957 [14.32]. Lederberg was very active in advocating for planetary protection, for instance, through the West Coast Committee on Extraterrestrial Life (WESTEX), which in addition to Lederberg counted several other very famous scientists among its members, including Nobel Laureate Melvin Calvin and Carl Sagan, whose influence on planetary protection we will say more about shortly [14.30].

On NAS's initiative, the International Council of Scientific Unions (ICSU) set up an ad hoc committee called the Committee on Contamination by Extraterrestrial Exploration (CETEX) specifically tasked with the construction of guidelines for avoiding contamination [14.17] [14.30] [14.32]. CETEX reported their work in a paper in *Nature* the coming year where they also stated: "The need for sterilization is only temporary. Mars and possibly Venus need to remain uncontaminated only until study by manned ships becomes possible" [14.8]. It was also CETEX that recommended that the responsibility for formulating future planetary protection policies should fall on COSPAR [14.32].

COSPAR's first resolution [14.14] [14.17] followed the same lines as the NAS statement but with a slight but significant change of the time constraint. NAS wanted to keep Mars uncontaminated until manned missions become possible. COSPAR wanted to keep the planet uncontaminated "until such time as this search can have been satisfactorily carried out."

In 1963, the United Nations General Assembly unanimously accepted a suggestion from the United Nations Committee on the Peaceful Uses of Outer Space (COPUOS) founded four years earlier, for a Declaration of Legal Principles Governing the Activities of States in the Exploration and Use of Outer Space [14.17]. This declaration was later to become the basis for the OST. COPUOS did not manage to reach an agreement regarding the part of the suggested declaration most relevant for planetary protection, the regulation of harmful experiments in space, and it was never included in the declaration. Instead, COPUOS acknowledged and came "close to outright endorsement of COSPAR's standards" [14.17]. The UN General Assembly, however, only took note of COSPAR's work but it was not mentioned in the declaration or in the OST, and the COSPAR guidelines are therefore still not legally binding in a formal sense [14.17].

The most elaborate early suggestion in the academic literature for an axiological basis for planetary protection, here in the case of Mars, as well as the most detailed suggestion for how to operationalize these values, was presented in an article by Carl Sagan and Sidney Coleman in *Astronautics and Aeronautics* in 1965. In this paper, the authors provide some concrete numbers as well as the calculations and the epistemic as well as the axiological assumptions behind the calculations [14.51]. They are also clear that "[t]he type and duration of sterilization procedure must depend on some estimate of what constitutes an acceptable risk of planetary contamination." They suggest "a probability very close to unity that N biological experiments be successfully completed on Mars before biological contamination occurs... [where N stands for] the desired number of experiments for a thorough survey of Martian biology." They explain that N needs to be very high but do not wish to take a definite stand on the exact number. In their own calculations they set N to 1000 and aim for a probability of 99.9% that 1000 experiments can be performed before contamination [14.51].

As Greenberg points out, the calculations of Sagan and Coleman are based on "pre-space-age" knowledge [14.25]. This is of course unavoidable, and Sagan and Coleman are aware of this weakness. They therefore emphasize the need to continuously update the numbers [14.51]. In that way, their paper is a good role model. They do not suggest any opening for a continuous update of the axiological basis, however.

Regarding the axiological basis for their calculations, there is not much room for doubt from Sagan and Coleman's formulations that the only value they acknowledge for the Martian environment and possible Martian life, is as study objects. They do not mention any other value but they do set a time limit for protection that is only based on the value these environments and possible life have as study objects [14.51].

Overall, it seems safe to conclude that the four axiological principles that guide planetary protection today have been the same from the start, long before the Outer Space Treaty. It also seems clear that these values came directly from the science community. This is also the conclusion of Cypser [14.17] [14.22].

This seems like a good explanation for how the present axiological basis for planetary protection has come about, though it does not fully explain or justify why these concerns still make up the entire axiological basis for planetary protection, and it does, of course, not show that this will always be the case, in particular since the composition of the players on the space arena is quickly widening. More about this point later.

What does it mean that the practical guidelines as well as the axiological basis for planetary protection clearly predates OST and that none of these are included in the OST or the declaration behind the OST? In international law, "practice makes principle." That is, if a practice is followed by a sufficient number of sufficiently influential players, it eventually gets the status of a binding legal principle. This means that as long as the most influential players in the form of space agencies and space researchers in space-faring nations accept that following COSPAR's planetary protection policy is what it takes to live up to Article IX in OST, then COSPAR's guidelines have a fairly strong de facto legal status.

Does this mean that COSPAR can widen the axiological basis for planetary protection and following that amend their guidelines to accommodate for the new values (as would have to be the case if, for example, the time limit has to go)? This is a tricky question. A potentially important fact is that something of the kind has actually happened. In 1982, the COSPAR guidelines were quite substantially recast. The most salient change was the establishment of five categories or mission type/target body combinations and the connection of measures to these categories. This included a special category for sample return missions, including guidelines for how to avoid contamination of Earth by extraterrestrial biological material, something that had previously only been handled ad hoc by NASA in connection with the moon landings [14.22]. This means that a non-epistemic value was in fact added to the axiological basis for planetary protection, namely.

This is a value that was clearly stated in OST, Article IX, which means this update can justifiably be considered as merely an adaption to the OST. On the other hand, it also shows that non-epistemic values are not banned in principle from being part of the axiological basis for planetary protection. It also means that Cypser's conclusion [14.17] that only scientific experiments connected with ongoing space missions can be covered by the principle of respect for the other parties of the OST, has to be wrong.

14.7   Ethics and Planetary Protection

I have in other publications discussed different approaches to the question of moral status for extraterrestrial life. I am not going to repeat that discussion here but just summarize the main conclusions in very few words and refer the interested reader to these earlier discussions [14.35] [14.36] [14.37] [14.38] [14.39].

There are different theories regarding who or what has moral standing on Earth and why. The historically most influential theory, anthropocentrism, claims that only humans can have moral standing [14.7] [14.56]. This theory is losing ground the more we learn about other life forms.

An alternative theory called sentientism states that all and only sentient beings can have moral standing [14.1] [14.24] [14.40] [14.41] [14.54] [14.55]. This theory is based on the idea that ethics is really about considering the interests of others. Therefore, if someone has interests, they automatically qualify as moral objects. If they do not have interests, they automatically disqualify.

The theory about moral standing that initially seems most promising for granting moral status to any extraterrestrial life we might find in our own solar system would be biocentrism, a theory that grants moral standing to all life [14.23] [14.53]. This theory is considerably more controversial, however. Also, and maybe initially unexpectedly, even though it is the only theory that grants moral standing to all extraterrestrial life, it may actually call for a weakening of planetary protection. The reason is that bioload reduction in the form of decontamination kills large numbers of Earth microbes. Biocentrism may, therefore, all things considered, not favor mass extermination of actual Earth microbes to protect merely possible extraterrestrial microbes.

This latter problem would be avoided if we instead accept the ecocentric theory of moral standing. This theory focuses primarily on species instead of individuals [14.3] [14.4] [14.5] [14.6] [14.26] [14.27] [14.44] [14.45] [14.46] [14.47] [14.48]. Killing large numbers of individuals from common species is not a problem according to this theory. On the other hand, it is even more controversial than biocentrism. It is generally considered highly implausible that species can have interests in a morally relevant sense. It also has other peculiarities. According to one version of this theory, extraterrestrial species do not count morally since they do not belong to the same biota as Earth life [14.5].

Sentientism is clearly the most plausible of these theories. The probability that we will find sentient life in our solar system outside Earth is considered extremely low, however. It therefore seems implausible that we would have any moral duties to any extraterrestrial life in our solar system. This does, however, not mean that there are no ethical considerations to be made. Greenberg expresses concern over how present standards of planetary protection affect future generations of astrobiologists. If we follow COSPAR and others and decide that planetary protection will only be necessary for a limited period of time, we will rob future generations of astrobiologists of their chance to continue the exploration of extraterrestrial life [14.25].

Greenberg clearly has a point here. Considering that we have studied the biology of our own planet for countless generations and we are not close to being finished understanding Earth biology, we can probably be sure that it will take a long time to understand the biology of the solar system (if there is any biology off Earth) to a degree that would make the science community say, "yes we are finished, no more to see here." In fact, a seemingly inherent property of scientific research is that the more questions we answer, the more questions turn up.

Concern for future astrobiologists is not the only moral concern, however. We also need to consider the interests of the people of planet Earth outside the relatively small community of astrobiologists. If the future existence of extraterrestrial life after it has been discovered and studied has a sufficiently high positive value to a sufficiently high number of people outside the astrobiology community, it seems there is a moral prerogative to take this into account. That the interests of people outside the scientific community have to be taken into consideration is in fact pointed out, for example, by NAS and the National Research Council's Committee on Planetary and Lunar Exploration [14.11] [14.32]. There are also initiatives by NASA, for example, to include the wider society in dialogues about planetary protection [14.2] [14.49] [14.50]. I am not aware, however, of any large-scale attempts to really find out whether there is a sufficiently large interest among a sufficiently large number of people on planet Earth to make it morally required to include this in the axiological basis for planetary protection. Such surveys are, of course, difficult to perform since value questions are tricky to assess, but it may be worth the effort to get the ethics right. For any attempt to include the values of the wider population, it is also essential to make it clear that it is not just for show, but that additional values actually can be added to the axiological basis for planetary protection.

14.8   Competing Values – Planetary Protection and the Commercial Use of Space

In addition to the ethical implications, there is also a more pragmatic reason why astrobiologists should consider widening the axiological basis for planetary protection. Astrobiology is part of a bigger world where news regarding extraterrestrial life is met with great interest by the general public, where funding is ultimately dependent on political decisions, and not least where the "old" space-faring nations and space agencies are being followed by new space-faring nations as well as by private initiatives.

The emergence of private initiatives in space with the explicit aim of landing on and performing operations in other worlds is particularly relevant for three reasons. 1) The total number of spacecraft aimed at bodies of interest to astrobiology will increase. 2) The competitive pressure will incentivize private actors to look at all possible ways of saving money. The necessity of strict non-contamination rules might be questioned as a result. 3) The main aim of the private actors is not to do science. Protecting the science might therefore not be a strong motivational factor for them to maintain high standards of planetary protection. This also has political implications. Both the USA and Luxembourg have recently passed laws with the explicit purpose of encouraging private space initiatives. If the only reason for planetary protection that is accepted in the scientific community is strictly internal to astrobiology, it will be very difficult for the scientific community to maintain the importance of planetary protection against the lobbying from the commercial sector. In connection with a hearing held by the space subcommittee of the US Senate Commerce Committee, US Senator Ted Cruz stated: "As we look to the future of American free enterprise and settlement in space, we should also thoroughly review the United Nations' Outer Space Treaty, which was written and enacted in a very different time and era in 1967.... It's important that Congress evaluate how that treaty, enacted 50 years ago, will impact new and innovative activity within space" [14.20].

In this situation, the great interest in questions about extraterrestrial life among the general public can be a great ally in the quest to keep extraterrestrial environments unspoiled, but in order to mobilize that support, it will be necessary to also include other values of relevance to the general public. In fact, a mobilization of the general public may be the only way for the scientific community to achieve the strengthened legal status necessary to withstand the push from the commercial sector to weaken planetary protection, and an active inclusion of other values more readily embraced by the general public may be the only way of achieving this.

14.9 Conclusions

Planetary protection needs a solid epistemic basis as well as a solid axiological basis. The almost universally accepted rules for planetary protection are formulated by COSPAR. The axiological basis for planetary protection assumes that the knowledge of whether there is life outside Earth and understanding of that life has a very high end value, while the life as such has merely instrumental value in relation to the former. The Earth biota, on the other hand, seems to be assigned a very high end value.

COSPAR's guidelines have no formal legal status but have a fairly strong de facto legal status due to their general acceptance among space-faring nations. The axiological basis originates from the science community but is closely associated with the OST, the purpose of which is to protect the interests of the participating states. To be ethically sound, however, any planetary protection policy needs to include a wider set of relevant interests. This includes the interests of future scientists, other sectors of society and the general public. It seems implausible, however, that extraterrestrial organisms in our solar system are advanced enough to have interests of their own. There are strong reasons to believe that the business sector has an interest in weakening the guidelines for planetary protection but it is also plausible that the large interest in extraterrestrial life among the general public may be able to outweigh these interests.

References

[14.1]   Bentham, J., *An Introduction to the Principles of Morals and Legislation*, Adamant Media Corporation, Boston, MA, USA, 2005.

[14.2]   Billings, L., Public communication strategy for NASA's planetary protection program: Expanding the dialogue. *Adv. Space Res.*, 38, 10, 2225–2231, 2006.

[14.3]   Callicott, J.B., Animal liberation: a triangular affair. *Environ. Ethics*, 2, 4, 311–338, 1980.

[14.4]   Callicott, J.B., The conceptual foundations of the land ethic, in: *Companion to a Sand County Almanac—Interpretive Critical Essays*, J.B. Callicott (Ed.), pp. 186–217, The University of Wisconsin Press, Madison, WI, 1987.

[14.5]   Callicott, J.B., Moral considerability and extraterrestrial life, in: *The Animal Rights/Environmental Ethics Debate*, E.C. Hargrove (Ed.), pp. 137–150, State University of New York Press, Albany, USA, 1992.

[14.6]   Callicott, J.B., *Beyond the Land Ethic*, State University of New York Press, Albany, USA, 1999.

[14.7]   Carruthers, P., *The Animals Issue*, Cambridge University Press, Cambridge, UK, 1994.

[14.8]   CETEX, Contamination by extraterrestrial exploration. *Nature*, 183, 4666, 925–928, 1959.

[14.9]   Cockell, C., Ethics and Extraterrestrial Life, in: *Humans in Outer Space – Interdisciplinary Perspectives*, N.-L. Nina-Louisa Remuss, K.-U. Schrogl, J.-C. Jean-Claude Worms, U. Landfester (Eds.), pp. 80–101, Springer, New York, NY, USA, 2011.

[14.10] Cockell, C.S., Horneck, G., Planetary parks – formulating a wilderness policy for planetary bodies. *Space Policy*, 22, 4, 256–261, 2006.

[14.11] Committee on Planetary and Lunar Exploration, National Research Council, *A Science Strategy for the Exploration of Europa*, National Academy Press, Washington, D.C., USA, 1999.

[14.12] Conley, C.A., Rummel, J.D., Planetary protection for humans in space: Mars and the Moon. *Acta Astronaut.*, 63, 7, 1025–1030, 2008.

[14.13] Conley, C.A. and Rummel, J.D., Planetary protection for human exploration of Mars. *Acta Astronaut.*, 66, 5–6, 792–797, 2010.

[14.14] COSPAR, COSPAR Resolution No. 26. *COSPAR Inf. Bull.*, 20, 25–26, 1964.

[14.15] COSPAR, *Planetary Protection Policy*, COSPAR, Paris, 2002.

[14.16] Crawford, R.L., Microbial Diversity and Its Relationship to Planetary Protection. *Appl. Environ. Microbiol.*, 71, 8, 4163–4168, 2005.

[14.17] Cypser, D.A., International law and policy of extraterrestrial planetary protection. *Jurimetrics*, 33, 2, 315–229, 1993.

[14.18] Debus, A., Planetary protection: Elements for cost minimization. *Acta Astronaut.*, 59, 8–11, 1093–1100, 2006.

[14.19] Fairén, A.G. and Schulze-Makuch, D., The overprotection of Mars. *Nat. Geosci.*, 6, 510–511, 2013.

[14.20] Foust, J., Commercial space's policy wish list. *Space Rev.*, 2017. 1 May 2017.

[14.21] Glavin, D.P., Dworkin, J.P., Lupisella, M., Williams, D.R., Kminek, G., Rummel, J.D., In Situ Biological Contamination Studies of the Moon: Implications for Planetary Protection and Life Detection Missions. *Earth Moon Planets*, 107, 1, 87–93, 2010.

[14.22] Goh, G. and Kazeminejad, B., Mars through the looking glass: An interdisciplinary analysis of forward and backward contamination. *Space Policy*, 20, 3, 217–225, 2004.

[14.23] Goodpaster, K., On being morally considerable. *J. Philos.*, 75, 6, 308–325, 1978.

[14.24] de Grazia, D., *Taking Animals Seriously*, Cambridge University Press, Cambridge, UK, 1996.

[14.25] Greenberg, R., Exploration and Protection of Europa's Biosphere: Implications of Permeable Ice. *Astrobiology*, 11, 2, 183–191, 2011.

[14.26] Johnson, L.E., *A Morally Deep World*, Cambridge University Press, Cambridge, UK, 1991.

[14.27] Johnson, L.E., Toward the moral considerability of species and ecosystems. *Environ. Ethics*, 14, 2, 145–157, 1992.

[14.28] Kminek, G., Conley, C., Hipkin, V., Yano, H., *COSPAR's Planetary Protection Policy*, COSPAR, Paris, 2017, cosparhq.cnes.fr/assets/uploads/2019/12/PPPolicyDecember-2017.pdf.

[14.29] Levin, G.V., Detection of Metabolically Produced Labeled Gas: The Viking Mars Lander. *Icarus*, 16, 1, 153–166, 1972.

[14.30] Meltzer, M., *When biospheres collide: a history of NASA's planetary protection Programs*, NASA, Washington, D.C., USA, 2011.

[14.31] National Academy of Sciences, *Resolution adopted by the Council of the NAS*, National Academy of Sciences, Washington, DC, 1958, February 8, 1958.

[14.32] National Academy of Sciences, *Preventing the Forward Contamination of Mars*, National Academy of Sciences, Washington, DC, 2006.

[14.33] Nicholson, W.L., McCoy, L.E., Kerney, K.R., Ming, D.W., Golden, D.C., Schuerger, A.C., Aqueous extracts of a Mars analogue regolith that mimics the Phoenix landing site do not inhibit spore germination or growth of model spacecraft contaminants Bacillus subtilis 178 and Bacillus pumilus SAFR-032. *Icarus*, 220, 2, 904–910, 2012.

[14.34] Nicholson, W.L., Schuerberger, A.C., Race, M.S., Migrating microbes and planetary protection. *Trends Microbiol.*, 17, 9, 389–392, 2009.

[14.35] Persson, E., The Moral Status of Extraterrestrial Life. *Astrobiology*, 12, 10, 976–984, 2012.

[14.36] Persson, E., Interplanetär etik, in: *Extrema världar – Extremt liv*, D. Dunér (Ed.), Pufendorfinstitutet, Lund, Sweden, 123–132, 2013a.

[14.37] Persson, E., Philosophical aspects of astrobiology, in: *The History and Philosophy of Astrobiology*, D. Dunér, J. Parthemore, E. Persson, G. Holmberg (Eds.), pp. 29–48, Cambridge Scholar, Newcastle upon Tyne, England, 2013b.

[14.38] Persson, E., Ethics and the potential conflicts between astrobiology, planetary protection and commercial use of space. *Challenges*, 8, 1, 12, 2017.

[14.39] Persson, E., A philosophical outlook on potential conflicts between planetary protection, astrobiology and commercial use of space, in: *Our Common Cosmos*, Z. Lehmann-Imfeld and A. Losch (Eds.), pp. 141–160, Bloomsbury Publishing, London, UK, 2019.

[14.40] Regan, T., *Defending Animal Rights*, University of Illinois Press, Urbana, IL, USA, 2001.

[14.41] Regan, T., *The Case for Animal Rights*, University of California Press, Berkeley, CA, USA, 2004.

[14.42] Rettberg, P., Anesio, A.M., Baker, V.R., Baross, J.A., Cady, S.L., Detsis, E., Foreman, C.M., Hauber, E., Gabriele, O.G., Pearce, D.A., Renno, N.O., Ruvkun, G., Sattler, B., Saunders, M.S., Smith, D.H., Wagner, D., Westall, F., Planetary Protection and Mars Special Regions – A Suggestion for Updating the Definition. *Astrobiology*, 16, 2, 119–125, 2016.

[14.43] Rettberg, P., Fritze, D., Varbarg, S., Nellen, J., Horneck, G., Stackebrandt, E., Kminek, G., Determination of the microbial diversity of spacecraft assembly, testing and launch facilities: First results of the ESA project MiDiv. *Adv. Space Res.*, 38, 6, 1260–1265, 2006.

[14.44] Rolston, H. III., The preservation of natural value in the Solar System, in: *Beyond Spaceship Earth: Environmental Ethics and the Solar System*, E. Hargrove (Ed.), Sierra Club Books, San Francisco, CA, USA, 140–182, 1986.

[14.45] Rolston, H. III., Duties to ecosystems, in: *Companion to a Sand County Almanac—Interpretive Critical Essays*, J.B. Callicott (Ed.), pp. 246–274, The University of Wisconsin Press, Madison, WI, USA, 1987.

[14.46] Rolston, H. III., *Environmental Ethics—Duties to and Values in The Natural World*, Temple University Press, Philadelphia, PA, USA, 1988.

[14.47] Rolston, H. III., *Conserving Natural Value*, Columbia University Press, New York, NY, USA, 1994.

[14.48] Rolston, H. III., Ethics on the home planet, in: *An Invitation to Environmental Philosophy*, A. Weston (Ed.), pp. 107–139, Oxford University Press, New York, NY, USA, 1999.

[14.49] Rummel, J.D. and Billings, L., Issues in planetary protection: policy, protocol and implementation. *Space Policy*, 20, 1, 49–54, 2004.

[14.50] Rummel, J.D., Race, M., Horneck, G., Ethical Considerations for Planetary Protection in Space Exploration: A Workshop. *Astrobiology*, 12, 11, 1017–1023, 2012.

[14.51] Sagan, C. and Coleman, S., Spacecraft sterilization standards and contamination of Mars. *Astronautics Aeronautics*, 3, 5, 22–27, 1965.

[14.52] Schon, S.C., Reversible exploration not worth the cost. *Science*, 323, 5921, 1561, 2009.

[14.53] Schweitzer, A., The ethic of reverence for life, in: *Animal Rights and Human Obligations*, T. Regan and P. Singer (Eds.), pp. 133–138, Prentice-Hall, Englewood Cliffs, NJ, USA, 1976.

[14.54] Singer, P., *Practical Ethics*, Cambridge University Press, Cambridge, UK, 1993.

[14.55] Singer, P., *Animal Liberation*, Pimlico, London, UK, 1995.

[14.56] Smith, K.C., The trouble with intrinsic value: an ethical primer for astrobiology, in: *Exploring the Origin, Extent, and Future of Life*, C.M. Bertka (Ed.), pp. 261–280, Cambridge University Press, Cambridge, UK, 2009.

[14.57] Smith, S.A., Benardini, J.N. III, Anderi, D., Ford, M., Wear, E., Schrader, M., Schubert, W., DeVeaux, L., Paszczynski, A., Childers, S.E., Identification and Characterization of Early Mission Phase Microorganisms Residing on the Mars Science Laboratory and Assessment of Their Potential to Survive Mars-like Conditions. *Astrobiology*, 17, 3, 253–265, 2017.

[14.58] UNOOSA, *Treaty on Principles Governing the Activities of States in the Exploration and Use of Outer Space, including the Moon and Other Celestial Bodies*, UN General Assembly RES 2222 (XXI), New York, 1966.

[14.59] deVincenzi, D., Race, M.S., Klein, H.P., Planetary protection, sample return missions and Mars exploration: History, status, and future needs. *J. Geophys. Res.*, 103, E12, 28,577–28,585, 1998.

[14.60] Wikipedia, *Planetary protection*, 2020, http://en.wikipedia.org/wiki/Planetary_protection.

[14.61] Zubrin, R. and Wagner, R., *The Case For Mars*, Simon and Schuster, New York, NY, USA, 1996.

15

Who Speaks for Humanity?
The Need for a Single Political Voice

Ian A. Crawford

*Department of Earth and Planetary Sciences, Birkbeck College,
University of London, London, UK*

Abstract

Future astrobiological activities and discoveries, along with other human activities in the transnational domain of outer space, will require the development of political institutions able to legitimately speak for humanity as a whole. I identify a range of possibilities, including the formation of a world space agency and a strengthening of the UN system; but I argue that ultimately the logic points in the direction of bringing space exploration within the remit of a federal world government, the creation of which would also be desirable for other reasons. Although, at present, humanity lacks a sufficiently strong sense of global community for the formation of strong global political institutions, I argue that the cosmic and evolutionary perspectives provided by astrobiology and related disciplines can help lay the psychological foundations on which such institutions may be built.

Keywords: Astrobiology, space exploration, federalism, global governance, world government

15.1   Introduction

"[L]aws and institutions must go hand in hand with the progress of the human mind. As that becomes more developed, more enlightened, as new discoveries are made, new truths disclosed ... institutions must advance also, and keep pace with the times" [15.52].
                                                        – Thomas Jefferson, 1816

Email: i.crawford@bbk.ac.uk

Octavio Alfonso Chon Torres, Ted Peters, Joseph Seckbach and Richard Gordon (eds.) *Astrobiology: Science, Ethics, and Public Policy*, (313–338) © 2021 Scrivener Publishing LLC

The relatively new science of astrobiology is usually defined as the study of the origin, evolution, distribution and future of life in the universe. It follows that one of the main scientific objectives, indeed the holy grail, of astrobiology is the discovery of extraterrestrial life. Only then will we be able to extend our understanding of the complex phenomenon of life beyond the single example found on Earth.

That said, and as discussed elsewhere in this volume, there are also wider societal and philosophical aspects of astrobiology that extend beyond its purely scientific aspirations. Some of these, for example the intellectual enrichment arising out of what is necessarily a highly interdisciplinary research field, will manifest themselves regardless of whether extraterrestrial life is discovered or not [15.24] [15.30] [15.72]. Others, including a host of moral and ethical considerations relating to humanity's interactions with alien life, will only become apparent, or at least pressing, if alien life is actually encountered [15.36] [15.37]. In this chapter I want, perhaps provocatively, to extrapolate from these ethical and philosophical considerations to what I see as their political implications.

One could make the case that politics, here understood as the process of decision-making within and between groups of intelligent social animals, itself falls within the remit of astrobiology given that, like other aspects of culture, it is ultimately a result of biological evolution. As Aristotle [15.4] realized long ago, "man is by nature a political animal," one whose natural environment is a political community.[1] The same may be true of other intelligent life forms that have evolved, or may yet evolve, elsewhere in the universe.

Moreover, the definition of astrobiology includes a concern with the *future* of life, and it seems inevitable that the future of life in the universe will, at least in part, depend on the political decisions of intelligent technological species. Of course, as yet we know nothing about the prevalence, or otherwise, of other intelligent life in the universe, and still less about their political arrangements, but our own case is clear enough: *Homo sapiens* is currently the dominant technological species on the only known inhabited planet in the Universe. Unless or until astrobiology itself teaches us otherwise, it is possible that the whole future of life, not only on Earth but also in the wider Universe, will depend on the political choices of this one

---

[1]The context of this famous quote makes clear that Aristotle was making the case that humanity's natural environment is actually a *polis*, or city-state. However, I shall argue below that today our *polis* has effectively expanded to encompass the whole planet and may one day extend beyond it.

technological species. Needless to say, this places an enormous responsibility on human political institutions.

To my mind, there are two broad, at first sight distinct (although, as I shall argue below, ultimately synergistic) political aspects of astrobiology. On the one hand, there are essentially practical political issues regarding human decision-making in the context of astrobiological activities and discoveries, and, on the other, there are the psychological and social implications of these activities, especially their attendant *perspectives*, for the evolution of human political institutions. We will consider these two broad categories in turn.

15.2   The Need for Global Decision-Making in an Astrobiological Context

A moment's reflection will reveal that multiple political decisions will have to be made as humanity goes about searching for life in our Solar System and beyond. By definition, many of these decisions will affect, and to a small extent have already affected, transnational domains (e.g., the surface of Mars) where no existing human institutions can claim political legitimacy. As Margaret Race [15.71] asks in her excellent review of our institutional preparedness for encountering extraterrestrial life, "Who would be involved in decision making on behalf of humankind?"

It is true that, at the relatively modest level of current activities, there is a framework of internationally recognized policies to guide our astrobiological activities [15.71], including intergovernmental treaties (most notably the 1967 Outer Space Treaty[2]) and internationally accepted guidelines such as the COSPAR Planetary Protection Policy.[3] However, although these existing agreements provide an excellent foundation on which to build, it is all too obvious that the latter are entirely voluntary, and that even the former would be difficult to enforce in practice.

As our exploratory activities increase, and especially if extraterrestrial life is actually encountered, the essentially political question of "Who speaks for humanity?" in the transnational domains beyond Earth will

---

[2]Treaty on Principles Governing the Activities of States in the Exploration and Use of Outer Space, including the Moon and Other Celestial Bodies (herein the 'Outer Space Treaty,' OST); https://www.unoosa.org/oosa/en/ourwork/spacelaw/treaties/introouterspacetreaty. html

[3]Committee on Space Research (COSPAR): Planetary Protection Policy; https://cosparhq. cnes.fr/assets/uploads/2019/12/PPPolicyDecember-2017.pdf

become increasingly pressing.[4] To give a flavor of the kinds of astrobiologically-related political issues that will require some form of legitimate international decision-making for their resolution, consider the following non-exhaustive list:[5]

- In the early stages of exploration, it may not be possible to determine whether or not a given planetary environment is inhabited or not, yet many important decisions will depend on this determination. In the short term these might include decisions on planetary protection and, at least in the case of Mars, when or if it might be appropriate to send human missions to such an environment. In the longer term, it will have implications for colonization and/or terraforming schemes. It follows that at some point someone, or some institution, may have to decide that the evidence against the existence of an indigenous biology is *sufficiently* strong to enable these activities to go ahead, but who will decide on the criteria required for such judgements?
- If microbial life is discovered elsewhere in the Solar System (e.g., on Mars, Europa, Enceladus, etc.) then a range of important decisions will need to be taken fairly quickly. For example, should a moratorium be implemented on interaction with such life while scientists and policy makers consider the options? Who will decide on the subsequent policy? Should sampling such life be allowed? Should any such samples be returned to Earth for analysis? Should plans for human missions to inhabited (or apparently inhabited) extraterrestrial environments be initiated or should they be forbidden? If permitted, what protocols should govern their activities and who will decide what these should be?
- Suggestions for preserving alien environments, inhabited or not, in the context of human exploration or colonization include the establishments of "planetary parks" [15.23] or

---

[4]This question was perhaps most famously asked by Carl Sagan [15.79] in Chapter 13 of *Cosmos*: "Who speaks for Earth?" but was interestingly anticipated in another context by the international relations scholar John Herz [15.49] who, in the course of a discussion on global political institutions (to be discussed below), asked "Who speaks for Man? How can a planetary mind be developed?".

[5]A similar set of questions, more from an ethical than a political perspective, has been posed by Peters [15.69].

the implementation of a "one-eighth principle" [15.64]. At first sight, these appear sensible suggestions, but who will decide on the location of the parks or the areas to be preserved? And who will enforce these decisions?

- In the context of searching for life beyond our Solar System, there are important political issues related to the Search for Extraterrestrial Intelligence (SETI). In the event of a *bona fide* detection of an alien signal there will be immediate decisions to be made regarding how to proceed, and longer-term questions about the wisdom or otherwise of responding to such a signal [15.41] [15.62]. Currently, guidelines are provided by the International Academy of Astronautics (IAA) SETI Protocols,[6] but these are entirely voluntary and unenforceable. In the event of an actual detection there seems every likelihood that they would be swept aside by the governmental and security apparatuses of nation-states eager to try to secure some advantage for themselves, and this could turn into a source of international conflict here on Earth.

- Even if extraterrestrial signals of alien origin are never detected by SETI, there is an increasingly pressing political issue related to the deliberate sending of radio signals into space in the hope of them being detected by extraterrestrial intelligence (ETI), an activity known as Messaging Extraterrestrial Intelligence (METI). Transmissions of this kind are already taking place without any international political oversight [15.40]. As Michaud [15.63] has noted, "[w]hatever the consequences of such transmissions may be, our descendants will not be able to opt out of them. We might expose our species to risks we cannot calculate." This being so, it seems especially important that some process or institution that can legitimately speak for "our species" has a say on whether such transmissions should be permitted or not. As Michaud himself asks, "who speaks for Earth? Should we speak with one voice or many?"

- We also need to consider the possibility that at some point in the future humanity, or some sub-set of humanity, may decide that it would be desirable to artificially spread Earth-life to

---

[6]International Academy of Astronautics (IAA): Declaration of Principles Concerning the Conduct of the Search for Extraterrestrial Intelligence; http://resources.iaaseti.org/protocols_rev2010.pdf

locations elsewhere in the Solar System or to planets orbiting other stars. This would involve humanity engaging in a program of "directed panspermia" [15.34] and has been advocated by a number of recent authors [15.43] [15.44] [15.61] [15.106]. This would be a controversial step, fraught with all sorts of ethical issues, and completely contrary to current planetary protection policies [15.22]. However, if one accepts that the complexity and potential of life transcends that of non-life [15.55] [15.73] [15.90] then an ethical case for spreading life to places where it doesn't yet exist could be made.[7] Clearly, we already have the capability to transport life around the Solar System, and there are reasons for believing that human space-faring technology is nearing the point where even directed interstellar panspermia might be seriously contemplated [15.43] [15.61].[8] The question of who will take responsibility on behalf of humanity for such possibly very far-reaching activities will therefore need to be addressed.

These examples of astrobiologically-relevant political issues can be supplemented by others related to the future of space exploration but not directly relevant to astrobiology *per se* (except insofar as *any* human activity in space has the potential to affect the future of life, and will therefore fall within a broad definition of astrobiology). Foremost among these are political questions related to the exploitation and ownership of extraterrestrial resources. It seems clear that the utilization of space resources will

---

[7]For example, the "astrobiology ethic" proposed by Randolph and McKay [15.73] that "promotes the goal of protecting and expanding the richness and diversity of life." This implies that life should be protected where it exists (note their important concept of a "Cosmic Golden Rule"), but that it should proactively be introduced to places where it is absent.

[8]For the foreseeable future we would presumably only be talking about microbial life, possibly genetically engineered for the purpose, and interstellar travel times with current technology would be tens of thousands of years. However, provided that microbes or their spores can remain viable during transit, travel times of thousands of years are not really an issue in the context of seeding the Galaxy with life because they are essentially instantaneous relative to astronomical and evolutionary timescales; note also that some recent proposals for near-term rapid interstellar travel (e.g., *Project Starshot,* https://breakthroughinitiatives.org/initiative/3) could potentially enable much shorter transit times, and that microorganisms appear to be ideally suited for the very low payload masses envisaged by such concepts. For those interested, I have summarized various interstellar travel concepts elsewhere [15.31]; note that the implementation of any of these proposals will also raise important questions of political oversight and legitimacy, quite apart from any application to directed panspermia.

require the establishment of an international legal and political regime that will encourage investment in prospecting and extraction activities, while at the same time protecting scientifically (and ethically) important locations from interference and ensuring that these activities do not become a flashpoint for human conflict. Although the Outer Space Treaty provides an excellent foundation on which to build appropriate institutions,[9] its provisions are in urgent need of development to cope with likely 21st century developments [15.7] [15.56] [15.70] [15.87] [15.91]. In this context, it seems appropriate to draw attention to William Hartmann's "Golden Rule of Space Exploration" to the effect that "space exploration must be carried out in a way so as to reduce, not aggravate, tensions in human society" [15.46] [15.47]. However, realizing this in practice will surely require the development of an appropriate, internationally supported, legal and political framework.

Ted Peters [15.69] has argued that addressing these issues will require building "a single planetary community of moral deliberation." I agree, but I also think that such a moral "planetary community" will need to be underpinned by appropriate political institutions able legitimately to speak for humanity as a whole in the transnational domains beyond Earth. Fortunately, as we will discuss below, there is actually a wide spectrum of possibilities for future international political institutions which may be appropriate for achieving this overarching socio-political objective. Unfortunately, as will become painfully apparent, all of them will, to greater or lesser degrees, require a greater commitment to international cooperation and solidarity than is manifested in global human society at the present time. I will argue, however, that astrobiology, and space exploration more widely, can themselves help generate the required international solidarity by engendering a cosmic perspective on human affairs.

15.3   Some Socio-Political Implications of Astrobiological Perspectives

As noted above, there is a widespread recognition that, beyond its purely scientific focus of searching for extraterrestrial life, the study of astrobiology

---

[9]Key provisions include the concept that space activity should be considered "the province of all mankind" (Article I), that outer space is free for the "exploration and use" by all states (Article I), that celestial bodies cannot be appropriated by nation-states (Article II), and that international law, including the UN Charter, applies to outer space (Article III).

has the potential to convey a range of wider social and intellectual benefits. Many of these arise from the inherently interdisciplinary nature of astrobiology, and its consequent ability to bridge the intellectual gaps between different sciences and between the sciences and the humanities, e.g., [15.6] [15.17] [15.24] [15.30] [15.36] [15.37] [15.38] [15.39] [15.51] [15.72] [15.88].

I have argued previously [15.30] [15.32] that many of these societal benefits result from the cosmic and evolutionary *perspectives* that are the natural, and in fact unavoidable, companions of astrobiology. It is not possible to be engaged in searching for life on Mars, or on planets orbiting other stars, without moving away from the narrow Earth-centric perspectives that dominate the social and political lives of most people most of the time. Moreover, it is only by sending spacecraft to explore the Solar System, in part for astrobiological purposes, that we can gain a truly cosmic perspective on our own planet (Figure 15.1); see also White [15.98], Som [15.84].

Importantly, astrobiology also provides a temporal and evolutionary perspective on human affairs, helping to locate *Homo sapiens*, and human society, in time as well as space. As noted by Dick [15.37]; see also Crawford [15.30], there is a strong synergy here with the emerging discipline of "big history" [15.18] [15.19] [15.85], which, following the lead of earlier authors such as Chambers [15.15] [15.16], Humboldt [15.50] and Wells [15.95], aims to integrate human history into an evolutionary history of the Universe. It has long been recognized that exposure to these cosmic and evolutionary perspectives may help stimulate the development of

Figure 15.1 (a) Earthrise over the lunar surface, photographed by the crew of Apollo 8 in December 1968. (b) The Earth photographed from the surface of Mars by the Mars Exploration Rover Spirit in March 2004. Such images powerfully reinforce a "cosmic perspective" that can help build a sense of human community. (Images courtesy of NASA.)

cosmopolitan worldviews.[10] Indeed, this was the explicit hope of several of the authors listed above,[11] as well as of more recent thinkers [15.8] [15.17] [15.42] [15.59] [15.83][12] [15.93][13] [15.101]. I have provided a more extensive discussion of this argument elsewhere [15.30] [15.32], so a couple of examples will have to suffice here.

In 1844, Robert Chambers [15.15] anonymously published his *Vestiges of the Natural History of Creation*, which is perhaps the first scientifically grounded attempt to provide an evolutionary history of the Universe and humanity's place within it. The publication of *Vestiges* caused a sensation at the time [15.81], and the following year Chambers felt the need to offer some "Explanations" in the course of which he drew the ethical implication that his "new view of nature" would assist in

> "establishing the universal brotherhood and social communion of man. And not only this, but it extends the principle of humanity to the other meaner creatures also. Life is everywhere ONE" [15.16].

This quotation is especially significant when it comes to considering the ethical implications of astrobiological perspectives: it shows that Chambers was concerned not only with laying a foundation for "the universal brotherhood and social communion of man," but also his expectation that a proper understanding of cosmic and evolutionary perspectives would have ethical implications for our relations with other living things.

---

[10]Here, I adopt the definition of a worldview given by Aerts *et al.* [15.1]: "A world view is a ... frame of reference in which everything presented to us by our diverse experiences can be placed. It is a symbolic system of representation that allows us to integrate everything we know about the world and ourselves into a global picture." Although I have here intended the word "cosmopolitan" to refer to the wider human community, we could take it more literally: for example, Lupisella [15.59] and Dick [15.37] have persuasively argued that cosmic and evolutionary perspectives may lead to a literally "cosmocentric" worldview, and associated cosmocentric ethics, which in principle might be shared by all intelligent entities that evolve in the Universe.

[11]H.G. Wells (1866-1946), in particular, was a life-long advocate of developing cosmopolitan political institutions, and this was one of his main motivations for popularizing evolutionary and historical perspectives in works like *The Outline of History* (1920); for a comprehensive review of Wells' political thought, see Partington [15.68].

[12]The astronomer Harlow Shapley (1885-1972) dedicated much of his career to popularizing the cultural benefits of a cosmic perspective; see [15.67].

[13]The economist Barbara Ward (aka Baroness Jackson, 1914-1981) was much taken by the planetary perspective provided by early space missions; her slim book *Spaceship Earth* (Ward 1966) contains much of interest to the present discussion, especially her insistence on the need to build global institutions for planetary management.

A century and a half later, the biologist Ursula Goodenough advanced essentially the same argument, writing

"Any global tradition needs to begin with a shared worldview: a culture-independent, globally accepted consensus as to how things are. ... our scientific account of nature, an account that can be called The Epic of Evolution. The Big Bang, the formation of stars and planets, the origin and evolution of life on this planet, the advent of human consciousness and the resultant evolution of cultures—this is the story, the one story, that has the potential to unite us, because it happens to be true" [15.42].[14]

The suggestion that these cosmic and evolutionary perspectives may have specifically political benefits rests on the realization that by encouraging more cosmopolitan *worldviews* they can help pave the way towards building more cosmopolitan political *institutions*. I think we have to accept that tribalism is probably instinctive in *Homo sapiens*, possibly as a result of group selection during our evolutionary past [15.92] [15.99] [15.101] [15.102],[15] and that this tribalism gets in the way of developing the kind of global, cosmopolitan, institutions that the world increasingly needs. As Kwame Appiah put it in his influential essay on modern cosmopolitanism:

"The challenge, then, is to take minds and hearts formed over long millennia of living in local groups and equip them with ideas and institutions that allow us to live together as the global tribe we have become" [15.3].

Fortunately, there are grounds for hope when we realize that throughout human history the size of tribes to which we feel allegiance has been expanding, and has now almost, but sadly not quite, reached a global scale.

---

[14]Goodenough's phrase "The Epic of Evolution" neatly captures the big historical and astrobiological perspectives. To my knowledge, the first person to write of an "evolutionary epic," and to view it as a kind of "origin myth" that comes as close to truth as science can make it, was E.O. Wilson [15.100] (see also Segerstråle [15.82]). James Malazita [15.60] has recently expanded on the role of astrobiology in creating a modern "origin myth" that can help "answer material-cultural questions about modern humanity's origins, identity, ethics and future." David Christian [15.19] has done something similar for big history. My own view is that the term "worldview," in the sense developed by Aerts *et al.* [15.1], is preferable to "myth" in this context.

[15]For a scholarly discussion of the various controversies associated with the concept of group selection, and other evolutionary influences on human behavior, see Segerstråle [15.82].

Today, the dominant political tribes are the 200 or so nation-states into which the world is divided, but as recently as a few thousand years ago there were probably hundreds of thousands of such independent political units [15.14]. As Benedict Anderson [15.2] pointed out, political communities such as nations are essentially "imagined communities" because "the members of even the smallest nation will never know most of their fellow members...yet in the minds of each lives the image of their communion."

It follows that if we wish to build global political institutions to speak for humanity in a cosmic context, as well as to deal with many other pressing global problems, it will be helpful, and perhaps essential, to strengthen feelings of an "imagined" global community, or, as Wilson [15.101] put it, to "globalize the tribe" (see also Appiah [15.3], p. xi). Sagan [15.79] put it most starkly (p. 371): "If we are to survive, our loyalties must be expanded further, to include the whole human community, the entire planet Earth" (see also [15.80]). Other authors who have come to essentially the same conclusion include Wells [15.95], Shapley [15.83], Ward [15.93], Aerts *et al.* [15.1], White [15.98], Burke *et al.* [15.11], Leinen and Bummel [15.58] and, most recently, Som [15.84].

We can therefore identify a symbiotic (strictly mutualistic) relationship between the cosmic and evolutionary perspectives provided by astrobiology and related disciplines on the one hand, and the development of the global institutions needed for the long-term management of astrobiology-related political issues on the other. That is, by helping to lay the psychological foundations for enhanced global cooperation, astrobiological perspectives may stimulate the development of appropriate international political institutions which, in turn, may stimulate and enable greater astrobiological research activity.

Conceivably, the greatest contribution astrobiology could make in this respect would be the discovery of ETI, because then humanity would, for the first time, have an "Other" against whom we could define ourselves as a community. For example, Sagan [15.78] suggested that following such a discovery "the animosities which divide the peoples of the Earth may wither. The differences among human beings of separate races and nationalities, religions and sexes are likely to be insignificant compared to the differences between all humans and all extraterrestrial intelligent beings."

Andre Novoa [15.66] has developed this argument and concludes that "our internal frontiers will hardly be dissolved until we encounter one Other, an outsider from which difference may be built and opposition constructed...cosmopolitanism, until then, cannot be a political project." I agree that what we know of human tribalism suggests that the discovery of ETI would be very helpful in this context, but I don't think we can rely on it actually happening. My own view [15.27] is that the Fermi Paradox[16] already indicates that ETI is probably very rare in the Universe. Rather, my argument here is that the cosmic and evolutionary perspectives engendered by astrobiology, perhaps stimulated by the discovery of non-intelligent life in our Solar System or beyond, will help push human society towards more cosmopolitan outlooks even in the absence of the discovery of ETI.

15.4   Who Speaks for Humanity? Building Appropriate Political Institutions for Space Activities

As noted above, current institutional arrangements appear inadequate for the management of the kinds of political issues that are likely to arise as a result of future astrobiological activities and discoveries. The same is true for other, not directly astrobiology-related, human activities in space (e.g., the governance of space resources), where international governance structures are currently weak or non-existent. In order to properly manage human activities in the Solar System (and eventually beyond), the various high-sounding statements to the effect that space exploration is "the province of all mankind"[17] will need to be underpinned by political institutions able to speak for, and to take responsibility on behalf of, humanity as a whole.

Because we are interested in finding genuinely cosmopolitan solutions to the governance of human activities beyond Earth, I am not here going to consider limited, albeit easier to implement, near-term measures such as domestic national legislation (e.g., the 2015 US Commercial Space Launch Competitiveness Act[18]), agreements brokered by non-governmental organizations (e.g., COSPAR, IAA, etc.), or bi-lateral or multi-lateral agreements between a limited number of nation-states (e.g., the International Space Station Agreements[19] and the recently proposed Artemis Accords[20]). Some such initiatives may play important roles in the near future, and some (e.g., the COSPAR Planetary Protection Policy) are clearly beneficial [15.71].

---

[16]For reviews, see Webb [15.94] and Ćirković [15.20].
[17]Outer Space Treaty, Article I.
[18]https://www.congress.gov/bill/114th-congress/house-bill/2262
[19]https://www.state.gov/wp-content/uploads/2019/02/12927-Multilateral-Space-Space-Station-1.29.1998.pdf
[20]https://www.nasa.gov/specials/artemis-accords/img/Artemis-Accords_v7_print.pdf

However, they all fall well short of legitimately "speaking for humanity" in outer space affairs, and some, by extending concepts of national sovereignty beyond Earth, may actually work counter to the direction in which, as I have argued above, we should aspire to go.

I considered some more ambitious proposals in an earlier article [15.26], where I identified a hierarchy of possible global institutional developments that might improve the governance of future space activities. These ranged from suggestions for creating a world space agency and/or strengthening the United Nations, to proposing that in the longer-term space activities would most logically fall within the remit of a future federal world government. I still believe that these suggestions adequately delineate the spectrum of desirable possibilities, so will briefly reiterate the arguments here.

15.4.1   A World Space Agency

In an important, and sadly rather overlooked, article from the mid-1970s, Seyom Brown and Larry Fabian [15.10] advocated the creation of much stronger international institutions to govern human activities in what they called the "nonterrestrial realms," which they then took to include Earth's oceans and climate as well as outer space. Clearly, all of these areas would benefit from stronger international governance, but it is their suggestions for coordinating global activities in space that are relevant here. In order to give institutional support to the provisions of the Outer Space Treaty, especially the provisions in Article I that space activities are "the province of all mankind" and "shall be carried out for the benefit and in the interests of all countries," Brown and Fabian [15.10] advocated the creation of what they called an "Outer Space Projects Agency." They envisaged that all countries in the world would belong to this agency, and that, among other responsibilities, it would be "empowered to give final approval to all... outer space exploration projects for civilian purposes, under guidelines requiring international participation and the international dissemination of all data and results." Similar proposals for a world space agency/authority have independently been advanced by Crawford [15.25], Tronchetti [15.87], Pinault [15.70], and Koch [15.56].[21]

---

[21]The proposals by Tronchetti [15.87] and Koch [15.56] were made specifically in the context of space resource utilization, although the proposed institutional structures could in principle be expanded to cover other aspects of space exploration, including those relevant to astrobiology.

The success of the European Space Agency (ESA), established in 1975 and now comprising 22 member states, clearly shows that large international space agencies are workable in practice and can result in many scientific and cultural benefits [15.9]. There has not yet been any serious attempt made at expanding the concept to a global scale, although a positive start was perhaps made in this direction in 2007 when fourteen of the world's space agencies developed the Global Exploration Strategy.[22] This initiative resulted in the formation of the International Space Exploration Coordination Group (ISECG),[23] which now consists of 22 national space agencies (including the multinational ESA), and which could perhaps be viewed as a tentative step towards a global space agency. Among the first fruits of ISECG was the formulation of *The Global Exploration Roadmap*, now in its third edition,[24] which outlines an international collaborative framework for the robotic and human exploration of the Solar System, focusing on destinations where humans may one day live and work. Clearly this focus is relevant to some of the astrobiological issues raised in Section 15.2.

However, although the creation of a world space agency would be desirable for coordinating global space exploration activities, it would not in itself be able legitimately to "speak for humanity" in a cosmic context. As a purely functional agency, formed through intergovernmental agreements, its political authority and legitimacy would ultimately be derived from the governments of its participating nation-states.[25] Some of these will be much more powerful than others, and, at least as the world is currently constituted, not all of them are likely to be democratically accountable. Thus, just as NASA obtains its political authority from the US federal government, which is ultimately answerable to US citizens through elections, any future world space agency would need to take its political direction from a higher-level political structure able to represent the world's citizens. This logic points inescapably towards some form of planetary government.

---

[22]https://www.globalspaceexploration.org/wordpress/wp-content/uploads/2013/10/Global-Exploration-Strategy-framework-for-coordination.pdf

[23]https://www.globalspaceexploration.org

[24]https://www.globalspaceexploration.org/wordpress/wp-content/isecg/GER_2018_small_mobile.pdf

[25]A tentative start has also been made in the direction of intergovernmental coordination of global space policy through the International Space Exploration Forum (ISEF; https://www.globalspaceexploration.org/?p=792); however, as for the agency-level ISECG, ISEF can only claim to represent national governments and not humanity as a whole.

15.4.2 Strengthening the United Nations for the Governance of Space Activities

Although falling well short of a true world government, the United Nations (UN) is arguably the closest approximation to one that has yet been attempted, and its very existence is an implicit recognition by national governments that *some* kind of global political institution is desirable for the management of global affairs. Given that space is a transnational domain, it would appear to be especially appropriate that human activities in space, including those of relevance to astrobiology, should fall under UN jurisdiction. Indeed, this was recognized at the very dawn of the space age by the creation of the UN Office of Outer Space Affairs (UNOOSA)[26] and the UN General Assembly's Committee on the Peaceful Uses of Outer Space (UNCOPUOS)[27] in 1958. Since then, the UN has been instrumental in negotiating the current legal regime that governs human activities in space (i.e., the 1967 Outer Space Treaty and its successors[28]), and continues to act as a valuable global forum for coordination, decision-making, and information-sharing related to international space activities.[29]

Relatively modest reforms of the UN structure could further strengthen international oversight of human activities in outer space and other transnational domains. One possibility, suggested by the 1995 Report of the Commission on Global Governance [15.13], would be to repurpose the now defunct UN Trusteeship Council for this purpose. The Trusteeship Council was established in 1946 to supervise the administration of former colonies as they transitioned to independent nation-states, and suspended its activities in 1994 when decolonization was essentially complete. In principle, therefore, this major UN organ, which has its own chamber at UN Headquarters in New York, is available to take on new functions such as overseeing human activities in transnational domains, including outer space. This would elevate oversight of human activities in these domains to one of the six principal organs in the UN system, placing them on a par, in principle if not initially in practice, with the deliberations of the Security Council. Indeed, as noted by Carlsson *et al.* [15.13] "the time has come to

---

[26]https://www.unoosa.org

[27]https://www.unoosa.org/oosa/en/ourwork/copuos/

[28]https://www.unoosa.org/oosa/en/ourwork/spacelaw/treaties.html

[29]E.g., the 2018 UNCOPUOS Guidelines for the Long-term Sustainability of Outer Space Activities (https://www.unoosa.org/res/oosadoc/data/documents/2018/aac_1052018crp/aac_1052018crp_20_0_html/AC105_2018_CRP20E.pdf) and UNOOSA's Thematic Priorities for space development (https://www.unoosa.org/documents/pdf/unispace/plus50/thematic_priorities_booklet.pdf).

acknowledge that the security of the planet is a universal need to which the UN system must cater," and several of the astrobiologically-related issues discussed in Section 15.2 have the potential to fit into this category.

Unfortunately, at present the UN suffers from a lack of political legitimacy because the world's citizens are not represented in its decision-making processes. Moreover, it is predicated on the (increasingly outmoded) concept of nation-state sovereignty, which means that, ultimately, it cannot enforce any decisions it may take. Rather, in practice, the UN is just one more forum within which nation-states are free to exercise their own sovereignty in their own perceived national interests. As Fremont Rider predicted just a year after its creation, the UN has, like its predecessor the League of Nations, been treated by national governments "as merely another piece to be moved about on the international board in the game for national power—and as not a very important piece at that" [15.75]. One way to increase the democratic accountability of the UN, and to at least reduce its subservience to the whims of its member governments (not all of which have any meaningful democratic legitimacy of their own), would be to add an elected Parliamentary Assembly to its governing organs [15.58].[30] Of course, deciding on the franchise of such a parliament would be fraught with problems [15.58] [15.75], but some such innovation is likely to be necessary if the UN is to play a more meaningful role in representing humanity in the governance of global, and extra-global, affairs.

15.4.3   Space Activities in the Context of a Future World Government

If the UN evolves to the point where more of its authority is derived from the world's citizens, as represented in a parliament of some kind, and less from the governments of nation-states, then it will be evolving in the direction of a federal world government. There is a large body of literature on the desirability or otherwise of establishing some form of world government, and a great many different forms that such a government might take.[31]

---

[30]See also the Campaign for a UN Parliamentary Assembly: https://en.unpacampaign.org
[31]E.g., Kant [15.53], Russell [15.77], Wells [15.95], Laski [15.57], Reves [15.74], Rider [15.75], Toynbee [15.86], Kerr [15.54], Hamer [15.45], Wendt [15.96] [15.97], Yunker [15.104] [15.105], Cabrera [15.12] and Leinen and Bummel [15.58]. Comprehensive historical overviews of world government proposals have been provided by Wynner and Lloyd [15.103], Heater [15.48] and Baratta [15.5]; interested readers may also wish to follow the contemporary online discussions at the *World Government Research Network* (https://www.wgresearch.org/).

My own view [15.28] is that dealing effectively with planetary scale problems will eventually require a federal system of planetary governance able to implement the principle of subsidiarity on a global scale (i.e., a world government responsible solely for global matters that cannot be addressed effectively at a local or national level). Examples of such planetary scale problems include: (i) the currently anarchic international environment where heavily armed nation-states act as judges in their own cause (making military confrontation all but inevitable); (ii) global environmental pollution, including anthropogenic contributions to climate change; (iii) habitat destruction and biodiversity loss; (iv) global-scale natural threats to human society (e.g., pandemics, mega-volcanoes and asteroid impacts); (v) long-term development challenges (e.g., provision of sufficient food and water, and the satisfaction of aspirations for higher living standards, for a growing world population); and (vi) inefficient, and often irresponsible, management of the global commons.

It seems clear that many of the astrobiologically-related political issues identified in Section 15.2 would also most appropriately fit within the remit of future federal world government. Indeed, the fact that there is already a general acceptance that such matters should ideally be referred to the UN, and the very existence of UNOOSA and UNCOPUOS, is a recognition that global-level governance of these issues is considered desirable. However, a genuine world government would be likely to have far greater legitimacy and effectiveness in managing extraterrestrial activities on behalf of humanity as a whole. To my knowledge, this connection between world government and space exploration has only occasionally been noted in the professional international relations literature, although it is explored more frequently in science fiction.[32] One international relations scholar who has explicitly made the connection is James Yunker in his book *Political Globalization: A New Vision of Federal World Government,* where he observes that a world government might need a "Ministry of External Development" to coordinate human activities beyond Earth [15.104], and speculates (p. 87) that a world government might one day be required to protect Earth from extraterrestrial threats.

If we consider that the future of humanity may involve a significant human presence elsewhere in the Solar System, contemplation of which falls within the remit of astrobiology as usually understood, then a careful

---

[32]Perhaps most notably in the *Star Trek* universe created by Gene Roddenberry (TV fiction, first broadcast in the United States on 8 September 1966), where a federal government exists not only on Earth but has been extended to include non-human civilizations on other planets.

consideration of federal forms of government becomes even more pertinent. This is because federal systems of government are inherently expand-
able, limited only by the speed and reliability of communication and
transportation technologies. This is perhaps demonstrated most clearly
by the history of the United States' federal constitution, drafted over the
summer of 1787 and which, within little more than a century, had enabled
a form of government designed to ensure cooperation between thirteen
former English colonies on the Atlantic coast of North America to expand
across the entire continent.[33] As I have argued elsewhere [15.28], a federal
form of government, employing the principle of subsidiarity on interplanetary scales, may be the only form of government able to maintain diversity
among human colonies elsewhere in the Solar System while at same time
minimizing the risk of conflict between them.[34]

However, although we might agree that a world (and later interplanetary)
government would be desirable in principle, the practical implementation
of such a government in the near future would be a daunting task, and perhaps politically infeasible. It is important to understand that the obstacles
are not technological (in terms of travel and communication timescales,
the whole planet today is far more compact than were the original thirteen North American colonies in 1787[35]), but psychological. Although in
1787 there were many Americans opposed to the proposed federal constitution,[36] there was at least a sufficiently strong sense of community to

---

[33]And beyond: consider, in the present context, that the islands of Hawaii in the Pacific
Ocean, admitted as a State of the United States in 1959, could just as well be a colony on
Mars as far as the federal institutions are concerned.

[34]Creating robust political institutions to prevent interplanetary conflict will be essential
given the biosphere-destroying energies that any interplanetary society will have at its disposal [15.33] [15.35]. Some dystopian fictional representations of what may happen in the
absence of appropriate interplanetary government are provided by Robinson [15.76] and
the TV series *The Expanse* (https://en.wikipedia.org/wiki/The_Expanse_(TV_series)).

[35]In this context, it is worth recalling Arthur C. Clarke's words at the signing of the
International Telecommunications Satellite Organization (INTELSAT) Agreement in
1971: "What the railroads and the telegraph did here [in the USA] a century ago, the jets
and communications satellites are doing now to all the world. ...You have just signed a
first draft of the Articles of Federation of the United States of Earth" [15.21].

[36]There are many histories available detailing the contemporary arguments for and against
the US federal constitution, but I recommend especially that by Carl Van Doren [15.89].
Originally published in 1948, Van Doren's book is notable both for its scholarship and
its clear-eyed sense of the relevance of the US constitution for future developments in
international governance; as he writes in his Preface (p. viii): "it is impossible to read the
story of the making and ratifying of the Constitution of the United States without finding there all the arguments in favour of a general government for the United Nations, as
well as all the arguments now raised in opposition to it."

make it a politically realistic project. As the leading (realist) international relations scholar Hans Morgenthau put it while contemplating the infeasibility of a world government, just as "the community of the American people antedated the American State...a world community must antedate a world state" [15.65]. Thus, however desirable in principle, a world government is likely to remain politically impractical unless or until humanity is able to overcome the innate tribalism of our species and develop a sufficient sense of Anderson's [15.2] "imagined community," Herz's [15.49] "planetary mind," or what Barbara Ward [15.93] called "a patriotism for the world itself."

As argued in Section 15.3, it is in overcoming these psychological obstacles to global political unification that astrobiology may be able to help. Specifically, I suggest that the cosmic and evolutionary perspectives engendered by astrobiology (together with related disciplines such as big history and bolstered by on-going space exploration activities) may play a valuable role in laying the psychological foundations for the political unification of our world. Moreover, this will not be a one-way street: astrobiology and space exploration are themselves likely to benefit from the creation of a federal world government, partly because of the extra resources such a government would have at its disposal, but mainly because any world government would have strong political incentives for strengthening a sense of global community by leveraging the perspectives provided by astrobiology and space exploration. I have developed this argument elsewhere [15.29], but it was also glimpsed by Yunker [15.104] where he noted that a world government is likely to be especially interested in space exploration because this would place it "at the center of attention in this exciting and inspiring area of human endeavour."

15.5 Conclusions

I have argued that future astrobiological activities and discoveries, along with other human activities in the Solar System (and perhaps one day beyond), will require the development of political institutions able to legitimately speak for humanity as a whole. I have identified a hierarchy of possibilities, including the formation of a world space agency and a strengthening of the UN system in the context of outer space affairs.

However, ultimately, I believe that the logic points in the direction of bringing human activities in space within the remit of a future (federal) world government. The creation of such a government would in any case be desirable to oversee and coordinate other activities that affect humanity as a whole. Eventually, such a federal form of government might be extended to include human colonies and outposts beyond Earth.

Unfortunately, at present, humanity appears to lack a sufficiently strong sense of global community for the formation of strong global political institutions, let alone a genuine world government, to be politically realistic. Creating a stronger sense of global identity will, at least in part, depend on strengthening our sense of humanity's place in the Universe. It is my thesis here that the cosmic and evolutionary perspectives provided by astrobiology (and attendant activities and disciplines such as space exploration and big history) can play a valuable role in laying the psychological foundations for the political unification of our species. Moreover, I have argued that a virtuous circle may develop between developing institutions for global governance and opportunities for future space exploration and development, from which astrobiology as a discipline would surely benefit.

References

[15.1]   Aerts, D., Apostel, L., de Moor, B., Hellemans, S., Maex, E., Van Belle, H., Van der Veken, J., *World Views: From Fragmentation to Integration*, VUB Press, Brussels, 1994, Available in an on-line edition produced by Clément Vidal and Alexander Riegler (2007) at: http://www.vub.ac.be/CLEA/pub/books/worldviews.pdf.

[15.2]   Anderson, B., *Imagined Communities*, Revised ed, Verso, London, UK, 1991.

[15.3]   Appiah, K.A., *Cosmopolitanism: Ethics in a World of Strangers*, Penguin, London, UK, 2006.

[15.4]   Aristotle, *Politics*, c.350 BCE, 1253a, lines 2-3.

[15.5]   Baratta, J.P., *The Politics of World Federation*, Praeger Publishers, Westport, CT, USA, 2004.

[15.6]   Bertka, C.M. (Ed.), *Exploring the Origin, Extent and Nature of Life: Philosophical, Ethical and Theological Perspectives*, Cambridge University Press, Cambridge, UK, 2009.

[15.7]   Bittencourt Neto, O. de O., Hofmann, M., Masson-Zwaan, T., Stefoudi, D., *Building Blocks for the Development of an International Framework for the Governance of Space Resource Activities: A Commentary*, Eleven International Publishing, The Hague, The Netherlands, 2020, (available on-line at: https://boeken.rechtsgebieden.boomportaal.nl/publicaties/9789462361218#0).

[15.8]   Bohan, E., How big history could change the world for the better. *J. Big History*, III, 3, 37–45, 2019.

[15.9] Bonnet, R.M. and Manno, V., *International Cooperation in Space: The Example of the European Space Agency*, Harvard University Press, Cambridge, MA, USA, 1994.

[15.10] Brown, S. and Fabian, L.L., Toward mutual accountability in the nonterrestrial realms. *Int. Organ.*, 29, 877–892, 1975.

[15.11] Burke, A., Fishel, S., Mitchell, A., Dalby, S., Levine, D.J., Planet politics: A manifesto from the end of IR. *Millennium*, 44, 499–523, 2016.

[15.12] Cabrera, L., *Global Governance, Global Government: Institutional Visions for an Evolving World System*, State University of New York Press, Albany, NY, USA, 2011.

[15.13] Carlsson, I., *Our Global Neighbourhood: The Report of the Commission on Global Governance*, Oxford University Press, Oxford, UK, 1995.

[15.14] Carneiro, R.L., The political unification of the world: Whether, when and how – some speculations. *Cross Cult. Res.*, 38, 162–177, 2004.

[15.15] Chambers, R., *Vestiges of the Natural History of Creation*, John Churchill, London, UK, 1844.

[15.16] Chambers, R., *Explanations: A Sequel to Vestiges of the Natural History of Creation*, John Churchill, London, UK, 1845.

[15.17] Chon-Torres, O.A., Astrobioethics: A brief discussion from the epistemological, religious and societal dimension. *Int. J. Astrobiology*, 19, 61–67, 2020.

[15.18] Christian, D., The case for big history. *J. World Hist.*, 2, 223–238, 1991.

[15.19] Christian, D., *Origin Story: A Big History of Everything*, Penguin, London, UK, 2018.

[15.20] Ćirković, M.M., *The Great Silence: The Science and Philosophy of Fermi's Paradox*, Oxford University Press, Oxford, UK, 2018.

[15.21] Clarke, A.C., *Profiles of the Future: An Inquiry into the Limits of the Possible*, Pan Books, London, UK, 1973.

[15.22] Cockell, C.S., Interstellar planetary protection. *Adv. Space Res.*, 42, 1161–1165, 2008.

[15.23] Cockell, C.S. and Horneck, G., Planetary parks: Formulating a wilderness policy for planetary bodies. *Space Policy*, 22, 256–261, 2006.

[15.24] Connell, K., Dick, S.J., Rose, K., *Workshop on the Societal Implications of Astrobiology: Final Report. NASA Technical Memorandum*, NASA Ames Research Center, Moffett Field, CA, 2000, Available at http://www.astrosociology.org/Library/PDF/NASA-Workshop-Report-Societal-Implications-of-Astrobiology.pdf.

[15.25] Crawford, I.A., On the formation of a global space agency. *Spaceflight*, 23, 316–317, 1981.

[15.26] Crawford, I.A., Space development: Social and political implications. *Space Policy*, 11, 219–225, 1995.

[15.27] Crawford, I.A., Where are they? Maybe we are alone in the Galaxy after all. *Sci. Am.*, 283, 1, 28–33, 2000.

[15.28] Crawford, I.A., Interplanetary federalism: Maximising the chances of extraterrestrial peace, diversity and liberty, in: *The Meaning of Liberty Beyond Earth*, C.S. Cockell (Ed.), pp. 199–218, Springer, Cham, Switzerland, 2015.

[15.29] Crawford, I.A., Space, World Government, and a 'Vast Future' for Humanity', in: *World Orders Forum*, 2017, 4th August 2017. Available on-line at: http://wgresearch.org/space-world-government-vast-future-humanity/.

[15.30] Crawford, I.A., Widening perspectives: The intellectual and social benefits of astrobiology (regardless of whether extraterrestrial life is discovered or not). *Int. J. Astrobiology*, 17, 57–60, 2018a.

[15.31] Crawford, I.A., Direct exoplanet investigation using interstellar space probes, in: *Handbook of Exoplanets*, H.J. Deeg and J.A. Belmonte (Eds.), Springer, Cham, Switzerland, 2018b.

[15.32] Crawford, I.A., Widening perspectives: The intellectual and social benefits of astrobiology, big history, and the exploration of space. *J. Big History*, III, 3, 205–224, 2019.

[15.33] Crawford, I.A. and Baxter, S., The lethality of interplanetary warfare: A fundamental constraint on extraterrestrial liberty, in: *The Meaning of Liberty Beyond Earth*, C.S. Cockell (Ed.), pp. 187–198, Springer, Cham, Switzerland, 2015.

[15.34] Crick, F.H.C. and Orgel, L.E., Directed panspermia. *Icarus*, 19, 341–346, 1973.

[15.35] Deudney, D.H., *Dark Skies: Space Expansionism, Planetary Geopolitics and the Ends of Humanity*, Oxford University Press, Oxford, UK, 2020.

[15.36] Dick, S.J. (Ed.), *The Impact of Discovering Life Beyond Earth*, Cambridge University Press, Cambridge, UK, 2015.

[15.37] Dick, S.J., *Astrobiology, Discovery, and Societal Impact*, Cambridge University Press, Cambridge, UK, 2018.

[15.38] Dick, S.J. and Lupisella, M.L., *Cosmos & Culture: Cultural Evolution in a Cosmic Context*, 2009, NASA SP-2009-4802.

[15.39] Finney, B., SETI and the two terrestrial cultures. *Acta Astronaut.*, 26, 263–265, 1992.

[15.40] Gertz, J., Post-detection SETI protocols and METI: The time has come to regulate them both. *J. Br. Interplanet. Soc*, 69, 263–270, 2016.

[15.41] Goldsmith, D., Who will speak for Earth? Possible structures for shaping a response to a signal detected from an extraterrestrial civilisation. *Acta Astronaut.*, 21, 149–151, 1990.

[15.42] Goodenough, U., *The Sacred Depths of Nature*, Oxford University Press, Oxford, UK, 1998.

[15.43] Gros, C., Developing ecospheres on transiently habitable planets: The genesis project. *Astrophys. Space Sci.*, 361, 324, 2016.

[15.44] Gros, C., Why planetary and exoplanetary protection differ: The case of long duration genesis missions to habitable but sterile M-dwarf oxygen planets. *Acta Astronaut.*, 157, 263–267, 2019.

[15.45] Hamer, C.J., *A Global Parliament: Principles of World Federation*, CreateSpace Publishing, UK, 1998.

[15.46] Hartmann, W.K., The resource base in our solar system, in: *Interstellar Migration and the Human Experience*, B.R. Finney and E.M. Jones (Eds.), pp. 26–41, University of California Press, Berkeley, CA, USA, 1985.

[15.47] Hartmann, W.K., Miller, R., Lee, P., *Out of the Cradle: Exploring The Frontiers Beyond Earth*, Workman Publishing, New York, NY, USA, 1984.

[15.48] Heater, D., *World Citizenship and Government*, Macmillan Press, Basingstoke, UK, 1996.

[15.49] Herz, J.H., *International Politics in the Atomic Age*, Columbia University Press, New York, NY, USA, 1962.

[15.50] Humboldt, A.v., *Cosmos: A Sketch of a Physical Description of the Universe*, vol. I, II, E.C. Otté (trans.), (1850) Harper & Brothers, New York, NY, USA, 1859.

[15.51] Impey, C., Spitz, A.H., Stoeger, W., *Encountering Life in the Universe: Ethical Foundations and Social Implications of Astrobiology*, University of Arizona Press, Tucson, AZ, USA, 2013.

[15.52] Jefferson, T., Letter to Samuel Kercheval, in: *The Portable Jefferson*, M.D. Peterson (Ed.) (1975), Penguin, New York, NY, USA, 1816.

[15.53] Kant, I., To Perpetual Peace: A Philosophical Sketch, in: *Perpetual Peace and Other Essays*, T. Humphrey (Ed.) (1983), pp. 107–143, Hackett Publishing Company, Indianapolis, USA, 1795.

[15.54] Kerr, P. and Lothian, L., *Pacifism Is Not Enough*, J. Pinder and A. Bosco (Eds.), Lothian Foundation Press, London, UK, 1990.

[15.55] Ketcham, C., Towards an ethics of life. *Space Policy*, 38, 48–56, 2016.

[15.56] Koch, J.S., Institutional Framework for the province of all mankind: Lessons from the International Seabed Authority for the governance of commercial space mining. *Astropolitics*, 16, 1–27, 2018.

[15.57] Laski, H.J., *A Grammar of Politics*, Allen and Unwin, London, UK, 1925.

[15.58] Leinen, J. and Bummel, A., *A World Parliament: Governance and Democracy in the 21st Century*, Democracy Without Borders, Berlin, Germany, 2018.

[15.59] Lupisella, M.L., Cosmocultural evolution: The co-evolution of culture and cosmos and the creation of cosmic value, in: *Cosmos and Culture: Cultural Evolution in a Cosmic Context*, S.J. Dick and M.L. Lupisella (Eds.), pp. 321–359, 2009, NASA: SP-2009-4802.

[15.60] Malazita, J.W., Astrobiology's cosmopolitics and the search for an origin myth for the Anthropocene. *Biol. Theory*, 13, 111–120, 2018.

[15.61] Mautner, M.N., *Seeding the Universe with Life: Securing Our Cosmological Future*, Legacy Books, Christchurch, New Zealand, 2004.

[15.62] Michaud, M.A.G., *Contact with Alien Civilisations: Our Hopes and Fears about Encountering Extraterrestrials*, Copernicus Books, New York, NY, USA, 2007.

[15.63] Michaud, M.A.G., Searching for extraterrestrial intelligence: Preparing for an expected paradigm break, in: *The Impact of Discovering Life Beyond Earth*, S.J. Dick (Ed.), pp. 286–298, Cambridge University Press, Cambridge, UK, 2015.

[15.64] Milligan, T. and Elvis, M., Mars environmental protection: An application of the 1/8 principle, in: *The Human Factor in a Mission to Mars*, K. Szocik (Ed.), pp. 167–183, Springer, Cham, Switzerland, 2019.

[15.65] Morgenthau, H.J., *Politics Among Nations: The Struggle for Power and Peace*, Alfred Knopf, New York, NY, USA, 1948.

[15.66] Novoa, A., Alien life matters: Reflections on cosmopolitanism, otherness and astrobiology. *Cosmopolitan Civil Societies J.*, 8, 1, 1–26, 2016.

[15.67] Palmeri, J., Bringing Cosmos to Culture: Harlow Shapley and the Uses of Cosmic Evolution, in: *Cosmos and Culture: Cultural Evolution in a Cosmic Context*, S.J. Dick and M.L. Lupisella (Eds.), pp. 489–521, 2009, NASA: SP-2009-4802.

[15.68] Partington, J.S., *Building Cosmopolis: The Political Thought of H.G. Wells*, Routledge, London, UK, 2016.

[15.69] Peters, T., Astroethics and Microbial Life in the Solar Ghetto, in: *Astrotheology: Science and Theology Meet Extraterrestrial Life*, T. Peters, M. Hewlett, J.M. Moritz, R.J. Russell (Eds.), pp. 391–415, Cascade, Eugene OR, USA, 2018.

[15.70] Pinault, L., Towards a world space agency: Operational successes of the International Seabed Authority as models for commercial-national partnering under an international space authority, in: *Human Governance Beyond Earth*, C.S. Cockell (Ed.), pp. 173–196, Springer, Cham, Switzerland, 2015.

[15.71] Race, M., Preparing for discovery of extraterrestrial life: Are we ready?", in: *The Impact of Discovering Life Beyond Earth*, S.J. Dick (Ed.), pp. 263–285, Cambridge University Press, Cambridge, UK, 2015.

[15.72] Race, M., Denning, K., Bertka, C.M., Dick, S.J., Harrison, A.A., Impey, C., Mancinelli, R., Astrobiology and society: building an interdisciplinary research community. *Int. J. Astrobiology*, 12, 958–965, 2012.

[15.73] Randolph, R.O. and McKay, C.P., Protecting and expanding the richness and diversity of life, an ethic for astrobiology research and space exploration. *Int. J. Astrobiology*, 13, 28–34, 2014.

[15.74] Reves, E., *The Anatomy of Peace*, Harper, New York, USA, 1946.

[15.75] Rider, F., *The Great Dilemma of World Organisation*, Reynal and Hitchcock, New York, USA, 1946.

[15.76] Robinson, K.S., *2312*, Orbit books, London, UK, 2012.

[15.77] Russell, B., *Principles of Social Reconstruction*, Allen and Unwin, London, UK, 1916.

[15.78] Sagan, C., *The quest for intelligent life in space is just beginning*, pp. 38–47, Smithsonian Magazine, Washington DC, USA, 1978, Reproduced on-line in Cosmic Search Magazine, vol. 1, No. 2, (2004) at: http://www.bigear. org/vol1no2/sagan.htm.

[15.79] Sagan, C., *Cosmos*, Abacus, London, UK, 1980.

[15.80] Sagan, C., *Pale Blue Dot: A Vision of the Human Future in Space*, Random House, New York, NY, USA, 1994.

[15.81] Secord, J.A., *Victorian Sensation: The Extraordinary Publication, Reception, and Secret Authorship of 'Vestiges of the Natural History of Creation'*, University of Chicago Press, Chicago, IL, USA, 2000.

[15.82] Segerstråle, U., *Defenders of the Truth: The Battle for Science in the Sociobiology Debate and Beyond*, Oxford University Press, Oxford, UK, 2000.

[15.83] Shapley, H., *The View from a Distant Star: Man's Future in the Universe*, Dell Publishing, New York, NY, USA, 1963.

[15.84] Som, S.M., Common identity as a step to civilization longevity. *Futures*, 106, 37–43, 2019.

[15.85] Spier, F., *Big History and the Future of Humanity*, 2nd, Wiley-Blackwell, Chichester, UK, 2015.

[15.86] Toynbee, A., *A Study of History*, One-volume edition, Thames and Hudson, London, UK, 1972.

[15.87] Tronchetti, F., *The Exploitation of Natural Resources of the Moon and Other Celestial Bodies: A Proposal for a Legal Regime*, Martinus Nijhoff Publishers, Leiden, The Netherlands, 2009.

[15.88] Vakoch, D.A. (Ed.), *Astrobiology, History, and Society*, Springer-Verlag, Berlin, Germany, 2013.

[15.89] Van Doren, C., *The Great Rehearsal: The Story of the Making and Ratifying of the Constitution of the United States*, Greenwood Press, Westport, CT, USA, 1982.

[15.90] Vidal, C. and Delahaye, J.-P., Universal ethics: Organized complexity as an intrinsic value, in: *Evolution, Development and Complexity*, G.Y. Georgiev, J.M. Smart, C.L. Flores Martinez, M.E. Price (Eds.), pp. 135–154, Springer, Cham, Switzerland, 2019.

[15.91] Viikari, L., Natural resources of the Moon and legal regulation, in: *Moon: Prospective Energy and Material Resources*, V. Badescu (Ed.), pp. 519–552, Springer, Heidelberg, Germany, 2012.

[15.92] Wallace, A.R., *Contributions to the Theory of Natural Selection*, 2nd, Macmillan, New York, NY, USA, 1871, Available on-line at: http://www. gutenberg.org/files/22428/22428-h/22428-h.htm.

[15.93] Ward, B., *Spaceship Earth*, Columbia University Press, New York, NY, USA, 1966.

[15.94] Webb, S., *If the Universe is Teeming with Aliens Where is Everybody? Seventy-Five Solutions to the Fermi Paradox and the Problem of Extraterrestrial Life*, Springer, Cham, Switzerland, 2015.

[15.95] Wells, H.G., *The Outline of History: Being a Plain History of Life and Mankind*, Waverley Book Company, London, UK, 1920.

[15.96] Wendt, A., Why a world state is inevitable. *Eur. J. Int. Relat.*, 9, 491–542, 2003.

[15.97] Wendt, A., Why a world state is democratically necessary, in: *World Orders Forum*, 2015, 2 July, 2015; available on-line at http://wgresearch. org/why-a-world-state-is-democratically-necessary.

[15.98] White, F., *The Overview Effect: Space Exploration and Human Evolution*, American Institute of Aeronautics and Astronautics, Reston, VA, USA, 2014.

[15.99] Wilson, D.S. and Wilson, E.O., Rethinking the theoretical foundation of sociobiology. *Q. Rev. Biol.*, 82, 327–348, 2007.

[15.100] Wilson, E.O., *On Human Nature*, Harvard University Press, Cambridge, MA, USA, 1978.

[15.101] Wilson, E.O., *Consilience: The Unity of Knowledge*, Abacus, London, UK, 1998.

[15.102] Wilson, E.O., *The Social Conquest of Earth*, Liveright Publishing, New York, NY, USA, 2012.

[15.103] Wynner, E. and Lloyd, G., *Searchlight on Peace Plans: Choose Your Road to World Government*, Dutton and Company, New York, NY, USA, 1944.

[15.104] Yunker, J.A., *Political Globalization: A New Vision of Federal World Government*, University Press of America, Lanham, MD, USA, 2007.

[15.105] Yunker, J.A., *Evolutionary World Government: A Pragmatic Approach to Global Federation*, Hamilton Books, Lantham, MD, USA, 2018.

[15.106] Zubrin, R., Interstellar communication using microbial data storage: Implications for SETI. *J. Br. Interplanet. Soc.*, 70, 163–174, 2017.

16

Interstellar Ethics and the Goldilocks Evolutionary Sequence: Can We Expect ETI to Be Moral?

Margaret Boone Rappaport[1]*, Christopher Corbally[2] and Konrad Szocik[3]

[1]*The Human Sentience Project, LLC, 400 E. Deer's Rest Place, Tucson, AZ, USA*
[2]*Vatican Observatory, Department of Astronomy, University of Arizona, Tucson, AZ, USA*
[3]*Department of Social Sciences, University of Information Technology and Management in Rzeszów, Rzeszów, Poland*

Abstract

Interpretations have been made of a currently narrow range of signal types that humans can record from interstellar sources, to determine if distant life forms exist. This chapter briefly surveys the types of signals now available and what they and others could reveal about distant life forms in the future. The chapter then explores the assessment of intelligent species if they visit Earth. The authors ask which observable signs could indicate self-awareness and ethical thinking, and how humans can test for the presence of ethics. A "Goldilocks Evolutionary Sequence" of neurocognitive features is presented, which can be operationalized as a plan to vet interstellar visitors and a protocol for negotiating with them. Important distinctions may point toward ethics, but if a visit to Earth occurs, will humans have the time to make them?

Keywords: Goldilocks zone, intelligence, sociality, ethics, theory of mind, first contact, protocol, Goldilocks Evolutionary Sequence

16.1 Introduction

Unlike today's remote explorations of the stars using ever finer astronomical instrumentation, the ruminations of poets and philosophers seem to

Corresponding author: MSBRappaport@aol.com

Octavio Alfonso Chon Torres, Ted Peters, Joseph Seckbach and Richard Gordon (eds.) Astrobiology: Science, Ethics, and Public Policy, (339–360) © 2021 Scrivener Publishing LLC

suggest a different scenario for first meeting a species from a distant star system—a visit, right here on Earth. Indeed, in some nations there are little-known protocols for conveying a chain of information when some-one on Earth finally and firmly detects the existence of another intelligence in the universe. When it happens, if it happens, it will be a momentous event full of both hope and dread. It will instantaneously become politically and personally important. Will this species be peaceful or warlike? Will humans be able to communicate with them, or will they be so different in cognition that discussion will be almost impossible?

Answers to these questions hinge on the careful planning of how First Contact is managed, how information on the new species is gathered, and how humans deduce the way they should react. The nations on Earth need a joint, effective vetting protocol and a coordination plan to be used if species visit Earth. In this chapter, we hope to address methodological issues that will encourage new approaches toward those ends.

16.1.1   The Little Broached Question of Ethics

One issue that has only been indirectly broached, to date, is ethics. There are elaborate systems of ethical constructions available to us, but we shall reduce them all to straightforward questions for the sake of discussion. We ask: Can a species from a faraway star system distinguish "right" from "wrong"? This does not imply a specific human culture's definition of right and wrong, but a discrimination between correct and appropriate vs. incorrect and inappropriate for all intelligent species. Does the new species have a scale of valuation and a thought process that can assess others' behavior? We shall delve into the cognition of ethical thinking in later sections and attempt to discern a "test" of sorts that may be useful when humans encounter a species from another star system.

The issue remains as to whether we shall have time to apply it. However, we must try because our very lives may depend upon it. We could simply plan to be overwhelmed. Our view is that that option is not a good one. It will be essential to know whether a visiting species discerns "right" from "wrong," because all known human cultures make these valuations. Ethical thinking is part of our species' biology and its absence causes alarm when detected in some unfortunate humans. Will ethics be part of the way another species thinks and behaves, and how do we find out? That is the question if a species visits Earth.

However, there are methods for the long-distance assessment of stars and planets that could harbor life, and that is where we shall begin.

16.2   Astronomical Detection of Possible Life

Before we address an Earth-based vetting protocol, we turn to the detection of signals that have begun to point the way toward identification of distant stars with planets that could harbor life.

16.2.1   The Complex Relationship Between Signals and Ethics

Astronomy can reveal only so much about stars and planets that are light-years distant. At the present there is an enormous gap between the specificity of signal data, and details about planets and species that may inhabit them. However, that gap is closing rapidly and it will continue to do so as instruments have greater capacity and scientists learn to analyze information better. Let us first look at the types of signals available to humans today, and the logic behind their interpretation. There has been enormous progress in identifying stars that may have Earth-like planets, and among those planets, the ones in the Goldilocks Zone—neither too hot nor too cold. These places are likely to have liquid water, and so a chemistry, a biology, like ours.

It is important to acknowledge that no signal data available today answer questions about any specific species or its ability to form and use ethics. However, if the planet has liquid water, then there may be plant cover, and then, a chance it may have animal species like Earth's. If humans receive a signal broadcast from another star system, that would surely change the state of our knowledge, but so far, we have detected no purposeful signals. If humans were to ever receive a purposeful message, it is possible that we could determine if a species lives socially, and maybe, if it thinks ethically. Social life is the first step toward determining ethical capacity, but not all social animals on Earth have ethics, and therefore "sociality," broadly speaking, is not an absolute indicator of ethics. Ethics requires agency, intentionality, and self-consciousness in humans, as well as a neurologically based decision-making capability that weighs "good" vs. "bad" [16.12], so we cannot apply ethical assessments to social insects, which show no evidence of these features. Furthermore, there is disagreement as to when the human line began to think in a way that might include ethical decision-making. Still, the fields of paleoneurology and cognitive archaeology are beginning to identify clues as to when the human line began to have ethics [16.21]. Before we get to sociality and then to ethics, we must first get to water and plant cover, and those are treated next.

16.2.2   Astronomical Signal Detection, the Goldilocks Zone, Habitation, and Ethics

16.2.2.1   Exoplanets

There was great excitement on October 6, 1995, when astronomers Michel Mayor and Didier Queloz made the momentous announcement of discovering a planet in orbit around 51 Pegasi, a Sol-type star located about 50 light-years away, in the direction of the constellation Pegasus, the "Flying Horse." The astronomers' approach used the fact that a star with a planet will not be stationary but will move slightly around a common center of gravity, i.e., the balance point between the star and planet. If that movement is in the plane of the line of sight between us and the star, then the colors of the star's spectrum will oscillate very, very slightly from red to blue, as the star moves back and forth. This is called the "Doppler effect." Using the Doppler effect is an important way to determine if a star has a planet orbiting around it. However, this method is biased toward finding larger planets, like Jupiter and even more massive gas giants.

There is another way of detecting planets around stars, which is more sensitive to smaller planets. This method was used when the Kepler Space Telescope (which orbited around the Earth) was trained on a single patch of the night sky with about 150,000 stars. The goal was to monitor any tiny, regular decreases in light that would indicate a planet had crossed in front of or transited its host star. The Kepler telescope was very successful, detecting 2,662 planets around stars over its nine and a half years of service. A good number of the stars had multiple planets, and a good number of those were of a size approaching Earth's.

16.2.2.2   Exoplanets in the Goldilocks Zone

As exciting as these extrasolar planets are, we are now just at the beginning of the process of detecting intelligent life beyond the Earth. Although the universe is always ready to surprise us, we think that the best chance for the development of intelligent life is on a planet similar to Earth. It would have a solid surface and water in liquid form on some of the surface. This would exclude a gas-giant planet like the ones in the outer part of our solar system. We have a few candidate planets in this so-called habitable zone—the Goldilocks Zone—but none are very promising.

The easiest stars around which to detect Earth-like planets are the lowest mass stars, called the M-dwarfs (our Sun is a G-dwarf, two-to-twelve times more massive). The lower mass of the M-dwarf stars means that their surface temperatures are much lower than that of the Sun. Therefore, for water to remain liquid on their planet's surface, it must be so close to the M-dwarf that one face of the planet is "locked" toward the star (just as one surface of the Moon stays facing Earth). That "locking" may mean life is only possible, or comfortable, on the daylight-darkness interface—a relatively narrow strip around the planet's circumference. Further, there is often intense radiation from an M-dwarf close to a planet. M-dwarfs are more active than our Sun, so they may not be the best places, after all. Astronomers expect to be most successful in finding life on the "Goldilocks planets"—neither too hot nor too cold—around stars more like our Sun, but that is difficult to do. Nevertheless, from Earth-based and orbital telescopes the hunt is on for planets like ours. On Earth, life exists all over, in fact, in some very unlikely places like in deep ocean trenches.

16.2.2.3   Exoplanets, Oxygen, and the 'Red Edge'

The evidence for intelligent life (for an "ETI" or extra-terrestrial intelligence) on a Goldilocks planet is going to be much more elusive than finding the planet itself. A radio or visual signal, like the ones we have included on the Voyager probes we have sent out, would be a clear sign of intelligence. This has been the basis for an important idea from the SETI Institute (Search for Extra-Terrestrial Intelligence Institute).

Astronomer Jill Tarter and her colleagues advocate searching for complex "astroengineering," high levels of energy expenditure, or electronic communications emanating from a planet, which would indicate an advanced civilization [16.10] [16.11]. Another approach is to use sensitive spectroscopy of the planet's atmosphere, which is expected to reveal if there is free oxygen there. If the oxygen is accompanied by methane, which competes with it, then the continuous generation of methane molecules would be a very good indication of an Earth-like biology, although not necessarily intelligence. Chlorophyll in vegetation reflects near-infrared light, so finding a "red edge" in infrared spectroscopy would also be an excellent result, if indeed the forests on the planet are green rather than another color, like red. In our Earth's biota, red and green tend to function in a structurally analogous manner in some large molecules. For example, a chlorophyll molecule has a magnesium ion at the center of a porphyrin ring, and a hemoglobin molecule has an iron ion at the center of a porphyrin ring. The result is life on Earth that includes worms with red blood and also green blood! Biochemical evolution can take alternative routes.

16.2.2.4   *The Great Leap from Plant Cover to Ethics*

Let us assume that in the future, the field of infrared spectroscopy produces clear and irrefutable evidence of a "red edge," and so, very likely plant cover on an exoplanet. That evidence is still not enough to presume animal life, especially animal life that is intelligent. However, proof of plant life is approaching a state of knowledge that suggests habitation by life forms. With this possibility, the linkage of evidence now ends and we wait for the instrumentation to catch up with theory, and vice versa.

16.3   Operationalizing Human Neurological Features for an ETI Vetting Protocol

Let us now examine the preparedness of Earth's populations to meet a species from another star system, right here on our planet or elsewhere in our star system. Perhaps without knowing it, much of our anxiety about this meeting devolves into ethics. Do the visitors have a way to discern "right" from "wrong" actions? If a species is sufficiently powerful to travel from a distant star system, then they are likely to be socially and technologically advanced, because both social and technical expertise are needed for large-scale constructions. With understandable anxiety, we wonder whether humanity will be in peril or if the visitors will behave in such a way that signals a desire for mutual understanding.

We now switch focus and explore the mechanisms for determining whether an ETI who visits Earth has moral thinking, i.e., a sense of "right" and "wrong," a method for moral adjudication, or a lexicon that addresses, or even implies moral and ethical issues. Make no mistake: There is a risk that a species from another star system will act aggressively. At present, we do not have a basis on which to estimate that risk, except to note whether the visitor starts "shooting." The question about risk lies at the heart of the following discussion. To address that question, we first suggest an approach for operationalizing indicators that could guide our decision-making about contact with a visiting species.

16.3.1   Parallel Moral Assessments by Host and Visitor

However, broader issues haunt our discussion of "moral assessments" of other species. We can ask ourselves which behavioral features would allow us to ascertain whether a species has ethics. We can speculate about the types of evidence that would indicate thought and behavioral systems for determining appropriate behavior. Oddly enough, clues may come in a visiting species' interest in *our* moral capacity. Therefore, whether we behave in a manner sensitive to the other species' welfare may eventually tell us much about the species. *Their* reactions to *our* priorities may provide clues about *their* moral capacity. Those priorities may well include our concern for the welfare of other species on Earth. However, a contrarian possibility also exists: The status of the human species may fall quite short of the ideals the visitor holds. A visiting species might see humans as inferior. All these possibilities point toward the importance of behaving toward unknown species with circumspection, respect, and concern. What humans call "humanity" is a quality that must ultimately transcend the species on Earth and include species from other star systems, whether they are able to make moral determinations, or not [16.19].

While we cannot be certain that a visiting species would have a facility like human ethics, we conclude later that they must have something close to it, because of their inevitably social nature. Ethics derive ultimately from the need for individuals of a species to associate reliably. The other option is genetically-based social behavior as in Earth's insects. It is useful and it does work for them, but it is highly inflexible. It cannot build spaceships or navigate them. Logically, because there must be rules to design, construct, and travel in an interstellar spacecraft, it is likely that a visitor will have social rules in some form. How can we determine this? Unsurprisingly, we derive clues on our level of preparedness to vet ETIs from our own human neurology.

16.3.2   Anthropocene or 'Adolescence'?

Identification of intelligent life forms (ETIs) should be simple, and it is on Earth. We know the signs, we understand the parameters, and we have many creatures to compare that do not show the self-awareness, creativity, or social problem-solving skills of intelligent humans in groups. We belong to a single subspecies, *Homo sapiens sapiens*, the only one in our genus and species to survive into the Holocene geological epoch. Some analysts have concluded we are in a new era, the Anthropocene, but we three authors are not so sure. Our view is that we remain in a geological world reflecting humankind's "adolescence," not its maturity. It is an era when we may destroy the livability of our own planet and ourselves. This so-called human "adolescence" may be hindering other intelligent life forms from making themselves known to us. Indeed, it does make sense that they would stay away in order to determine our progress before contact.

Nevertheless, it is possible that visitors to Earth have hostile intentions. If we are in potential danger, it will be important to learn how we can "vet" other life forms, whether up close or from afar, and be prepared. On Earth, extraordinary caution is needed when evaluating beings from other star systems that are self-aware and intelligent. Safety comes first, but then what? What do we do? We begin to analyze the behavior we observe. Humans have evolved to be very good at this, and social sensitivities are genetically grounded [16.1] [16.26] [16.4]. We conclude that humans should be able to extrapolate this fine-tuned capability to a new species.

16.3.3   Vetting ETIs: Friend or Foe? Right vs. Wrong

Our goal is to ease the way forward in the task of assessing ETIs on Earth, so that we are more confident in our interpretation of their cognitive and behavioral features. We begin by operationalizing some aspects of the type of intelligence *Homo sapiens* represents. In truth, humans also want to be able to identify types of intelligence that humans do *not* possess. If the ETI is extremely different from us, they may not necessarily be friend or foe, but communication with them may be a very difficult process, as seen in our case study of the film *Arrival* [16.3] in Section 16.4, below.

Determination of friend vs. foe has always been an important social process for humans, even before their prehistoric interactions with Neanderthals, Denisovans, and other early human species in Eurasia. Humans (both early and modern) are deeply sensitive to the need to identify individuals and species that can help them, or that pose a danger. We have written elsewhere that there may be a special neurology, i.e., part of a brain organ called the putamen, which allows or even encourages humans to be sensitive and wary about unknown others. There is early evidence that this feature is genetically based [16.14]. There is also substantial evidence that there are many forms of human sensitivity, some types related to known others (family and local group members) and some types related to unknown others (strangers). These types of sensitivity are genetically grounded and have been researched in laboratories [16.1] [16.26] [16.4].

There is no guarantee that humans represent the only type of intelligence in the universe, and much reason to doubt it, in particular, factors like these: (1) the random chance of mutations in nuclear DNA or some other, non-terrestrial large proteins that carry information about the construction of still other large proteins; (2) examples such as the evolutionary reorganization of the lateral-medial cerebellum in three mammalian orders, not just higher Primates, laying foundations for innovation and higher intelligence in more than one taxonomic grouping, and so, parallel evolution of cognition [16.23] [16.25]; and (3) the fact that similar chemical species can substitute for each other, and so we can envisage different biochemical pathways taken toward intelligence, in faraway star systems. Because biology and chemistry vary and large proteins are subject to both random and non-random forces, it is quite conceivable that there are alternative forms of intelligence in the universe. It is also quite possible that some are achieved using biochemical and evolutionary pathways that are more likely than others, whose joint action would give rise to parallel features in different astrobiologies. The theory of evolutionary science suggests a conservation of energy may determine more likely pathways, but there are other determinants like the structural stability of genes and the stability of chromosome segments.

The hypothesis of a single evolutionary path toward full, self- and other-awareness, innovation, problem-solving in social groups, and intelligence, is very appealing, and there are aspects of this thesis that are reflected in the progression in Table 16.1. In some ways, it is a very complex progression, perhaps one *not* easily duplicated, and for that reason we have labeled it the "Goldilocks Evolutionary Sequence" (after the relatively low probability of a planet being in a "Goldilocks Zone" around a star).

The Goldilocks Evolutionary Sequence would not be impossible to duplicate, given the existence of billions of star systems, but it might be difficult because it would rely on unlikely events, as it did on Earth. The odds against the sequence in Table 16.1 were heavy and probably remain so elsewhere in the universe. On the other hand, the sequence of evolutionary innovations does have its own logic and momentum, because each innovation sets the stage for all the ones that follow. All evolutionary innovations in Table 16.1 did not simply encourage eventual human cognition. They encouraged the emergence of each other. Cognition as a system evolved.

16.3.4   Rationale and Approach: Operationalizing Human Neurology to Assess ETIs

Because of the possibility of a different type of intelligence in an ETI that visits Earth, it will be useful to identify the features that make information exchange easiest for humans. Then, we can improvise an accommodation, as needed. Communication will be the first task in vetting ETIs, after making sure they do not pose an immediate danger.

Table 16.1 The Goldilocks Evolutionary Sequence and implications for vetting extra-terrestrial ethics.

Sequence element and timing[a]	What are the signs to look for in ETIs?
Sociality all primates, 65–55 million years ago [mya]	Social life in groups.
Reorganization of the lateral-medial cerebellum in anthropoid apes and two other mammalian orders	Capacity for intelligent behavior. Capacity for innovation.
Basic ape model with species Proconsul, the first true ape, 19 mya in the Miocene	Relatively large body size and large brain. Behaviorally demonstrative at times.
Realignment of the senses occurred on the lines to humans and some modern apes	Good vision and hearing, which facilitates rapid social and linguistic interaction.
Lengthening developmental trajectory and secondary altriciality, 8–10 mya in some Miocene apes	Lengthened adolescence allows learning of advanced neurocognitive skills, and longevity allows them to be taught anew.
Down-regulation of aggression and greater social tolerance among adults, some late Miocene apes	Physical and behavioral signs of a modified "domestication complex" (as in humans)
Upgrades in intellect to manage aggression in the social group, in some groups of late Miocene apes	Previous tendencies for large brains continue and are accentuated.
Greater variety of genetically-based sensitivities: general sensitivity and sensitivity engaging the emotions; to insiders and/or strangers	Social sensitivity and Theory of Mind. Sensitivity to others not in the social group.
Some biological foundations for culture in the ancient apes leading to the genus *Homo* and the genus *Pan*, 8–10 mya	Cultural learning of arbitrary beliefs and symbols, which deeply strengthen group bonds and create a shared world view.

(Continued)

Table 16.1 The Goldilocks Evolutionary Sequence and implications for vetting extra-terrestrial ethics. (*Continued*)

Sequence element and timing[a]	What are the signs to look for in ETIs?
Aggressive scavenging of meat to feed the enlarging brain of *Homo habilis*, from 2.8 mya in the Pliocene	Hypertrophy of the brain, as on the line to modern humans. Exaptation of neural networks for new functions.
Moral capacity, in *Homo erectus*, 1–1.5 mya, after learning to control fire, and a learning context, "The "Human Hearth," develops around a communal fire	Evaluation of everything along scales of morally "good" and "bad," within cultural contexts that vary by social group (although there is a general similarity in all humans).
Religious capacity emerged in *Homo sapiens*, stabilizing 150,000–120,000 years ago, based on findings of the globular shape of fossil skulls and the brain organs that created that shape	Religion emerges to support the social group. Religious concepts are culturally defined and arbitrary, which renders them deeply held.

[a]Research findings and theory supporting each of these evolutionary steps in human evolution are documented extensively in *The Emergence of Religion in Human Evolution* by Rappaport, M.B. and Corbally, C.J. 2020, Routledge.

16.3.4.1 Theory of Mind

Our approach relies on the latest findings in paleoneurology and neuroscience [16.5] [16.6]. We know that modern humans are extremely expert at a type of mentalizing called "theory of mind," which attempts to understand others' motivations and perspectives, and gauge a response to them carefully. Without such a fine-tuned facility, social life as we know it would not be possible [16.15]. As part of the emerging field of network neuroscience, a "theory of mind network" is also recognized [16.2].

Theory of mind is an example of a cognitive capacity that would be extremely useful in an exchange with a species from a distant star system. It would also be helpful if that species had the facility, too, or something like it. Evidence of a "theory of mind" would be critical in any assessment of an ETI. It is logical that we find clues to its existence in ETIs from our own features, the ones that make us innovative, sensitive, self-aware, and other-aware, too. However, we cannot assume that there are other features that attend full, intelligent self-awareness in ETIs. That possibility must loom large, while we attempt to ask the right questions. We possess a theory of mind about each other human [16.20] [16.16], and we will attempt to do the same with ETIs, and hope they do the same with us.

16.3.4.2 Sequence of Evolutionary Innovations: Logical, Determinate, Systemic

The neurological features in Table 16.1 illustrate the long evolutionary climb that the human line accomplished, with major innovations at successive periods. The model represented in this table is published in *The Emergence of Religion in Human Evolution* [16.21], and we have found it useful for a variety of analyses. Here, it will guide the questions we ask in vetting ETIs. Our assumption is that the model can suggest features that we want to operationalize while we determine if ETIs are friendly, bellicose, just explorers, or have some other purpose. It is important to remember that the sequence of evolutionary innovations in Table 16.1 resulted in us, a self- and other-critical species stabilizing 300,000 and 400,000 years ago all over Africa [16.17]. We would not have important physical, emotional, and neurocognitive capacities if these evolutionary innovations had not occurred. They make us what we are today and form a basis for what we can be in the future.

We can be a worrisome species because we tend to evaluate everything as "good" or "bad" morally. We may protest that it is not true, and it *may not* be true in specific instances. However, humans do have an inborn ability, and a social tendency to view every person, action, and thought as morally good or morally bad, according to their own culture. This truth suggests one question about ETIs: Should all of them be able to do the same? Must they reason morally in order to be intelligent? Philosophers have been speculating on the answer to that question for many years, but now we need to know because we may have to determine the intentions, cognitive abilities, and capacity to do harm of an ETI, perhaps many of them. The philosophical question has taken on a new urgency.

In Table 16.1, we detect a lingering question: Do all ETIs go through the same, or a similar progression of evolutionary innovations? Our view is that to achieve high intelligence, innovation, and both self- and other-awareness, an evolutionary line would need to go through almost all of the steps in Table 16.1, perhaps with modifications. The steps might also occur in a different order and there might be differences in details, but we conclude that the end result, moral evaluation, is a key to higher life forms. While logical, it is only a tentative conclusion and will remain untested until we encounter an ETI.

It is difficult to imagine that a species who can design and manufacture starships would be able to function without almost all the features in Table 16.1, with the exception of the last, religious capacity. Cooperation, group problem-solving, learning, reproduction, nurturance, and control of aggression are all dependent on the evolutionary innovations listed in this table, and they, in turn, allow a striving for the stars and its accomplishment. Specific features in Table 16.1 encourage cooperative endeavors, for example, the genetic and physiological down-regulation of aggression in adults, and its further social and intellectual management (which may well have evolved later, because down-regulation of aggression appears to have occurred in "bouts" [16.18]). Almost all of these innovations encourage group problem-solving, which is obvious even in 3- and 4-year-old human children; comparison groups of chimpanzees and capuchin monkeys could not accomplish the same because problem solving remained an individual matter for these other species [16.7]. Our view is that no species that succeeds in space travel could do so without group problem-solving—an essential and critical feature of the human experience that is linked to ethics. Is it linked to ethics for the ETI before us?

16.3.4.3   Cultural, Moral, and Religious Capacities – How Important and in What Order?

It is important to keep in mind the order in which the basic human traits that encouraged and enabled moral capacity (and so, ethics) emerged. Its determinants and components did not evolve all at one time. Sociality emerged 55–65 million years ago with the first primates. Cultural capacity emerged some time before the split between humans and the anthropoid apes, now estimated at 8–10 million years ago. Moral capacity, we contend, emerged in *Homo erectus* after the species learned to control fire, and a learning context called "The Human Hearth" was available, around 1.0–1.5 million years ago. Religious capacity in a basic form was available with the stabilization of the human globular skull form (and the brain organs creating that shape, including the pre-frontal cortex, the precuneus, and the cerebellum) around 150,000 to 120,000 years ago. In a certain light, anatomy of this type does indeed suggest morality. The human genome evolved in certain directions because the results benefited social life—an essential adaptation that had been around since the first primates.

Our expectation is that moral evaluation may be more commonly found in ETIs than religious capacity because moral capacity is simpler neurologically and indeed may be programmable in artificial intelligences [16.12]. The sequence in which moral capacity arrived for ETIs could be different, but the results could not have been much different. We conclude that moral capacity uses fewer brain networks than religious capacity (Rappaport and Corbally [16.21], which describes the archaeological and paleobiological findings that support this thesis). Religious thinking and experience use a variety of different brain networks, probably with a great deal of overlap for different types of religious experience and numinous states.

The endowment of either moral adjudication or religious capacity with cultural meaning is an entirely separate question. While moral capacity and religious capacity emerged at different times, the latter incorporates, populates, and "makes cultural" moral capacity in virtually all known societies and cultures. They function together. The models for these capacities are also provided in detail by Rappaport and Corbally [16.21].

Our questions about biologically-based cultural capacity and religious capacity in ETIs persist because the two traits are relatively complex and unusual in comparison to "sociality," which emerged so much earlier, 55–65 million years ago. Still, if an ETI has a "system of beliefs" that guides sociality, it would be difficult to distinguish that from a "culture." Cultural capacity supports humans primarily at the level of group organization, helping to hold the group together. It is displayed in individual humans, but it is played out ultimately and, we contend, comes under the pressure of evolutionary forces, only in the social group.

16.3.4.4   Assessing ETIs for Culture

The previous discussion suggests the need for a "test for culture" that could be applied to ETIs. Culture is group-based, complex neurologically, and can be joined with other capacities like morality and religion to produce arbitrary results. That quality of arbitrariness in associating concepts with objects or events or thoughts, is a very useful facility. It allows cultures to construct new and elaborate systems that can be mentally manipulated, even if they are counterfactual and exist only in the future, which, of course, also enables planning. Culture's arbitrary quality and its ability to facilitate planning are especially important, because if an ETI attaches meaning to an object or activity that bears no clear relation to its obvious importance, then a capacity for culture might be suspected of operating. Adult humans are extremely adept at sensing when cultural beliefs differ among groups, although they often do not recognize the source of their malaise. Vetting ETIs for culture may become humans' greatest challenge, or their easiest! Nevertheless, it is one that needs bright, sensitive negotiators who are not bound by the conviction that there is "only one answer" to seemingly obvious questions. A possible alternate scenario is that an

ETI belief system is much more mechanical and only an expression of the genetics of sociality. If this is so, the belief system of an ETI from another star system would likely be derived much differently at a biological level. Still, it could be viewed as comparable to a "culture," and like our capacity for culture, some of the genomic and developmental origins of ETI belief systems may have come late in their evolution.

The arbitrariness of culture, which enhances the flexibility of human innovation for the creation of both physical and mental systems, possibly stands as one of the most important sources for misunderstanding between humans and an ETI. (For an example, see *Babylon 5*, in Section 16.4, below). Misinterpretation is also a problem among groups of humans that have different cultures. At first, a belief system appears to make little sense, but then, as more is learned about it, it can suddenly appear internally consistent and utterly logical. An important goal for a negotiator in assessing an ETI would be to attain sufficient knowledge to sense an internally consistent set of concepts and processes. This is the ultimate benefit of implementing a "theory of mind" over a period of time: A completely different set of ideas can eventually make utterly good sense, when initially, they did not. An important question is whether an ETI negotiator will have sufficient time.

16.3.5   A Test for Neuroplasticity: The Clincher if We Have Time

The high level of human neuroplasticity (which is related to the capacity for culture, according to Gómez-Robles and Sherwood [16.13]), may suggest an answer to questions about a possible ETI culture. Human neuroplasticity must surely have originally emerged to assist in food getting, fending off predation, and neurological recovery from accidents and attacks. The high level of neuroplasticity that eventually allowed human cultural learning may not have been important in higher cognitive functioning—initially. Did a similar sequence occur elsewhere? Organic evolution does not plan for one feature to be helpful for one purpose now, but a different purpose in an evolved species, even if this is eventually what happens [16.9] [16.27] [16.22].

Questions about the coherence of a specific culture suggest that the sequence in Table 16.1 may have its own momentum and direction, and so, be replicated elsewhere, at least in part. Intense social life kicks off a variety of evolutionary features that allow it, support it, and even shelter people from its intensity, so, the consequences may be more predictable than we imagine at first. The feature (the key innovation, in the language of the Extended Evolutionary Synthesis) that could enable the Goldilocks

Evolutionary Sequence is neuroplasticity. Therefore, if there is a way to test for its presence in ETIs, then there would be good evidence of the type of flexibility that characterizes the human species, and both its cultural and moral capacities.

However, it might take an unfortunate accident to "prove" the neuro-plasticity of an ETI, who might recover from an accident because of it. Alternatively, it might take a great deal of research on the species, for which there simply may not be enough time. Scientists will not be the only stake-holders in vetting a visiting ETI. There will be government and military officials who may forestall efforts to determine the characteristics of the visitors gradually. Some world leaders may suddenly decide on an aggres-sive approach, even a military one. This scenario was played out in the film *Arrival*, described below.

16.4   Fictional Case Studies of Vetting ETIs

16.4.1   Examples from Film and Television

Popular science fiction literature and film have been used to explore a variety of intelligent species from other star systems. Indeed, the original *Star Trek* television series [16.24] describes interaction of several different intelligent species as crew on a starship in the 23rd century. The series was known for its treatment of current social issues, which often merged into questions of ethics and ethical decision-making. The latter involved inter-esting comparisons between the Captain, a human, and his first officer, a member of the Vulcan species, and both the ethical differences and sim-ilarities of the two characters. Conflict often arose concerning the role of emotion in ethical decisions.

The charming and popular movie *E.T. the Extra-Terrestrial* [16.8] did not broach ethics per se, but the story had a strong theme of a social and emotional allegiance to one's group. Thus, the phrase "E.T., call home" became a popular phrase among moviegoers. That type of allegiance is at the foundation of human ethics, which only make sense in a social context where individuals share common values or try to.

Babylon 5 was an American television series, 1993–1998, set again in the 23rd century, on a space-based orbital station that hosted a half-dozen intelligent species who looked and acted quite differently and who were often in conflict over a variety of issues. The setting was likened to a small "United Nations" of the future. One episode from the series is germane to our discussion of the detection of "cultural and ethical differences" between rational and intelligent species, and how "arbitrariness of customs" can lead to misunderstanding and danger. The story goes this way: When a vessel from EarthForce, which was essentially a warship, contacted a species called the "Minbari," they greeted EarthForce with open gunports. The practice of opening gunports was not a signal of hostility to the Minbari, but a sign of respect for the visitors' strength. Nevertheless, it was misunderstood as an act of hostility. EarthForce opened fire on Minbari cruisers, which led to a three-year war and almost led to the extinction of humanity.

16.4.2   Case Study of the Film *Arrival*

Let us now examine in more depth a commercial film example of identification and communication with an ETI who comes to Earth, in the movie, *Arrival* [16.3]. How can our own evolutionary sequence in Table 16.1 inform our conclusions about this film and the action it depicts? In the film, ETIs arrive in 12 starships hovering over 12 Earth locations. The ship over the United States is met by a military presence. The primary question is "What is the ships' purpose?" Efforts to determine the answer come from the following American sectors: (1) the Military, (2) Science (represented by the character of a Physicist), (3) Linguistics (represented by the character of a Linguist), and (4) there is some Political background throughout the film. We will now ask a series of questions and use our discussion above in vetting the ETIs, who appear as large "heptapods." Online descriptions state they are 9.8 meters in height.

> *Question*: How do the Americans know the ETIs are not initially hostile?
> *Answer*: The heptapods have not started "shooting."

> *Question*: How do they know the ETIs have culture?
> *Answer:* The arbitrariness of their written language, which is displayed for the humans on a translucent screen between humans and heptapods. The cursive is circular in form with patterns of figures that repeat and represent different concepts and items, as the linguist discovers. She catalogues this on a handheld screen, and eventually, in a book. It is a true language of re-combinable units of meaning (morphemes).

> *Question:* How do you know that the ETIs have morality?
> *Answer:* After several exchanges between the Linguist and the heptapods, a rogue Military person plants a bomb on the

starship. The heptapods forcefully eject the humans just before the bomb detonates, saving the humans. The explosion results in the injury and death of one of the heptapods. They saved the humans and sacrificed themselves, demonstrating a respect for life, as well as an acknowledgment of the importance of their mission.

Question: How do you know the ETIs are social?

Answer: (1) They open the door to their vessel regularly every 18 hours—an invitation. (2) They continue communicating with the humans through their written language, which the Linguist is learning. They continue the relationship. (3) When the linguist goes by herself to converse with them, they allow her to come alone, and provide transport. (4) Later, we learn the purpose of the visit of the 12 ships, which is more broadly social. The heptapods bring the humans a gift (their language, which allows humans to conceive of events on a timeline stretching from past into the future), in exchange for something that they will need from the humans three thousand years in the future. This last "social" act represents a very broad conception of interspecies exchange. It is important for the same reason that archaeological remains of "traded items" between early humans suggest and represent a social unit larger than the family or local band. Trade represents a network structure of mutual exchanges. The heptapods in *Arrival* are requesting one such exchange.

This is a simplified vetting process of asking questions of a movie plot, but we can see how a deeper understanding of human qualities can help in determining the intentions of ETIs. An entire vetting protocol could be constructed using the items in Table 16.1.

16.5   Conclusion

Humanity is making excellent progress in developing instrumentation that can identify the conditions of life from light-years distant. There is no special signal for "ethics," but if a purposeful message were received, there might be a way to ascertain at least whether an ETI were social. If Earth is visited by a species from a distant star system, humans need to be prepared with a set of procedures and a set of questions that will reveal the intentions of the visitors. This assessment must be rapid. Unless humans can devise a protocol that can quickly determine whether the visiting species has a sense of ethics, there may be great danger. On the other hand, if the species responds to questions deriving from Table 16.1 in a manner that suggests a system of assessing social behavior according to group values, there may be hope for an interesting and fruitful exchange. The key will be to assess the species' status and avoid misinterpretation of the arbitrary elements of culture—ours and theirs.

References

[16.1]   Acevedo, B.P., Aron, E.N., Aron, A., Sangster, M.-D., Collins, N., Brown, L.L., The Highly Sensitive Brain: An fMRI Study of Sensory Processing Sensitivity and Response to Others' Emotions". *Brain Behav.*, 4, 4, 580–594, 2014.

[16.2]   Amodio, D.M. and Frith, C.D., Meeting of Minds: The Medial Frontal Cortex and Social Cognition". *Nat. Rev. Neurosci.*, 7, 4, 268–277, 2006.

[16.3]   *Arrival*, Paramount Pictures, Hollywood, California USA, 2016.

[16.4]   Brookshire, B., A Vivid Emotional Experience Requires the Right Genetics". *Scicurious (Neuroscience)*, 2015. Online, May 8, 2015. www.sciencenews. org/blog/scicurious/vivid-emotional-experience-requires-right-genetics.

[16.5]   Bruner, E. and Iriki, A., Extending Mind, Visuospatial Integration, and the Evolution of the Parietal Lobes in the Human Genus". *Quat. Int.*, 405, 98–110, 2016.

[16.6]   Bruner, E., Preuss, T.M., Chen, X., Rilling, J.K., Evidence for Expansion of the Precuneus in Human Evolution". *Brain Struct. Funct.*, 222, 1053–1060, 2017.

[16.7]   Dean, L.G., Kendal, R.L., Schapiro, S.J., Thierry, B., Laland, K.N., Identification of the Social and Cognitive Processes Underlying Human Cumulative Culture". *Science*, 335, 1114–1118, 2012.

[16.8]   *E.T. the Extra-Terrestrial*, Universal Studios, Los Angeles County, California USA, 1982.

[16.9]   Fiddick, L., Barrett, H.C., Smelser, N.J., Baltes, P.B., Evolution of Cognition: An Adaptationist Perspective, in: *International Encyclopedia of the Social and Behavioral Sciences*, pp. 4996–5000, Elsevier Science, Amsterdam, Netherlands, 2001.

[16.10] Forgan, D., Wright, J., Tarter, J., Korpela, E., Siemion, A., Almár, I., Piotelat, E., Rio 2.0: Revising the Rio Scale for SETI Detections. *Int. J. Astrobiology*, 18, 4, 336–344, 2018.

[16.11] Rebuttal to: Deconstructing the Rio Scale: Problems of Subjectivity and Generalization. *Int. J. Astrobiology*, 18, 5, 492–493, 2019.

[16.12] Glimcher, P., Gazzaniga, M.S., Mangun, G.R., The Emerging Standard Model of the Human Decision-Making Apparatus, in: *The Cognitive Neurosciences*, 5th ed, pp. 683–91, MIT Press, Cambridge, MA, 2014.

[16.13] Gómez-Robles, A., Sherwood, C.C., Human Brain Evolution; How the Increase of Brain Plasticity Made Us a Cultural Species. *MÉTODE Sci. Stud. J.*, 7, 35–43, 2016, https://doi.org/10.7203/metode.7.7602.

[16.14] Gunz, P., Tilot, A.K., Wittfeld, K., Teumer, A., Shapland, C.Y., van Erp, T.G.M., Dannemann, M., *et al.*, Neandertal Introgression Sheds Light on Modern Human Endocranial Globularity. *Curr. Biol.*, 29, 120–127, 2019.

[16.15] Hagoort, P., Levinson, S.C., Gazzaniga, M.S., Mangun, G.R., Neuropragmatics, in: *The Cognitive Neurosciences*, 5th ed, pp. 667–674, MIT Press, Cambridge, MA, 2014.

[16.16] Hicks, J.M., Coolidge, F.L., On the Role of Precuneal Expansion in the Evolution of Cognition. *Presentation at the Conference of the European Society for the Study of Human Evolution Annual Meeting*, Madrid, n.d. 2016, https://doi.org/10.13140/RG.2.2.17353.95841.

[16.17] Hublin, J.-J., Ben-Ncer, A., Bailey, S.E., Freidline, S.E., Neubauer, S., Skinner, M.M., Bergmann, I., Le Cabec, A., Benazzi, S., Harvati, K., Gunz, P., New Fossils from Jebel Irhoud, Morocco and the Pan-African Origin of *Homo sapiens*. *Nature*, 546, 289–292, 2017.

[16.18] MacLean, E.L., Unraveling the Evolution of Uniquely Human Cognition. *Proc. Natl. Acad. Sci.*, 113, 23, 6348–6354, 2016.

[16.19] Persson, E., The Moral Status of Extraterrestrial Life. *Astrobiology*, 12, 10, 976–984, 2012.

[16.20] Premack, D. and Woodruff, G., Does the Chimpanzee Have a Theory of Mind?. *Behav. Brain Sci.*, 1, 4, 515–526, 1978. Special Issue: Cognition and Consciousness in Nonhuman Species.

[16.21] Rappaport, M.B. and Corbally, C., *The Emergence of Religion in Human Evolution*, Routledge, Abingdon, Oxfordshire, 2020.

[16.22] Rappaport, M.B., Szocik, K., Corbally, C., Neuroplasticity as a Foundation for Human Enhancements in Space. *Acta Astronaut.*, 175, 438–446, 2020.

[16.23] Smaers, J.B., Turner, A.H., Gómez-Robles, A., Sherwood, C.C., A Cerebellar Substrate for Cognition Evolved Multiple Times Independently in Mammals. *eLife Sci.*, 7, e35696, 2018.

[16.24] *Star Trek; The Original Series*, 1966-1969, Produced from 1966 to 1967 by Norway Productions and Desilu Productions, and from 1968 to 1969 by Paramount Television, Los Angeles, California USA.

[16.25] Tanabe, H.C., Kubo, D., Hasegawa, K., Kochiyama, T., Kondo, O., Bruner, E., Ogihara, N., Tanabe, H., Cerebellum: Anatomy, Physiology, Function, and Evolution, in: *Digital Endocasts, Replacement of Neanderthals by Modern Humans Series*, pp. 275–289, Springer, Tokyo, Japan, 2018.

[16.26] Todd, R.M., Müller, D.J., Palombo, D.J., Robertson, A., Eaton, T., Freeman, N., Levine, B., Anderson, A.K., Deletion Variant in the ADRA2B Gene Increases Coupling between Emotional Responses at Encoding and Later

Retrieval of Emotional Memories. *Neurobiol. Learn. Mem.*, 112, 222–229, 2014.

[16.27] Whiten, A. and Erdal, D., The Human Socio-cognitive Niche and Its Evolutionary Origins. *Philos. Trans. R. Soc. Lond., B, Biol. Sci.*, 367, 2119–29, 2012.

17

Intrinsic Value, American Buddhism, and Potential Life on Saturn's Moon Titan

Daniel Capper

School of Humanities, University of Southern Mississippi,
Hattiesburg, Mississippi, USA

Abstract

This chapter concerns the process of how we develop astrobiological morals by examining some of the compromises within ethical argumentation. I illustrate these compromises by turning to Saturn's moon Titan to provide an ethical theory challenge in terms of protecting Titan life, should life be found there, as well as protecting the habitats of that life and enabling the scientific study of that life. Through analysis I find that an intrinsic value approach to astrobiological ethics may, in this case, provide better absolute protection for Titan life. However, an American Buddhist approach, which arises from alternative, deontological ethical presumptions, in this analysis may provide better protection for the habitats of life as well as stronger arguments for the scientific study of extraterrestrial life. In the end we find that the ethical models that we bring to our work strongly color the ethical outcomes that we realize because of the limited, yet still valuable, nature of all forms of ethical argumentation.

Keywords: American Buddhism, astrobiology, bioethics, Buddhist ethics, intrinsic value, life on Titan, planetary protection

17.1 Introduction

The computer that I use to write this essay emerges from a series of compromises. In a perfect world, it would be infinitely fast, have unlimited

Email: Daniel.Capper@usm.edu

Daniel Capper: https://www.buddhismandspace.org/, https://www.researchgate.net/profile/Daniel_Capper, https://southernmiss.academia.edu/DanielCapper

Octavio Alfonso Chon Torres, Ted Peters, Joseph Seckbach and Richard Gordon (eds.) Astrobiology: Science, Ethics, and Public Policy, (361–376) © 2021 Scrivener Publishing LLC

storage, and make my lunch for me, but I do not live in a perfect world. Therefore, I choose to abide a digital device that in glass-half-empty perspective possesses negative dimensions, because the net positive of the computer in my life makes accepting a compromise situation worth it. Indeed, such compromises surround us all, for often we rightly value situations of trade-off above circumstances of utter absence. My point is this: Just because a state of affairs involves a compromise does not mean that it necessarily lacks value to us. I, in fact, hope that my reader enjoys this essay despite its background in a number of compromises like computer microprocessor operations.

As with computers, so it is with astrobiological ethical argumentation. Compromises of some kind always are required, whether one argues from the perspective of Aristotelian virtue ethics, John Stuart Mill's utilitarianism, Wang Yang Ming's Neo-Confucianism, Kant's intrinsic value concept, American Buddhism, or some other methodological viewpoint. Ethics work innately entails locating not an unambiguous formulation but rather an ideal balance of gains and deficits and then supporting the implementation of that multifaceted position with strong rational arguments.

But this trade-off nature does not mean that ethics work is arbitrary or meaningless, as some people appear to imply [17.6]. Finding the right compromise position poses quite a challenge because inherently it means arguing, even if implicitly so, for some type of moral loss, not just moral gain. This compromise-imbued quality makes ethics work in the eyes of the environmental ethicist Dale Jamieson a "problem" with a "difficult nature" [17.9]. The challenging character of ethical argumentation supplies one reason why varying schools of ethical thought have developed and then remained active, since they offer helpful guidance down time-tested paths.

To date the literature in astrobiological ethics does not reflect a fulsome discussion of the compromise-laden character of ethics work, though, thus diminishing effervescence within the ethical conversations among space scholars. This lack of discussion also may increase confusion among those scholars without much training in ethics. Hence, as a central goal this chapter delineates ethical compromises within a comparative performance evaluation between the oft-employed ethical concept of intrinsic value and, alternatively, Buddhist ethics as expressed by American Buddhists in the ethnographic field. I put these two methods of ethical discourse to work in directing the search for life on Saturn's moon Titan so that we can appreciate similarities and differences emerging from diverse modes of ethical argumentation most starkly.

In order to add clarity, here I define "intrinsic value" as a form of ethical argumentation in which value intentionally is accorded to something in itself rather than for its instrumental use. I illuminate what I mean by American Buddhist ethics at some length below but for now mention that they reject the concept of intrinsic value in favor of deontological, or rule-based, ethical arguments.

Wielding intrinsic value and Buddhist moral approaches, I will show that both schools offer some protection for possible living beings that may dwell on Titan. Given the different premises of these modes of ethical thought, though, I will find that a thoroughgoing intrinsic value argument may provide more complete protection for the lives of organisms themselves than do many American Buddhists. At the same time, however, I will demonstrate how a Buddhist approach may create greater protection for habitats than some arguments that are based on the concept of intrinsic value, while the Buddhist method also may enable science more readily than an intrinsic value argument. This examination thereby teaches us how two divergent ethical tools can direct different yet desirable moral outcomes in terms of protecting living beings, preserving their habitats, and supporting science.

I alert my reader in advance that because this metaethical essay is about the process of how we do ethics work, not a deep exploration of one ethical viewpoint, my intentionally limited portrayals of intrinsic value and Buddhist ethical arguments are for illustration only, being inconclusive in themselves. There in fact exist many types of intrinsic value and Buddhist arguments beyond those that I investigate. For instance, within intrinsic value studies the space ethicists Charles Cockell and James S. J. Schwartz affirm establishing different degrees of intrinsic value [17.5] [17.15]. I have no room here to enter the controversy that accrues to these potential degrees of intrinsic value, however, which anyway fundamentally do not change my main point about the compromises that inevitably reside within ethics work.

In the next section I will outline my focus on the search for life on Titan, because potential Titan life provides a good test for discerning differences in ethical systems. Then, since intrinsic value arguments remain so familiar within astrobiological ethics literature, I explore their role on Titan only briefly. Afterwards, Buddhist ethical perspectives, being understudied in astrobiological literatures, receive a greater share of my attention before I discuss what we learn from comparing these two ethical approaches.

17.2   Titan and Possible Weird Life

The moral task of protecting the life on Saturn's moon Titan, should the chance of life there prove to be true, poses a vigorous test for the ethical theory challenge that is featured in this essay. Titan, the target of the European Space Agency's 2005 Huygens mission as well as NASA's future Dragonfly mission, remains intriguing to astrobiologists because, despite a surface temperature of only 94 K or −179 C, it nonetheless appears in many ways like Earth [17.10] [17.14]. Quite like Earth, Titan possesses an atmosphere that can shield living beings from radiation and has a pressure of a relatively comfortable 1.5 bar [17.13]. Yet, unlike Earth, organic molecules and hydrocarbon aerosols fill Titan's hazy air. As on Earth, rain falls from Titan's skies, drains into river systems, and then converges to form lakes, but on Titan the rain consists of methane and ethane, not water [17.10]. Titan, like Earth, appears to house volcanoes, but on Titan they should be made of water ice, not stone, and spew not magma but liquid water [17.13].

It is possible that Titan life exists in liquid water pools within volcano systems or in the underground water that supplies them, and, being water based, this life may not be that different from Earth life [17.13]. But, along with some other astrobiologists, Christopher McKay opines that seriously weird life by Earth standards may inhabit Titan's hydrocarbon ponds. Large but simple organisms dwelling in liquid methane lakes, McKay relates, could appear as flat membranes while metabolizing hydrogen, acetylene, or ethane for energy [17.12].

Of course, methane-based life on Titan like this could be so unfamiliar to humans that it remains indistinguishable from its habitat, and under such conditions humans may be prone unintentionally to killing living beings due to unmindfully utilizing their home ecologies. This peril creates a need for a prior ethical agreement to protect life's potential habitats as well as life forms themselves, so that we may inhibit accidental harm, arising from a lack of discernment, to life forms.

In addition, because of the weirdness that I describe, strong arguments to study thoroughly the genetics or molecular structure of such life would appear, even though such efforts would mean killing some of these living beings. Since the theme of this article involves compromises, please note now the trade-off between absolute protection of life and the sacrifice of life in the name of science. Absolute preservation of life inhibits the invasive but complete study of genetic or molecular structure, while allowing such study limits the absolute protection of life.

Hence, I pose this bruising ethical challenge to the moral theories in this chapter: Provide necessary ethical protection for possible Titan life forms, even the weirdest ones, as well as for the habitats that support those life forms. Along the way, enable robust scientific study within appropriate ethical boundaries.

In the face of this task, as we will see, intrinsic value arguments appear relatively to excel with protecting living beings themselves, without concern for preserving habitats. Alternatively, American Buddhists from the ethnographic field, following deontological ethical reasoning, in this essay appear superior in achieving the prevention of harm to habitats as well as enabling science. In light of these distinctions, situating this study vicariously on Titan allows more clear visibility of possible divergences in ethical results between the use of intrinsic value theory and Buddhist ethical arguments. We get a better sense of how these trade-off differences can emerge if we look more closely at the concept of intrinsic value as well as at its opponents.

17.3   Some Strengths and Limitations of the Intrinsic Value Concept

So that I am not misunderstood as an enemy to the concept of intrinsic value, I highlight that I have used it myself in an argument to establish nature reserves on Mars [17.3]. The ethicist Alan Marshall also discusses Mars, rather than Titan like this essay, yet his intrinsic value argument regarding the protection of possible life on Mars retains application to Titan and nicely exhibits some of the qualities of intrinsic value arguments that inform the argument of this paper. Thus, I turn now to Marshall's intrinsic value argument.

For Marshall, any life that is found on Mars must be preserved absolutely, because "Martian life is intrinsically valuable" [17.11]. If microbial life is found on the Red Planet, Marshall claims, Mars "should be left alone, with exploration only via sterilized automated spacecraft" [17.11]. Instead, the entirety of Mars should be turned into a "World Park," devoid of physical humans and their science, in order to preserve the planet's heuristically-presumed intrinsically valuable microorganisms [17.11]. According to Marshall, even sterilized exploratory robots cannot learn scientifically about this life in any way that may be intrusive, since intrinsically valuable "Martians have no duty to contribute to the knowledge of humanity" [17.11]. In this way Marshall's intrinsic value ethical formulation provides any possible life on Mars with stringent, thoroughgoing protection. Yet, because of the ways that it inhibits invasive research, Marshall's ethic does not flex to meet the needs of scientists who wish to utilize microscopes to understand the inner mechanisms of extraterrestrial organisms.

Additionally, Marshall's intrinsic value presentation recognizes that habitats of life should be protected, but Marshall musters no real intrinsic value argument that can result in this end. Feebly, for instance, Marshall spotlights that stony habitat entities may have intrinsic value, because, for all we know, Martian stones may enjoy a "blissful state of *satori* (the experience of nirvana in Japanese Zen Buddhism) only afforded to non-living entities" [17.11]. While perhaps stones have greater meditative capacities than currently recognized, examples from Earth indicate that a hypothesized and unproven belief in the spiritual capacities of rocks will not preserve environments from the rapacious capacities of human desires, commerce, and conflicts. Therefore, in the end, Marshall's argument from the standpoint of intrinsic value provides absolute protection to extraterrestrial life but without much of an ethic to preserve the ecologies upon which life depends or the adaptability to meet possible desirables like empowering genetic or molecular science.

Although intrinsic value arguments can vary from Marshall's approach, Marshall's presentation exhibits common themes in applying intrinsic value arguments to extraterrestrial realities, such as absolutely protecting life but not so much preserving habitats or enabling science for human benefit. Naturally, Marshall's argument has detractors, such as the space ethical philosopher Kelly C. Smith. Smith recognizes that the "intrinsic value justification is certainly central to modern environmental ethics" because of its attractiveness in assuring a limited ethical outcome, like, in this essay, absolutely preserving life [17.16]. Smith rightly indicates, "Something with intrinsic value simply should not be treated in certain ways, irrespective of the possible benefits to others of doing so" [17.16].

But Smith finds an intrinsic value approach to be flawed for several reasons. In concert with other theorists, for instance, Smith highlights the diminishing returns that remain inherent in the concept of intrinsic value. Taking an extreme scenario as instructive of principle, if we grant intrinsic value to everything in the natural world, then nothing functionally has value more than anything else, leaving us with no basis for making choices based upon comparative moral valuations. The more that the concept of intrinsic value is invoked, the less it ethically protects, because the harder it is to focus upon what truly possesses worth. On this note, Smith also asserts that the more that different things enjoy intrinsic value, likewise the more we create pointless moral dilemmas for ourselves [17.16]. Put differently, like rich foods or potent liquors the concept of intrinsic value should be used "sparingly," as Smith says, not liberally [17.16].

Reminding my reader that all ethical arguments involve compromises, so that the contributions of many divergent voices is desirable in the big picture, one can appreciate that it remains advantageous to have ethical voices be capable of entering conversations with each other. Unfortunately, another drawback to an intrinsic value argument beyond those mentioned by Smith concerns its sometimes deficient abilities as a partner in conversation with other ethical modalities.

Take, for instance, the poor fit between the concept of intrinsic value and Buddhist ethical systems. A distinctive mark of Buddhist philosophies concerns precisely their rejection of notions of intrinsic qualities or essences. The Buddhist doctrine of *anattā*, or no-self, denies that any empirical entity exists independently in time or space, since everything phenomenal arises as an effect of prior causes and in turn serves as a cause for other effects [17.20]. For instance, without the Big Bang 13.7 billion years ago, this essay presumably would not exist. In this light, despite ordinary human temporal perceptions, this essay is not separate from the birth of our universe. Therefore, in Buddhist perspective this essay, rather than arising independently as we may wish to think, is the product of ever-changing conditions that began long ago, leaving it with no final essence or intrinsic existence of its own. In Buddhist philosophies, moreover, the same argument applies to the concept of "value" itself, since value as a notion, being conditioned by various causes, lacks inherent or intrinsic existence. Needless to say, if in Buddhism nothing intrinsically exists, including the concept of value itself, there can be no such thing as intrinsic value in Buddhist ethics. This metaphysic is why the concept of intrinsic value remains little used by Buddhist ethicists: its unfortunate philosophical fit with treasured Buddhist premises [17.8].[1]

Alternatively, Buddhism commonly asserts deontological ethical arguments following established rules in the Buddhist scriptures, such as the central *Vinaya* code of rules of behavior for monks and nuns. Deontological ethical arguments ground themselves in the importance of following tested guidelines or rules. In different deontological ethical systems deities can declare these guidelines, for instance, as with the rules within the Ten Commandments; or, in the case of Buddhists, scriptures that contain the respected words of the Buddha, rather than those of a deity, supply a set of directions that is robust enough to support and maintain a large international monastic community for 2,500 years. While deontological and intrinsic value ethical arguments can be integrated, as one sees in the work of the philosopher Kant, for reasons that I have described, Buddhists intrinsic value and deontological approaches generally remain separate, thereby providing this chapter with useful ethical system contrasts.

---

[1]For a rare implementation of the concept of intrinsic value within Buddhist ethics see [17.8].

Now I turn to a Buddhist deontological approach to Titan life so that we can appreciate these contrasts more sharply.

17.4   Buddhist Scriptures and the Search for Extraterrestrial Life

If we consult the *Vinaya* code of behavioral rules for monastics as found in the scriptures of Theravāda Buddhism, we find four deontological ethical injunctions relevant to the search for life on Titan. For precision I add as an aside that these injunctions appear similarly in the scriptures of other forms of Buddhism, too. All four of these ethical injunctions are Pācitiyya offenses, meaning that violation of them requires confession and forfeiture of goods gained.

The first of these rules, Pācitiyya 142 in the rules for nuns or Pācitiyya 61 in the rules for monks, declares, "Should any *bhikkhunī* [nun] intentionally deprive an animal of life, it is to be confessed" [17.18]. With this injunction, which also features prominently as the First Precept of Buddhist rules for laypeople, Buddhism asserts protection for all living, sentient beings.

Along with this protection of life, Buddhist monastic rules also contain precepts for not harming the habitats of small living beings. Pācitiyya 20 for monks (nuns: Pācitiyya 116) specifies, "Should any *bhikkhu* (monk) knowingly pour water containing living beings—or have it poured—on grass or on clay, it is to be confessed," while Pācitiyya 62 (nuns: Pācitiyya 143) states, "Should any *bhikkhu* knowingly make use of water containing living beings, it is to be confessed" [17.19]. This deontological concern for preserving habitats provides a key difference from many intrinsic value arguments like Marshall's, as I will discuss further.

One more ethical injunction relevant for life on Titan concerns the propriety of killing those beings for the benefit of humanity through scientific research. Pācitiyya 107 for nuns or the similar Pācitiyya 11 for monks asserts that, "The damaging of a living plant is to be confessed" [17.18]. For some Buddhist monastics, this rule forbids their engaging in agriculture. Yet this apparently plant-friendly stricture has not always stopped Buddhist monastics from growing and harvesting plants, engaging in forestry themselves, or especially from encouraging lay people to pursue practices like agriculture or forestry. Otherwise, with no plants for food, wood for housing, etc., monastics cannot survive. Thus, the net practical effect of this injunction appears to be that the unreasonable destruction of living beings for human use should be avoided while reasonable human uses remain acceptable.

This sensibility of reasonable utilization opens the door for the scientific testing of organisms from Titan to answer the important research question of the potential possession by those organisms of DNA or molecular structure like those on Earth. Such study could reveal possible beneficial medical advances or provide evidence that life on Titan arose independently as a second genesis from that of Earth, which would be scientifically salient.

The guidelines that I have described are rules that the most ardent of Buddhists, the monastics, live by, and thus they represent a respected and time-tested authoritative moral code. But all forms of ethical argumentation face limits, and here Buddhist deontological ethics can struggle, too, because these ethical principles were promulgated long ago without any concern for their applicability to one of Saturn's moons. Do these rules, designed only for Earth, even work off-planet? Rather than let Buddhist ethics simply sputter with this question, I entered the ethnographic field among contemporary American Buddhists so that they can update our understanding. I share their input now.

17.5   American Buddhists and Life on Titan

Like people from Japan, India, Russia, China, and the European Space Agency countries, residents of the United States live in a spacefaring culture which generally promotes space exploration and therefore provides its citizenry with educational experiences to inspire support for space travel. This makes the United States among the best of locations for discovering informed grassroots ethical reactions to space exploration, including among American Buddhists, who therefore can offer us a capable contemporary Buddhist ethical perspective regarding Titan's possible life.

In order to understand Buddhist attitudes toward space exploration, between March and June of 2019 I visited seven important Buddhist centers in the southeastern United States. I surveyed practitioners at centers across all three Buddhist great sects of Theravāda (N=44), Mahāyāna (N=40), and Vajrayāna (N=37), gaining representative samples from each type as well as a balanced overall sample of N=121. Moreover, in order to highlight the distinctiveness of Buddhist voices from among those of the general American public, I collected a control data set from 78 randomly selected university undergraduate students. For the sake of economy in this piece, I refer my reader to another work of mine for understanding more of the demographic details of the ethnographic data that I present here [17.4].

Since American Buddhists tend not to be experts in space policy, in the field I avoided asking direct space policy questions and instead focused my interactions with Buddhists in terms of their extensions of ethical values. All informants took the same sixteen prompt survey about the application of Buddhist ethics to extraterrestrial environments. I also gathered qualitative comments both through my survey as well as through discussions with Buddhists and I share some of these qualitative data in this chapter. The quantitative prompts that are relevant for this chapter include:

1. I think that Buddhist principles should be utilized to guide our interactions with microbial life beyond Earth. (responses on a five-point scale from strongly agree to strongly disagree)
2. If we do use Buddhist principles to guide our interactions with microbial life beyond Earth, those principles should be? (choices offered but alternative responses welcomed)
3. We should protect from harm the extraterrestrial habitats of life, the ecologies on which life depends, whenever possible. (responses on a five-point scale)
4. If it intends to alleviate human suffering through the advancement of science, it is acceptable to take the lives of a small number of microbes from beyond Earth for the sake of their scientific study. (responses on a five-point scale)

American Buddhist responses to these survey prompts support the extension of deontological Buddhist rules as derived from the scriptures to guide the search for life on Titan, thereby creating a viable Buddhist ethical position that is alternative to outcomes from some intrinsic value arguments. To the question of whether Titan can represent an appropriate location for applying Buddhist scriptural ethics, 64% of Buddhists strongly agreed and another 25% agreed, making a notable 89% of Buddhists' asserting that Buddhist ethics should inform our actions on Titan. In this way, space-age savvy Buddhists help their tradition to overcome possibly being Earthbound by arguing for the application of its ethics on other worlds.

Further, in terms of what values to apply to the lives of organisms on other planets, 84% of Buddhists agreed or strongly agreed to extend the value of nonharm, also known as ahimsa from the Sanskrit, thus providing protection from harm for life forms. Of course, this result reflects the scriptural injunctions of Pācitiyyas 142 and 61 that I mentioned previously, so here we witness American Buddhists deontologically putting these rules into action. In so doing these Buddhists contrast with the control response of 59% who extend nonharm to Titan, so that Buddhists, as demonstrated by a Fisher's statistical test result of p=0.0001, express a distinctive voice on this count within American culture. With their application of Buddhist strictures to

Titan and extension of nonharm toward possible life on that moon, these American Buddhists influence an ethical outcome that is not unlike one that an intrinsic value argument could produce, spotlighting how different ethical argumentative forms can at times produce similar results.

But divergent ethical arguments, of course, also may produce contrasting results. Protection for the ecologies on which life depends rather than just for life itself appears little in many ethical arguments in the current astrobiological literature, including, as I have mentioned, those made from the standpoint of intrinsic value. This lack of protection thereby accentuates in importance the survey prompt regarding the protection of the habitats of life. To the survey prompt, "We should protect from harm the extraterrestrial habitats of life, the ecologies on which life depends, whenever possible," 75% of American Buddhists in this study strongly agreed and another 21% agreed, so that in total 96% of Buddhists approved. This protection remains important, for it makes little sense to avoid direct harm to an organism yet still eliminate that organism's way of making a living. Here, following provisions of their scriptures, these American Buddhists supply ethical safeguards for habitats, not just for living organisms, in a manner that is most ethically useful. A Fisher's statistical test reveals that these Buddhists in fact distinguish themselves on this count from the American general public. As I will discuss more fully, in a world of ethical theorizing compromises, in this instance the Buddhist protection of habitats can contribute an important gain.

But, like all forms of ethical theory, Buddhist ethics face compromises. We see this principle in action with Buddhist responses to the fourth survey prompt that I mentioned, "If it intends to alleviate human suffering through the advancement of science, it is acceptable to take the lives of a small number of microbes from beyond Earth for the sake of their scientific study." Several Buddhists described this prompt as the most difficult moral quandary of my survey. This dilemmatic character appears because, on one hand, Buddhists, following the deontological rules that I have examined, wish to protect life from harm. On the other hand, Buddhism, like all traditional religious forms, anthropocentrically favors humanity itself, including in this case support for science that benefits humans [17.2]. Indeed, without some anthropocentrism, Buddhists would have no food to eat or places of residence. In this specific situation, mobilizing anthropocentric support for scientific ventures that benefit humanity arguably can be a fully Buddhist action despite the countervailing obligation not to harm living beings.

Exhibiting this tension between protecting life and allowing the harvesting of life for human science, about 24% of Buddhists remained neutral with regard to the prompt concerning the ethical appropriateness of sacrificing extraterrestrial microbes for scientific testing. But only 20% of these Buddhists actually opposed this scientific testing. Overall, 56% of Buddhists agreed or strongly agreed that the limited harvesting of microbe lives for human benefit from science, even if those microbes must die, remains morally justified. Given the dilemmatic nature of the prompt, though, perhaps many of these Buddhists would agree with the Vietnamese American Buddhist who emphasized that "only a SMALL number of microbes" should be sacrificed.

With this support of science, American Buddhists flexibly make study-ing extraterrestrial life morally easier, within of course restraints, for we feel more justified ethically to learn from a tiny organism's genetics or molecular structure. In a world of ethical theory trade-offs, some people, like many of the Buddhists in this study, would consider empowering sci-ence in this way to be a moral benefit. Yet conflict remains, since other people wish, in the world of moral compromises, to protect life absolutely. This divergence in approaches brings me to discuss what these deonto-logical American Buddhist ethical positions regarding Titan life mean in comparison to intrinsic value views.

17.6   Discussion

These American Buddhist ethical perspectives could be framed differ-ently than I present them here, as could the intrinsic value argument that I offered above. In this essay the Buddhist and the intrinsic value arguments remain inconclusive in themselves and appear merely to illustrate the com-promise-requiring character of astrobiological ethical work.

Reflecting such compromises, in this essay an example intrinsic value argument, when stringently applied to extraterrestrial life like Alan Marshall does, avoids making the compromises for science in terms of protecting life that the Buddhists in this study collectively make. Marshall's intrinsic value argument completely preserves extraterrestrial life from intrusive study in contrast to the more liberal deontolological Buddhist approach. Therefore, in a world of ethical compromises, if we wish to protect life abso-lutely, because of the ethical reasoning used an intrinsic value argument like Marshall's may supply this protection more reliably than may the Buddhists in this study. Depending on our ethical goals and ideals, an intrinsic value argument may be more thoroughgoing when it comes to preventing harm to extraterrestrial life itself.

But an absolutist intrinsic value argument like Marshall's faces limits in terms of habitat protection and the practice of science, as we have seen.

For instance, Marshall's intrinsic value argument attempts to protect habitats, but does so by following the impractical path of preventing entry to planets and speculating about the spiritual capacities of rocks. Likewise, most other intrinsic value arguments protect only living beings, because in these formulations intrinsic value remains extended only to living beings, not to their abiotic habitats. The geologist Murray Gray relates that such situations arise because across times and places, humans have failed to accord intrinsic value to perceived inanimate habitat entities like mineral formations [17.2] [17.7]. Therefore, lacking in historical or geographic precedents for valuing nonliving things, intrinsic value arguments for abiotic entities like stones often remain quite difficult to make, leaving the task of habitat protection understudied in the intrinsic value astrobiological ethics literature. As the space ethicist James S. J. Schwartz helpfully indicates, "preoccupation with protecting extraterrestrial life for its own sake needlessly limits the scope of what should be said about the ethics of planetary protection" [17.15].

Conversely, as we have seen, a large 96% of American Buddhists insist on some protection for the abiotic habitats upon which extraterrestrial beings depend. This ethical advance may promote helpful moral outcomes not just for organisms but also for the ecologies that they inhabit. In the context of the search for life on Titan, in fact, this idea of preserving habitats takes on great importance. Because Titan life may be quite weird by Earth standards, it may be difficult to tell what is a living being and what is an abiotic habitat of that being, thereby complicating preservation.

Take, for example, the findings of the research team Svirčev et al. in terms of possible life in loess soils on Titan. Loess soil formations on Earth are formed through the action not just of abiotic minerals but also of microorganisms, since "the eco-physiological activities of microorganisms such as cyanobacteria, other bacteria, lichens, mosses, and fungi play an important role in the trapping of dust particles and formation of loess" [17.17]. Because Titan maintains "all geological factors needed for loess formation like on Earth," Svirčev et al. state that on Titan, "The search for extraterrestrial life should be expanded to loess-like deposits" [17.17].

If we encounter such biologically active loess-like soils on Titan like Svirčev et al. describe, are we certain that we could immediately recognize life in them, rather than just seeing lifeless regolith? Faulty discernment of what is biotic and what is not could lead to the tragic if unintentional destruction of life. If we cannot tell organism from habitat, along with saving life the protection of potential habitats, too, provides a desirable ethical outcome. And, on this issue, deontological, rather than intrinsic value, Buddhist ideals of habitat protection may offer the unexpected beneficial preservation of unfamiliar life forms, as we have seen.

Thus, by following the different path of deontological ethics, the American Buddhists in this study seem to ameliorate the problem of protecting ecologies exactly by focusing their ethical regard not just on living beings but also on their habitats. Within the compromises that these Buddhists face, they therefore decisively supply astrobiological ethics with a significant benefit, the ability to protect ecologies. On the question of habitat conservation and preservation, we can see that these Buddhist responses, relative to other models, may inspire advances in our collective astrobiological ethics.

Nonetheless, when it comes to practicing science, one of the Buddhists in my study also strongly asserted, "I do not support the scientific search for microbial life. This is not a 'sanctity of life' response," thereby demonstrating the desirability of a different moral end than habitat preservation, the uncompromising protection of life in itself. If the strict preservation of life itself is our goal, with the question of habitat preservation set aside, the rest of these American Buddhists mostly fail us, as we have seen, given that by more than two-to-one Buddhists collectively assert the acceptability of sacrificing limited numbers of tiny beings for science that aims to benefit human beings. This is because Buddhist ethics retain numerous inbuilt anthropocentric elements, and in this case these anthropocentric dimensions result in the benefit of human science rather than in an ethic of inviolate harm toward extraterrestrial life [17.1].

As this analysis reveals, if protecting specific Titan life in itself absolutely represents our goal, an intrinsic value ethical argument may be superior. However, although intrinsic value models may result in better protections for life itself, Buddhism may provide a seemingly better platform for the ethical protection of life's habitats as well as for the scientific study of weird life. This compromise situation between intrinsic value theories and Buddhist deontological ethics thereby demonstrates the thesis of this chapter: Astrobiological ethics always demand compromises that are difficult to negotiate, limiting in form, yet still valuable in terms of creating a grounded and beneficial moral compass.

17.7   Conclusion

Ethical deliberation in astrobiology remains difficult because of the many compromises within the task. Intrinsic value arguments are popular in the literature, since as employed they enjoy a traditional Western philosophical pedigree and can produce some useful ethical fruit. But intrinsic value arguments face limits, as do the ethics that American Buddhists in this study extend to life beyond Earth. Thoroughgoing intrinsic value arguments appear to result in more complete protection for various specific forms of life in themselves than contestations supplied by American Buddhists, who make room for the limited harvesting of life for science if human benefit thereby prevails. Alternatively, American Buddhists in this specific study promote the protection of the habitats in a way that helpfully contributes to our astrobiological ethics while they better enable science. Further, Buddhists do so through deontological methods rather than through intrinsic value arguments, which often fall short on these counts. In a Titan weird life situation in which it is difficult to separate an organism from its habitat, this preservation of habitats can supply greater preservation of living beings, although, of course, not the ones that some Buddhists support as scientific sacrifices. By examining the results of intrinsic value approaches in tandem with American Buddhist methods, we therefore learn to appreciate that ethical arguments can offer welcome moral guidance despite their embodiment of imperfect compromises.

References

[17.1] Capper, D., Learning Love from a Tiger: Approaches to Nature in an American Buddhist Monastery. *J. Contemp. Relig.*, 30, 1, 53–69, 2015, https://doi.org/10.1080/13537903.2015.986976.

[17.2] Capper, D., *Learning Love from a Tiger: Religious Experiences with Nature*, University of California Press, Oakland, CA, USA, 2016.

[17.3] Capper, D., Preserving Mars Today using Baseline Ecologies. *Space Policy*, 49, 101325, 2019, https://doi.org/10.1016/j.spacepol.2019.05.003.

[17.4] Capper, D., American Buddhist Protection of Stones in terms of Climate Change on Mars and Earth. *Contemp. Buddhism*, 1–21, 2020, https://doi.org/10.1080/14639947.2020.1734733.

[17.5] Cockell, C.S., The Ethical Status of Microbial Life on Earth and Elsewhere: In Defense of Intrinsic Value, in: *The Ethics of Space Exploration*, J.S.J., Schwartz and T., Milligan, (Eds.). pp. 167–179, Springer, Cham, Switzerland, 2016.

[17.6] Fogg, M.J., The ethical dimensions of space settlement. *Space Policy*, 16, 205–211, 2000, https://doi.org/10.1016/S0265-9646(00)00024-2.

[17.7] Gray, M., *Geodiversity: Valuing and Conserving Abiotic Nature*, John Wiley and Sons, West Sussex, UK, 2004.

[17.8] James, S., *Zen Buddhism and Environmental Ethics*, Ashgate Publishers, Aldershot, Hampshire, UK, 2004.

[17.9] Jamieson, D., Oxford University Press, New York, NY, USA, 2014.

[17.10] Lorenz, R. and Mitton, J., *Titan Unveiled: Saturn's Mysterious Moon Explored*, Princeton University Press, Princeton, NJ, USA, 2008.

[17.11] Marshall, A., Ethics and the Extraterrestrial Environment. *J. Appl. Philos.*, 10, 2, 227–236, 1993, https://doi.org/10.1111/j.1468-5930.1993.tb00078.x.

[17.12] McKay, C.P., Titan as the Abode of Life. *Life*, 6, 8, 8–12, 2016, https://doi.org/10.3390/life6010008.

[17.13] Meltzer, M., *The Cassini-Huygens Visit to Saturn: An Historic Mission to the Ringed Planet*, Springer, Cham, Switzerland, 2015.

[17.14] NASA, *Dragonfly's Journey to Titan*, NASA, Washington, D.C., USA, 2020, https://www.nasa.gov/dragonfly/dragonfly-overview/index.html.

[17.15] Schwartz, J.S.J., *The Value of Science in Space Exploration*, Oxford University Press, New York, NY, USA, 2020.

[17.16] Smith, K.C., The Trouble with Intrinsic Value: An Ethical Primer for Astrobiology, in: *Exploring the Origin, Extent, and Future of Life: Philosophical, Ethical, and Theological Perspectives*, C.M., Bertka (Ed.), pp. 261–280, Cambridge University Press, Cambridge, UK, 2009.

[17.17] Svirčev, Z., Nikolić, B., Vukić, V., Marković, S.V., Gavrilov, M.B., Smalley, I.J., Obreht, I., Vukotić, B., Meriluoto, J., Loess and Life out of Earth? *Quat. Int.*, 399, 18, 208–217, 2016, https://doi.org/10.1016/j.quaint.2015.09.057.

[17.18] Thānissaro, B., *Bhikkhunī Pātimokkha*, Metta Forest Monastery, Valley Center, CA, USA, 2019, https://www.accesstoinsight.org/tipitaka/vin/sv/bhikkhuni-pati.html.

[17.19] Thānissaro, B., *The Buddhist Monastic Code I and II*, Metta Forest Monastery, Valley Center, CA, USA, 2013.

[17.20] Walshe, M. (trans.), *The Long Discourses of the Buddha*, Wisdom Publications, Boston, MA, USA, 1995.

18

A Space Settler's Bill of Rights[1]

Russell Greenall-Sharp, David Kobza, Courtney Houston, Mohammad Allabbad, Jamie Staggs and James S.J. Schwartz*

Dorothy and Bill Cohen Honors College and Department of Philosophy
Wichita State University, Wichita, Kansas, USA

Abstract

We propose a Bill of Rights for space settlers, which includes rights associated with basic physiological needs; physical and psychological well-being; freedom of expression; privacy; reproductive autonomy; vocational and educational liberty; communication; constrained dissent; and self-governance. This document aims to inform future discussions about the construction and governance of space societies. While not intended to provide an exhaustive list of space settlers' rights; nor to outline a well-defined legal framework for space settlement, nevertheless it comprises an attempt to describe a set of values to which any space settlement should aspire to exemplify.

Keywords: Space settlement, space colonization, rights, liberty, ethics, governance

18.1 Introduction

Space ethics—also known as astroethics—is a burgeoning subdiscipline of ethics that focuses on questions of normative significance to the wide and multifaceted realm of space exploration. As such, its primary function is not to legitimize or otherwise advance existing or proposed spaceflight objectives. Instead, its principal tasks are to provide much-needed critical discussion of (a) the worthiness of existing or proposed spaceflight objectives

Corresponding author: james.schwartz@wichita.edu

[1]We thank Sheri Wells-Jensen, Charles Cockell, Tony Milligan, Kelly Smith, Ted Peters, and Neal Allen for discussion.

Octavio Alfonso Chon Torres, Ted Peters, Joseph Seckbach and Richard Gordon (eds.) Astrobiology: Science, Ethics, and Public Policy, (377–388) © 2021 Scrivener Publishing LLC

and (b) the permissibility of pursuing these objectives by specific means.[2] Thus, it can and should be expected that space ethics will sometimes produce scholarship that is perceived, at least in the eyes of especially fervid space advocates, as antithetical to space exploration. Similarly, space ethics does not exist merely to provide ethical "checklists" for securing spaceflight objectives in an ethically permissible way. Just as the need for bioethics does not disappear but grows dramatically even long after the hospital is built, so too the need for space ethics does not disappear but grows dramatically even long after the mission is launched. Space ethics, like eating, drinking, or caring for one's health, simply is not the kind of thing that we will ever "finish" or "get over with." These are features, not bugs.

Space settlement provides a particularly salient example of the need for space ethics. If the current iteration of the space settlement movement keeps its momentum, then human settlement of the space environment seems inevitable. However, as has been explored previously, the space environment is extremely hostile to human life, which raises deep and vexing questions—technological, political, social—about how to balance the survival and security needs of a space society against the personal, psychological, and social needs of its denizens [18.2] [18.3] [18.4] [18.14] [18.21] [18.23] [18.16] [18.9] [18.10]. These questions are, at heart, ethical ones: To what conditions may we permissibly subject humans in the name of whichever ultimate purpose space settlement serves? Thus, one important task for space ethics is to engage in the proactive consideration of ethical issues surrounding space settlement.

As a small step among many to this end, we propose a Bill of Rights for space settlers, which should inform future discussions about the construction and governance of space societies. This document is not intended to provide an exhaustive list of space settlers' rights; nor is it meant to outline a well-defined legal framework for space settlement. It also does not recommend or specify mechanisms for securing the rights proposed herein. Instead, it comprises an attempt to describe a set of values to which any permanent space settlement should aspire to exemplify, with the specific extent and practice of these rights subject to ongoing discourse in the life of the settlement. Although we approve of the idea that these rights may be amended in the future by settlements (to suit their particular or unanticipated needs), nevertheless we aim to endorse a perspective on the *initial conditions* of any space settlement so that its denizens are (at least) given opportunities to lead worthwhile lives. These rights are discussed in the context of permanent settlements. It may be acceptable that,

---

[2]In no way do we suggest that (a) and (b) are *exhaustive* of the tasks of space ethics.

in early phases of exploration or construction, transient workers (scientists, engineers, construction workers) may be subject to different, less pervasive entitlements (as specified in employment contracts, for instance). While such persons are clearly deserving of protections, these fall outside the scope of our discussion.

We propose the following nine rights—ranked in anticipated importance to space settlers—which include basic survival rights, social rights, and political rights:

1. Basic Physiological Needs
2. Physical and Psychological Well-Being
3. Freedom of Expression
4. Privacy
5. Reproductive Autonomy
6. Vocational and Educational Liberty
7. Communication
8. Constrained Dissent
9. Self-Governance and Revisability

Each proposed right is described and discussed below. Some of these rights are rights in the "positive" sense—that is, they are rights to access, or to be provided, various resources or opportunities. We acknowledge that securing these rights may be especially burdensome—especially if they must be secured by the settlers themselves. However, it would be a mistake to view these as burdens that would always have to fall directly on settlers; we maintain that the majority of the burden falls on whichever entities sponsor or initiate space settlement. Thus, this "Bill of Rights" can and should be viewed as hypostasizing obligations primarily borne by anyone interested in establishing space settlements. Whether any of these obligations should be backed by treaty, national or local law, or some other guarantee, we leave to further discussion, but we insist that initiating space settlement entails a serious ethical responsibility to the settlers by those supporting and funding the endeavor.

While space settlement will be an arduous undertaking, we should nevertheless aim to maintain basic rights for the people who will be living in these difficult conditions. These people are not merely construction materials for building an interplanetary future, they are human beings with dignity who are owed respect, and deserve worthwhile lives. As such, their welfare, rights, and living conditions should factor into our thinking, technological development, and cost evaluations for space settlement. Their ability to enjoy lives worth living is not a disposable extra, a luxury—it is a bare minimum. If we proceed with space settlement without recognizing this, we will not succeed in bringing humanity to the stars, but merely human bodies.

18.2   Basic Physiological Needs

Space settlers have a right to have their basic physiological needs met. These include access to adequate air, water, food, and shelter. Since it is unlikely that individual settlers will be able to meet these needs on their own, they must instead be met by the settlement as a whole. This will almost certainly require that the sponsoring entity, whether state or private, provide substantial backing to enable this from the start. It would be utterly negligent and irresponsible to begin such an undertaking without the ability to meet the basic needs of settlers.

Nowhere on Earth has it been necessary to view access to breathable air as a basic right. But in space, where air must be manufactured (and thus produced and controlled by some entity), access to it must be safeguarded. Obviously, without air, the settlers will quickly die. Guaranteeing a right to adequate breathable air enables basic survival. It also mitigates the potential for the air production and distribution infrastructure to be abused by bad actors. If someone were to maliciously control the oxygen, they would be able to control the settlement [18.6]. That cannot be allowed. Though the need is less immediate, similar concerns motivate a right to access adequate quantities of water and food. Without endorsing any particular economic system, we maintain that access to breathing air should never be commodified. The second-to-second urgency of our biological need for oxygen grants anyone controlling its supply, or even its cost, immense coercive power over potential settlers. Such commodification lends itself easily to exploitation in myriad ways and so it should in all cases be avoided.

The right to adequate shelter takes on a novel immediacy in the space environment. Moreover, settlers most likely will not have the option to "move away" from the settlement without returning to Earth. Thus, for most, leaving the settlement would be leaving all basic resources and would equate to suicide. Settlers therefore are entitled to a large enough area to live, relax, and rest, and exercise their other rights (including, we maintain, a right to privacy). Since settlements will likely be rigidly structured and difficult to modify, adequate living space must be planned from the outset [18.1].

We include accessibility among the requirements of adequate living space, and advanced planning must invoke principles of Universal Design. Indeed, space settlement provides us with a unique opportunity to plan for inclusive, accessible places from the outset. As Sheri Wells-Jensen has argued [18.23], there is no good reason to exclude disabled persons from participating in space settlement. The idea that disabled persons are a liability to missions is built upon an unjustified ableism and exclusionary design philosophies. Inclusion isn't merely politically desirable; it is beneficial also as a means of utilizing the full inventory of our intellectual and technical resources. Moreover, the fact that the space environment is hazardous, both in the short-term and long-term, to human health, makes disability an inevitability for space-dwellers—an inevitability that can be anticipated and accommodated for by designing for and including disabled persons.

We cannot stress enough that denying any settler these rights would be to essentially kill them and possibly endanger the entire settlement. We cannot imagine a functioning, productive space society that does not meet the basic physiological needs of its denizens.

18.3   Physical and Psychological Well-Being

The unique hardships of the space environment will require that settlements have an extra duty of care to create livable societies. To this end, well-being here is endorsed in a holistic sense and concerns the basic maintenance of both the settlers' functioning and the quality of their lives.

Settlers must be provided access to adequate healthcare. Since all settlers will be vital to the settlement's growth and success, it is essential that everyone remains healthy and receives whichever treatments they require. Denying adequate healthcare to any settler would be detrimental not only to the individual, but also to the settlement which stands to lose a productive member. Much of the obligation to secure this right falls on the instigator or sponsor of a space settlement, since the possibility of *in-situ* manufacturing of medical devices, treatments, and pharmaceuticals cannot be assumed. Moreover, space settlers as medical patients have a right to autonomy, including a right to refuse medical treatment provided that doing so does not threaten the overall survival of the settlement. Thus, we admit that in principle it is possible for the needs of the settlement to outweigh the obligation to protect personal liberties. But we reject any blanket presumption that the settlement is always, generally, or often entitled to carry out decisions that result in diminution of personal liberties. The burden of proof should always be on the settlement to demonstrate the need to override a particular personal liberty in a particular case; and never on individual settlers to demonstrate the importance of retaining their basic rights. As mentioned above, maintaining basic rights should be seen as a priority; as such, these rights should be seen as defaults rather than privileges or luxuries that can be easily removed. If adequate healthcare for settlers is too expensive to provide, then space settlement is too expensive to instigate.

Precautions such as regular psychological evaluations should be supported. By catching disruptions to someone's mental health early on, potential threats such as self-harm or violence towards others can be mitigated [18.22]. Furthermore, this right will provide an entitlement to the availability of a variety of treatment options, ranging from medication to therapy, in order to address the diverse needs of the population. It is important that individuals have at least some choice in how they are being treated. For this reason, medical consultation services must always be offered. At no point should patients ever be forced to undergo "mood enhancement" treatments that would, against their will, transform them into docile individuals.

Personal time for recreation and social needs must also be offered to ensure psychological well-being through the availability of opportunities for personal satisfaction. Personal enrichment opportunities will have a particular importance to the well-being of settlers, especially as these individuals are likely to experience stress due to living in confined environments with few personal choices. Settlements should strive to ensure that their inhabitants are not forced to live impoverished lives. Space environments will be taxing for all, but the difficulty should not mean the sacrifice of the possibility of worthwhile lives. Ensuring the possibility of worthwhile lives for settlers should also encompass issues surrounding disability. While employing principles of Universal Design provides a starting point, the settlement should actively strive to become a place that is inclusive and that is capable of providing worthwhile lives to all of its inhabitants, including its disabled inhabitants. Key to ensuring this is executed successfully is the inclusion of disabled persons throughout the design process, because such persons are better situated to be aware of the abilities and design requirements for disabled persons.

Lastly, individuals suffering unmanageable psychological distress caused by the settlement environment must be allowed to leave wherever possible. Colonists mustn't be made hostages or slaves to the settlement, which requires allowing settlers the option to leave and/or to return to Earth. While this is likely to be difficult to guarantee, and while emigration opportunities will be limited, these opportunities should nevertheless be facilitated where possible. This will provide the settlers a sense of freedom in their choice to join and remain in the settlement, as well as a respite if living in a space settlement becomes too distressing for them (helping to realize the liberal ideal of governance by consent of the governed). A more robust right of return could also resolve ethical concerns surrounding one-way trips to space settlements [18.19] [18.18] [18.12].

18.4   Freedom of Expression

Settlers have a right to freedom of expression. This includes a right to free speech, press, artistic liberty, religion, as well as sexual and gender expression. We regard these as genuinely universal rights, and space settlements should not be established if they will become places where these rights are stifled for the sake of making the settlement simpler to plan or easier to govern. Moreover, people should never be reduced to mere units of labor for the settlement or simple vessels for their genes, whether for future generations or for "the species."

Freedom of expression is an essential part of leading a human life, and this is especially the case for humans living in space. The sorts of expression protected in this right are often used by humans on Earth as outlets, i.e., as personal and social "release valves," and the privation of these outlets can result in needless suffering. Moreover, space settlers will have many new emotions and experiences to work through. Thus, the need for these outlets is even more prevalent in space, where the range of everyday activities open to settlers will be very limited in scope [18.11].

To remove a settler's freedom of expression is to deny them the right to be human. Only when expression produces a tangible threat to other settlers should it be stopped, and even then, only with extreme caution. Overly broad rules, which might stifle expression extensively, should be forbidden.

18.5   Privacy

A right to privacy will be a delicate right that must be balanced against the security and safety of the settlement and those in it [18.20] [18.7]. It may be that pervasive surveillance is necessary to guarantee the well-being of the settlement, but this justification should never infringe on the legitimate privacy rights of settlers. Surveillance should not be permitted in private dwellings and spaces, and allowances should be made for private correspondence among the settlers. Settlers' private correspondence should not be examined without appropriate due process. It may also be advisable to further restrict the scope of surveillance in public places where there is

minimal risk to vital infrastructure, so that settlers are freer from observation during innocuous daily activity and conversation, whether political
or trivial. A right to privacy also facilitates a right to basic psychological
needs. Settlers will be confined within their habitat with no chance to
escape—this already stressful situation, conjoined with extensive monitoring, is likely to be a considerable drain on mental well-being.[3]

Future settlers may eventually decide, with good reason, to curtail or
adapt the extent or interpretation of the right to privacy. Nevertheless, it is
our belief that grounding the foundation of the settlement in liberal democratic norms and values gives it the best chance to thrive and establish itself
and its own values.

18.6   Reproductive Autonomy

Space settlers have a right to bodily integrity in matters of intimacy and
procreation. This encompasses who settlers choose as romantic and life
partners, who settlers choose to mate with, whether or not they choose to
reproduce, and protection of their sexual preference [18.19]. To protect
the sexual preferences and identities of settlers, it is prohibited to coerce
individuals to procreate for any reason, including for the maintenance of
adequate genetic diversity. The gender identities and sexual preferences of
settlers are to be respected and the settlement has a responsibility to enable
its settlers to meet their companionship needs.

In looking to continue human progress through space settlement, it
would be an immense failing if we allowed recent, hard-fought gains in
equality to be lost, discarded, or treated as an expendable luxury. In space
settlements, we should aim to improve the human condition (and not just
the physical spread of human bodies). It would reflect extremely poorly on
us if we were to allow backsliding on these important issues for the sake of
expediency.

Genetic diversity could be maintained through a combination of the
autonomous pairing of settlers and influxes of new settlers. Under no circumstances is it permissible to use coercion to maintain the genetic diversity of the settlement. Thus, the settlement sponsor has a duty to assist in the
maintenance of the settlement's genetic diversity, for instance, by sending
additional settlers to alleviate any genetic diversity concerns before they
become too serious. Hence, there are limits on what can permissibly be done
to ensure stable population minima. Meanwhile, preventing destabilizing

---

[3]In this regard it is worth appreciating prisoners' perspectives on space settlement [18.8].

population growth can be accomplished by comparatively uncontroversial means, e.g., through honest sexual education and free access to contraception. If such passive measures fail, then the settlement may wish to consider a "one-child" policy as a means of preventing additional population growth if a population crisis seems inevitable—but only with the ongoing consent of those directly affected by such a policy.

18.7   Vocational and Educational Liberty

Settlers have the right to a minimum standard of education and to choose their path in secondary and tertiary education that leads to a vocation of their choosing [18.11]. In order to maintain a well-educated and functioning society, primary and secondary education will be required for the children (ages 5–18) of space settlers. After a settler has completed secondary education, they will be entitled to the right of choice pertaining to their vocation, possibly including further education to that end. Securing a necessary minimum diversity of skills should not infringe on settlers' vocational autonomy. This may be accomplished through either encouraging immigration of skilled workers or through incentivizing necessary vocations—but it should never involve force or excessive coercion of settlers [18.19]. It is likely that many settlers will feel a sense of duty to the settlement, either instilled purposefully or naturally, and this will vitiate the need to pursue coercive strategies for maintaining a necessary skill base. As we have noted before, no one should be made to be a "slave" to the settlement.

18.8   Communication

A right to communication entitles every settler to the ability to communicate with whomever they choose to, including those living outside of the settlement. This right undergirds and enables rights to expression and to basic psychological welfare. Attempts to restrict communications networks, including long-distance networks, would negatively impact the social aspects of settlers' lives. Open communications are especially crucial to the well-being of small settlements, where there is little opportunity to interact with new people. Without the right to communicate, there is a very high risk of dissatisfaction and other negative feelings that could be detrimental to the success and growth of the settlement. Freedom of communication is also an important tool for exposing, mitigating, and preventing abuses of power.

18.9   Constrained Dissent

A limited right to dissent and protest should be protected. Activities not protected by this right include physically harming settlers and damaging or restricting access to vital systems (e.g., air and water production and distribution equipment). This right enables the right to revisability of the terms of governance. It also protects against unjust social and working conditions that might lead to an authoritarian regime [18.5] [18.13]. Special exceptions might exist in the case of essential service providers. For instance, a maintenance worker may not be permitted to strike when air production systems are in immediate need of maintenance [18.17]. More generally, anyone who has a responsibility for vital systems cannot protest in ways that compromise the provision of essential resources (air, water).

18.10   Self-Governance and Revisability

Settlers have a right to be free from external control and to amend this Bill of Rights. Space settlements will place humans in entirely novel environments, which means that at present we cannot predict which set of basic rights will best facilitate the welfare of settlers and the health of settlements. As settlements develop, they may encounter novel and unanticipated social and environmental factors, and the settlement as a whole must retain sufficient ingenuity and autonomy to adapt itself to these new factors. Revisions should be instigated and adopted via democratic processes [18.20] [18.15].

18.11   Conclusion

We find it troubling that many discussions of space settlement appear to ignore the human/experiential aspects of the endeavor. While questions concerning, e.g., minimum genetically viable populations, minimum vocationally viable populations, etc., are important, they are, in our view, secondary to questions about how to ensure that space settlers are able to lead meaningful lives—lives they regard as worth living. Our future in space should ideally be guided by expressions of values that prioritize our humanity rather than sacrifice it for the sake of expediency. While each right mentioned above warrants much more discussion than we provide here, nevertheless our goal has been to provide a brief and holistic overview of concerns and ethical principles that ought to guide the construction and management of space settlements.

We hope, then, that our proposed Bill of Rights for space settlers encourages scholars and potential sponsors to grant these issues the serious consideration they warrant. If one lesson here is clear, it is this: Space settlement is not the kind of endeavor that should be pursued at the earliest possible opportunity with the absolute minimum of personnel and resources. The "Faster, Better, Cheaper" strategy is not one that is likely to produce truly livable space settlements. If securing the above rights places space settlement at an apparently unreachable financial and technological distance, then we implore that this be taken as evidence that we simply are not yet ready to settle space. Only when we are prepared to accept the costs associated with providing settlers with worthwhile lives will we truly be ready to make humans an interplanetary species.

References

[18.1] Abood, S., Martian Environmental Psychology: The Choice Architecture of a Mars Mission and Colony, in: *The Human Factor in a Mission to Mars*, K. Szocik (Ed.), pp. 3–34, Springer, New York, 2019.

[18.2] Cockell, C. (Ed.), *Dissent, Revolution and Liberty Beyond Earth*, Springer, New York, 2016.

[18.3] Cockell, C. (Ed.), *Human Governance Beyond Earth*, Springer, New York, 2015.

[18.4] Cockell, C. (Ed.), *The Meaning of Liberty Beyond Earth*, Springer, New York, 2015.

[18.5] Cockell, C., Disobedience in Outer Space, in: *Dissent, Revolution and Liberty Beyond Earth*, C. Cockell (Ed.), pp. 21–40, Springer, New York, 2016.

[18.6] Cockell, C., *Extraterrestrial Liberty: An Enquiry into the Nature and Causes of Tyrannical Government beyond the Earth*, Shoving Leopard, Edinburgh, 2013.

[18.7] Cockell, C., Freedom in a Box: Paradoxes in the Structure of Extraterrestrial Liberty, in: *The Meaning of Liberty Beyond Earth*, C. Cockell (Ed.), pp. 47–68, Springer, New York, 2015.

[18.8] Cockell, C., *Life Beyond: From Prison to Mars*, The British Interplanetary Society, London, 2018.

[18.9] Cowley, R., *A Manifesto for Governing Life on Mars*, King's College London, London, 2019.

[18.10] Cowley, R., Yes, We Earthlings Should Colonize Mars if Martian Rights can be Upheld. *Theol. Sci.*, 17, 332–3405, 2019.

[18.11] de Vigne, J., Education and Liberty in Space, in: *The Meaning of Liberty Beyond Earth*, C. Cockell (Ed.), pp. 219–226, Springer, New York, 2015.

[18.12] Koepsell, D., Mars One: Human Subject Concerns? *Astropolitics*, 15, 97–111, 2017.

[18.13] Milligan, T., Constrained Dissent and the Rights of Future Generations, in: *Dissent, Revolution and Liberty Beyond Earth*, C. Cockell (Ed.), pp. 7–20, Springer, New York, 2016.

[18.14] Milligan, T., *Nobody Owns the Moon: The Ethics of Space Exploitation*, MacFarland, Jefferson, NC, 2015.

[18.15] Milligan, T., Rawlsian Deliberation About Space Settlement, in: *Human Governance Beyond Earth*, C. Cockell (Ed.), pp. 9–22, Springer, New York, 2015.

[18.16] Oman-Reagan, M., Politics of Planetary Reproduction and the Children of Other Worlds. *Futures*, 110, 19–23, 2019.

[18.17] Schwartz, J., Lunar Labor Relations, in: *Dissent, Revolution and Liberty Beyond Earth*, C. Cockell (Ed.), pp. 41–58, Springer, New York, 2016.

[18.18] Schwartz, J., *The Value of Science in Space Exploration*, Oxford University Press, New York, 2020.

[18.19] Schwartz, J., Worldship Ethics: Obligations to the Crew. *J. Br. Interplanet. Soc.*, 17, 53–64, 2018.

[18.20] Smith, K., Cultural Evolution and the Colonial Imperative, in: *Dissent, Revolution and Liberty Beyond Earth*, C. Cockell (Ed.), pp. 169–187, Springer, New York, 2016.

[18.21] Smith, K., Abney, K., Anderson, G. *et al.*, The Great Colonization Debate. *Futures*, 110, 4–14, 2019.

[18.22] Tachibana, K. and Hobbesian, A., Qualm with Space Settlement. *Futures*, 110, 28–30, 2019.

[18.23] Wells-Jensen, S., Miele, J., Bohney, B., An Alternate Vision for Colonization. *Futures*, 110, 50–53, 2019.

Index

Printed in the USA/Agawam, MA
October 4, 2021

782116.009